85

Fortschritte der Chemie organischer Naturstoffe

Progress in the Chemistry of Organic Natural Products

Founded by
L. Zechmeister

Edited by
W. Herz, H. Falk,
and G. W. Kirby

Authors:
D. P. Chakraborty, K. Krohn,
P. Messner, S. Roy, C. Schäffer

Springer-Verlag Wien GmbH

Prof. W. Herz, Department of Chemistry,
The Florida State University, Tallahassee, Florida, U.S.A.

Prof. H. Falk, Institut für Chemie,
Johannes-Kepler-Universität, Linz, Austria

Prof. G. W. Kirby, Chemistry Department,
The University of Glasgow, Glasgow, Scotland

Originally published by Springer- Verlag Wien New York in 2003
Softcover reprint of the hardcover 1st edition 2003

Library of Congress Catalog Card Number AC 39-1015

Typesetting: Thomson Press (India) Ltd., Noida

Printed on acid-free and chlorine-free bleached paper

SPIN: 10875643

With 12 Figures

ISSN 0071-7886
ISBN 978-3-7091-7290-2 ISBN 978-3-7091-6051-0 (eBook)
DOI 10.1007/978-3-7091-6051-0

Contents

Carbazole Alkaloids IV
D. P. Chakraborty and S. Roy

List of Contributors

Chakraborty, Prof. Dr. D. P. †, Institute of Natural Products, Satchasi Para Lane, Calcutta 700 036, India

Krohn, Prof. Dr. K., Fachbereich Chemie und Chemietechnik der Universität Paderborn, Warburger Str. 100, D-33098 Paderborn, Germany, e-mail: kk@chemie.uni-paderborn.de

Messner, Prof. Dr. P., Zentrum für Ultrastrukturforschung und Ludwig Boltzmann Institut für Molekulare Nanotechnologie, Universität für Bodenkultur Wien, Gregor-Mendel-Str. 33, A-1180 Wien, Austria, e-mail: paul.messner@boku.ac.at

Roy, Dr. S., 8, Jagadishnath Roy Lane, Calcutta 700 006, India, e-mail: shyamaliroy@yahoo.com

Schäffer, Dr. C., Zentrum für Ultrastrukturforschung und Ludwig Boltzmann Institut für Molekulare Nanotechnologie, Universität für Bodenkultur Wien, Gregor-Mendel-Str. 33, A-1180 Wien, e-mail: christina.schaeffer@boku.ac.at

Natural Products Derived from Naphthalenoid Precursors by Oxidative Dimerization

K. Krohn
Fachbereich Chemie und Chemietechnik,
Universität Paderborn, Germany

Contents

1. Introduction

In the last decade, a structurally diverse class of new bioactive natural products was isolated from fungi in which two naphthalene units are fused together via oxidative coupling. These compounds attracted attention because of their antifungal, antibacterial, and antitumoral activity. In addition, several enzymes, namely: phospholipase D, DNA gyrase, and Ras farnesyl-protein transferase, were also found to be inhibited with high selectivity.

It has been known for a long time that nature can dimerize simple naphthalene building blocks through one or more C-C bonds to construct a variety of secondary metabolites. Examples for these compounds include the neopodine glycosides (*1*), the sphaerolones (*2*), the stemphytriols (*1*), and the stemphyltoxins (*3*). More highly substituted compounds of this class are represented in the perylenequinones (*4, 5*) as exemplified by the elsinochromenes (*6*) (for a review on some C-C-dimeric naphthalenes see Gill and Steglich in (*7*)). The biosynthetic origin of this type of naphthalene "dimers" is well established (see for example (*8*)).

In this overview, we will focus on the more recently elucidated structures where the naphthalene units are linked by two or three oxygen atoms. In most of these "dimers" the naphthalene cores are partially modified (saturated or oxidized) and the oxygen atoms are also incorporated into the acetal bridges. As one naphthalene part is generally linked to a decalin system via a spiroketal bridge, this class of compounds is referred to as "spirobisnaphthalenes" (*9*), or, more specifically, "bisnaphthospiroketals" in case of the spiroxins, (*10*) or "spiroacetal-linked naphthodiepoxydecalinones" in case of the diepoxins (*11*). Reflecting on the relatively small number of carbon atoms, the spirobisnaphthalene type of natural products belong one of the most diverse classes of secondary metabolites. They are an excellent example of how flexible nature is in producing numerous diversified compounds from relatively small building blocks (monomeric naphthols) by employing a variety of simple chemical modifications like epoxidation, reduction, methylation, oxidation, halogenation, etc.

Three different types of oxygen-bridged spirobisnaphthalenes have thus far been isolated. Their molecular skeletons are represented in Chart 1 by spiroxin A (**1**), with **one** carbon and **two** oxygen bridges (*10*); palmarumycin CP_1 (**2**) (*12*), the parent compound of the spirobisnaphthalenes with **two** oxygen bridges; and preussomerin A (**3**), where the naphthalene units are linked over **three** oxygens, representing **two** spiroacetal groups (*13*).

All existing data for the naturally occurring members of each family of spirobisnaphthalenes are compiled in tables, in which the synonyms used in the original publications are listed as well (e.g. reference, source, melting point, specific optical rotation, CD/absolute configuration, X-ray analysis, placement of formulae in Charts/Schemes, and biological activity).

References, pp. 46–49

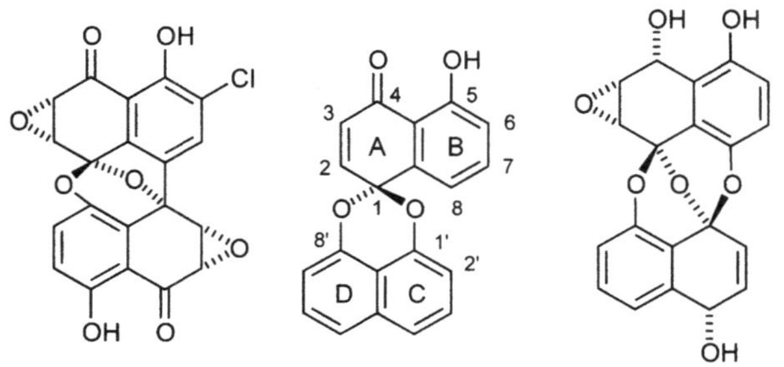

Spiroxin A (**1**) Palmarumycin CP_1 (**2**) Preussomerin A (**3**)

Chart 1. The three molecular skeletons of spirobisnaphthalenes

2. Isolation and Structure Elucidation

2.1. Spirobisnaphthalenes with Two Oxygen Bridges

The first reported spirobisnaphthalene (with two oxygens in the bridge) was the antibiotic MK 3018. It was the subject of a patent in which structure **4** (Scheme 1) was tentatively assigned (*14*). However, its relative or absolute stereochemistry has never been reported. The compound was apparently isolated from the fungus *Tetraploa aristata* I R 25 and showed a broad spectrum of antibacterial activity. Unfortunately, no similar or identical compound was subsequently isolated that could have confirmed the details of the substitution pattern and assigned a stereochemistry.

Shortly after this report, a metabolite called bipendensin (**5a**) (m.p. 254°C) was isolated from the trunk wood of *Afzelia bipendensis* (*15*). Despite the small amount of material available, its structure could be elucidated with the aid of modern NMR techniques. In particular, two-dimensional NMR proved to be useful to assign the signals for the carbon atoms and protons to the two fragments **A** and **B** and corroborate the connectivity of the carbon skeleton (Scheme 1). The dioxynaphthalene fragment **A** showed the expected shifts for aromatic protons and carbons. Notably, the two lowfield signals for C-1′ and 8′ at $\delta = 146.9$ and 146.8 ppm in the ^{13}C NMR spectrum were typical for aromatic carbon atoms bonded to oxygen atoms. The most significant signal at $\delta = 97.0$ ppm in the ^{13}C NMR spectrum of fragment **B** originated from the acetal carbon. These signals are characteristic for all spirobisnaphthalenes and mark a spectral feature that was common to all spirobisnaphthalenes subsequently isolated from the culture broth of

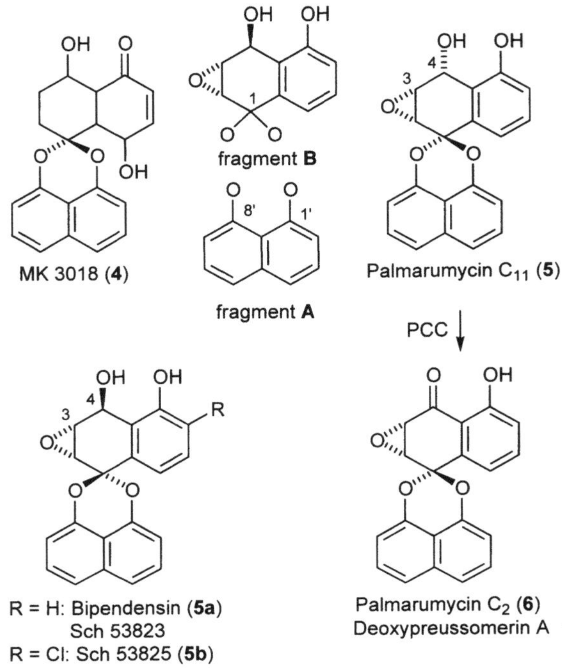

Scheme 1. Structure of MK 3018 (**4**), palmarumycin C_{11} (**5**), bipendensin and Sch 53823 (**5a**), Sch 53825 (**5b**) and PCC oxidation of palmarumycin C_{11} to palmarumyin C_2 (**6**)

fungal fermentations. The two fragments **A** and **B** of bipendensin can be assembled to structure **5a**. However, the relative stereochemistry of bipendensin (**5a**) currently remained unresolved because the observed small coupling constant of $J = 2.2\,\text{Hz}$ between H-3 and H-4 did not allow an unambiguous assignment. The author suggested that bipendensin is presumably derived from a naphthoquinone called juglone. However, later investigations of the biosynthesis showed that 1,8-dihydroxy-naphthalene (1,8-DHN) is the likely biosynthetic precursor ((*16*, *17*, *18*) *vide infra*). (Remark: as all of the other spirobisnaphthalenes were isolated from fungi, we have to assume that compound **5** is also produced by an endophytic fungus living in the tree *Afzelia bipendensis*).

Later on, more complete data of this metabolite **5a** were published in a full paper (*19*). The spectroscopic data deviate from those of palmarumycin C_{11} (m.p. 237–238°C), a metabolite with the same gross structure, isolated from *Coniothyrium palmarum* (*16*). The stereochemical relevant coupling constants in the ^{1}H NMR spectrum of bipendensin (**5a**) for $J_{3,4} = 2.2$ Hz differ slightly from those of palmarumycin C_{11} with $J_{3,4} = 2.7$ Hz. The gross structure of palmarumycin C_{11} (**5**) was confirmed by oxidation with pyridinium chlorochromate to palmarumycin C_2 (**6**) (Scheme 1) and its relative stereochemistry was later confirmed by total synthesis (*20*). Chu *et al.* (*21*) isolated a compound Sch 53823 of the same composition with a nearly identical melting point (235–240°C) but with positive specific optical rotation (see Table 1). The agreement of the physical and spectroscopic data (notably the NMR spectra) suggests the identity of Sch 53823 with bipendensin (**5a**) and, based on reduction experiments, Taylor *et al.* (*20*) suggested the *anti*-configuration **5a** for bipendensin and Sch 53823 and the *syn*-configuration for palmarumycin C_{11} (**5**). Considering the NMR data, a chlorinated product Sch 53825 probably also has the *anti*-configuration **5b**. The stereochemical questions of relative and absolute configuration have been resolved for a number of spirobisnaphthalenes and will be discussed later (*vide infra*).

In the year 1993 and in subsequent years, five different research groups from pharmaceutical industry as well as academy, including our group at the university of Paderborn, and Zeeck *et al.* at Göttingen, published a number of papers on the isolation, structure elucidation, and biological activity of spirobisnaphthalenes that were related to MK 3018 (**4**) or bipendensin (**5a**). Some of the information was originally communicated in patents (*34*) or as posters at meetings (see Ref. 7 in (*31*)) and reported new secondary fungal metabolites now known as spirobisnaphthalenes. These compounds were produced by different fungi, were given different names, and their structural formulae were drawn in different ways. We have now compiled their synonyms and relevant data in Table 1 and combined them with the formulae.

Schlingmann *et al.* at Lederle Laboratories (American Cyanamid Company) investigated the metabolites of a non-sporulating fungus isolated by MYCOsearch from a tree trunk growing in Panama. This endophytic fungus, subsequently grown only as a *mycelium sterilum*, also produced an antibiotic of the allenic polyacetylene family (*35*), but the major metabolites were spirobisnaphthalenes named diepoxins because they contained two epoxide groups at the decalin ring system (*24*). The absolute stereochemistry of the diepoxins was established later (*vide infra*) by employing the exciton coupled CD method on some of their bis-dimethylaminobenzoate derivatives (*11*). The formulae shown here reflect the absolute stereochemistry thus determined.

Table 1. Naturally Occurring Spirobisnapthalenes. Source (organism), melting point, specific optical rotation, absolute configuration, structural confirmation by X-ray analysis, and biological activity. M.p. mostly decomposition

Name	Ref.	Source	*M.p.* °C	$[\alpha]_D$	X-ray	CD/abs. conf.	Scheme/ Chart	Bioactivity
MK 3018 (**4**)	(*14*)	*Tetraploa aristata* I R 25	–	–	–	–	S 1	Broad antibacterial activity
Bipendensin (**5a**)	(*15, 19*)	*Afzelia bipendensis*	254	–	–	–	S 1	not tested
Palmarumycin C_{11} (**5**)	(*16*)	*Coniothyrium palmarum*	237–238	−153	–	–	S 1	weakly antibacterial
Sch 50676, identical with p. C_{11} (**5**)	(*23*)	*N. mangiferae*	235–238	−133.5	–	–	S 1, C 4	antitumor
Sch 53823 (**5a**)	(*21*)	*Endophyte*	235–240	+227	–	–	S 1	phospholipase D inhibitor
Sch 53825 (**5b**)	(*21*)	*Endophyte*	182–183	+74	–	–	S 1	phospholipase D inhibitor
Palmarumycin C_2 (**6**)	(*16*)	*C. palmarum*	228	−341	+	+(*22*)	S 1, C 4	antifungal, antibacterial
Diepoxin α (**7**)	(*24*)	nonsporulating fungus	–	+30	–	+	C 2	antifungal, antibacterial
Diepoxin η (**8**)	(*11, 24*)	nonsporulating fungus	250	+30, +23	–	+(*11*)	C 2, S 3	virtually inactive
Palmarumycin C_{14} (**8**) (mixture with **9**)	(*16*)	*C. palmarum*	–	–	–	–	C 2, S 3	not tested
Sch 53516 (**8**)	(*25*)	*N. mangiferae*	270–272				C 2, S 3	phospholipase D inhibitor
Sch 53517 (**8a**)	(*25*)	acetate of **8**	120–122	–	–	–	–	
Diepoxin ζ (**9**)	(*11, 24*)	nonsporulating fungus	–	+75	–	–	C 2, S 3	antifungal, antibacterial
Palmarumycin C_{13} (**9**)	(*16*)	*C. palmarum*	–	–	–	–	C 2, S 3	not tested
Sch 53514 (**9**)	(*25*)	*N. mangiferae*	152–154	–	–	–	C 2, S 3	antitumor, phospholipase D inhibitor
Sch 53515 (**9a**)	(*25*)	acetate of **9**	225–227	–	–	–	–	phospholipase D inhibitor
Cladospirone bisepoxide (**9**)	(*26*)	*Sphaeropsidales* sp.	>160	–	+	+	C 2, S 3	antibacterial, herbicidal
Diepoxin σ (**10**)	(*24*)	mycelia sterila	–	+67	–	+	C 2, S 2	
Sch 49209 (**10**)	(*27*)	*N. mangiferae*	144–146	+79.1	–	–	C 2	antitumor
Acetate (**10a**)	(*27*)	acetate of **10**	–	–	–	–	S 2	antitumor
Diepoxin γ (**11**)	(*11*)	nonsporulating fungus	–	–	+	+	C 3	antifungal, antibacterial
Diepoxin δ (**12**)	(*11*)	nonsporulating fungus	241	–	–	+	C 3	not tested
Diepoxin ϕ (**13**)	(*11*)	nonsporulating fungus	–	–	–	+	C 3	not tested
Palmarumycin C_{10} (**13**)	(*16*)	*C. palmarum*	236	–	–	–	C 3	antifungal, antibacterial
Diepoxin ι (**14**)	(*11*)	derivative	242	–	–	+	C 3	not tested

Table 1 (continued)

Name	Ref.	Source	*M.p.* °C	$[\alpha]_D$	X-ray	CD/abs. conf.	Scheme/ Chart	Bioactivity
Diepoxin κ (**15**)	*(11)*	derivative	158; 232	–	+	+	C 3	not tested
Sch 50674 (**16**)	*(23)*	derivative of **10**	245–247	+33.1	+	–	S 2	antitumor
Sch 49210 (**17**)	*(25)*	*N. mangiferae*	140–143		–	rel.	C 4	PLD inhibition
Sch 50673 (**18**)	*(23)*	*N. mangiferae*	164–166	−89.8	–	–	C 4	antitumor
Sch 49211 (**20**)	*(28)*	*N. mangiferae*					C 4	PLD inhibition
Sch 49212 (**21**)	*(28)*	*N. mangiferae*					C 4	PLD inhibition
CJ–12, 371 (**22**)	*(29)*	fungus	>265 dec.	−46.8	–	+/+	C 5	Gyrase inh.
CJ–12, 372 (**23**)	*(29)*	fungus	>238 dec.	−82.0	–	+/+	C 5	Gyrase inh.
4-Oxo-cladospirone-bisepoxide (**25**)	*(30, 26, 31)*	*Spaeropsidales* sp.	>200	–	–	–	S 3	not tested
Cladospirone B (**31**)	*(9)*	*S.* sp.	230	−270	+	+	C 6	not active
Cladospirone C (**32**)	*(9)*	*S.* sp.	164	−35	–	+	C 6	antibacterial
Cladospirone D (**33**)	*(9)*	*S.* sp.	127	+55	–	+	C 6	antibacterial, herbicidal
Cladospirone E (**34**)	*(9)*	*S.* sp.	236	−217	+	+	C 6	not active
Cladospirone F (**35**)	*(9)*	*S.* sp.	140	−150	–	+	C 6	not active
Cladospirone G (**36**)	*(9)*	*S.* sp.	135	+5	–	+	C 6	not active
Cladospirone H (**37**)	*(9)*	*S.* sp.	156	−22	–	+	C 6	not active
Cladospirone I (**38**)	*(9)*	*S.* sp.	138	+14.6	–	+	C 6	not active
(**39**)	*(9)*	derivative	68	+6.5	–	+	C 6	not tested
Palmarumycin CP_1 (**2**)	*(12)*	*C. palmarum*	170 dec.			+/+	C 1	weakly antibacterial, antifungal
Palmarumycin CP_2 (**40**)	*(12)*	*C. palmarum*	170 dec.	–	–	–	C 7	not active
Deoxypreus-somerin B (**40**)	*(32)*	coelomycetous fungus			–		C 7	inactive
Palmarumycin CP_3 (**41**)	*(12)*	*C. palmarum*	190 dec.	−102.8	+	+/+ *(22)*	C 7	antibacterial, antifungal
Palmarumycin CP_4 (**42**)	*(12)*	*C. palmarum*	193	+495	+		C 7	antibacterial, antifungal
Palmarumycin CP_{4a} (**43**)	*(33)*	*C. palmarum*	213	+70.6	+	+/+	C 7	not tested
Palmarumycin CP_5 (**44**)	*(33)*	*C. palmarum*	168	+45.5	+	+/+	C 7	not tested
Palmarumycin C_1 (**45**)	*(16)*	*Coniothyrium* sp.	>180	–	–	–	C 8	not tested
Palmarumycin C_2 (**6**)	*(16)*	*C.* sp.	228	−341	+	+/+	S 1	antibacterial, antifungal
Deoxypreus-somerin A (**6**)	*(32)*	unidentified coelomycetes	235–236	−300	–	–	S 1	not tested
Palmarumycin C_3 (**46**)	*(16)*	*C.* sp.	220	−300 $CHCl_3$	–	+	C 8	antibacterial, antifungal

Table 1 (continued)

Name	Ref.	Source	*M.p.* °C	$[\alpha]_D$	X-ray	CD/abs. conf.	Scheme/ Chart	Bioactivity
Palmarumycin C_4 (**47**)	(*16*)	*C.* sp.	(84)	−285.5	–	–	C 8	antibacterial, antifungal
Palmarumycin C_5 (**48**)	(*16*)	*C.* sp.	170	–	+	–	C 8	not tested
Palmarumycin C_6 (**49**)	(*16*)	*C.* sp.	191–192	–	–	–	C 8	not active
Palmarumycin C_7 (**50**) mixture with **51**	(*16*)	*C.* sp.	–	–	–	–	C 8	not tested
Palmarumycin C_8 (**51**)	(*16*)	*C.* sp.	–	–	–	–	C 8	not tested
Palmarumycin C_9 (**52**)	(*16*)	*C.* sp.	–	–	–	+/+ (*43*)	C 8	antibacterial, antifungal
Palmarumycin C_{10} (**13**)	(*16*)	*C.* sp.	236	−48.2	–	+/+ (*43*)	C 3	antibacterial, antifungal
Palmarumycin C_{11} (**5**)	(*16*)	*C.* sp.	237–238	−153	–	–	S 1	weakly antibacterial, antifungal
Palmarumycin C_{12} (**53**)	(*16*)	*C.* sp.	207–208	−179.6	–	+/+ (*43*)	C 8	antibacterial, antifungal
Palmarumycin C_{13} (**9**) mixture with **8**	(*16*)	*C.* sp.	–	–	–	–	C 2	not tested
Palmarumycin C_{14} (**8**)	(*16*)	*C.* sp.	–	–	–	–	C 2	not tested
Palmarumycin C_{15} (**54**)	(*16*)	*C.* sp.	148–149	−18.1	–	–	C 8	antifungal
Palmarumycin C_{16} (**17**)	(*16*)	*C.* sp.	187–188	−43.3	–	–	C 4	not tested
Pre-palmarumycin (**75**)	(*18*)	*C.* sp.	220	+43	+	+/+		not tested
Palmarumycin C_{17} (**76**)	(*18*)	semisynthetic	178	+55°C	–	+/+		not tested

The first reported metabolites of this type were the diepoxins α (**7**), η (**8**), ζ (**9**), and σ (**10**) (Chart 2) (*24*). The major component of the antibiotic complex was diepoxin ζ (**9**). The structure determination relied mainly on HRMS and NMR spectra. Similarly as described for bipendensin (**5a**), two fragments related to **A** and **B** (Scheme 1) could be discerned from the two-dimensional NMR spectra, notably the HMBC spectra. Proton resonances at $\delta = 5.22$, 3.58, and 3.55 ppm bonded to carbon signals at $\delta = 61.7$, 54.6, and 55.5 ppm were typical for the presence of epoxide groups. The quaternary carbon atoms at $\delta = 63.7$ and 70.6 ppm formed the bridgeheads connecting the two fragments by oxygen atoms. The fragments can only be combined as shown in structure **9** and considering the fourteen elements of unsaturation. The relative configuration of **9** was finally established by X-ray structure analysis, revealing the *cis* orientation of the two epoxide rings and the hydroxyl group at C-4 and its *trans* configuration to the 5-OH. The other related structures **7**, **8**, and **10**, differing only in the degree of desaturation of ring B and the oxidation state of carbon C-4, were easily deduced from their differences in the spectral data based on the X-ray structure analysis of **9**. The diols **8** and **9** predominated in the early stages of the fermentation, whereas the abundance of the ketones **7** and **10** increased with time. Most of these metabolites showed antibacterial and antifungal activities, whereas diepoxin η (**8**) was virtually inactive. Diepoxin σ (**10**) was found to be identical with Sch 49209 as disclosed by Chu *et al.* at Schering-Plough (*23*, *27*); diepoxin η (**8**) was identical with palmarumycin C_{14} (*16*) and Sch 53516 (*25*); and diepoxin ζ (**9**) was identical with palmarumycin C_{13} (*16*), cladospirone bisepoxide (*26*, *30*) and Sch 53514 (*25*). The hydroxyl group of cladospirone bisepoxide isolated by the Ciba Geigy group in Basel, was initially placed at C-8 (*30*), but was later reassigned to C-5 to reflect the correct structure as shown in (**9**) (*26*). Further, the absolute configuration of cladospirone bisepoxide was proven indirectly to be structure **9** by X-ray analysis of the iodine derivative of **9** (*26*) (*vide infra*).

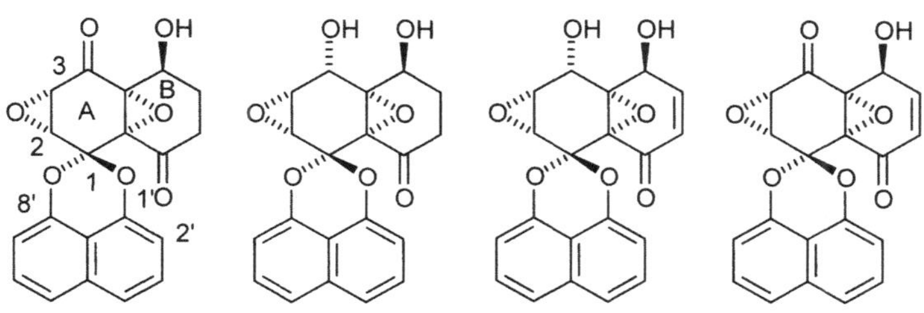

Diepoxin α (**7**)

Diepoxin η (**8**)
Palmarumycin C_{14}
Sch 53516

Diepoxin ζ (**9**)
Palmarumycin C_{13}
Sch 53514
Cladospirone bisepoxide

Diepoxin σ (**10**)
Sch 49209

Chart 2. Absolute structures of diepoxins α (**7**), η (**8**), ζ (**9**), and σ (**10**) and their synonyms

In a subsequent paper, the Lederle chemists were able to elucidate the absolute configuration of the diepoxins by employing the exciton coupled CD method on bis-dimethylaminobenzoate derivatives of the diepoxins η (**8**), ι (**14**), and κ (**15**) (*11*) (Chart 3). The bis-dimethyl-aminobenzoates (DMAB) were chosen as the chromophore because its red-shifted UV adsorption at $\lambda_{max} = 309$ nm would not couple with the naphthalene chromophore at $\lambda_{max} = 226$ nm. The (*S*)-configuration was assigned to both C-4 and C-5 on the basis of the **positive** exciton couplet between the two DMAB-chromophores in the corresponding bis-dimethylaminobenzoates **8a**, **14a**, and **15a** (Chart 3) of the diepoxins η (**8**), ι (**14**), and κ (**15**).

The *cis*- and *trans*-methoxylated products **14** and **15** were prepared by base-catalyzed addition of methanol to diepoxin ζ (**9**). Similarly, the epimeric 6-hydroxy-diepoxins γ (**11**) and δ (**12**) might be formed from **9** by addition of water (Chart 3). Diepoxin ϕ (**13**), with an enedione

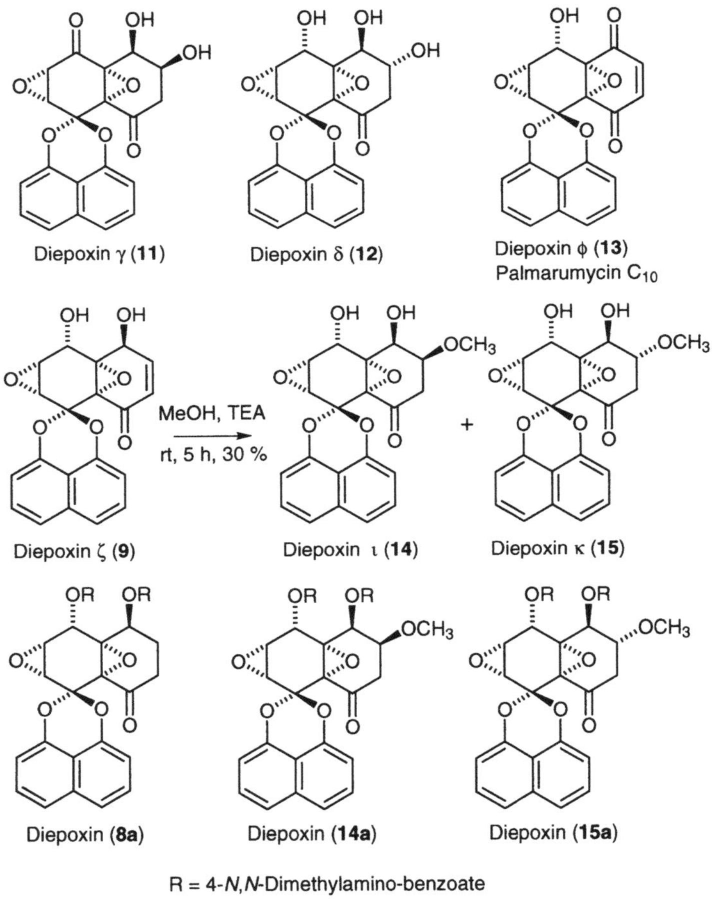

Chart 3. Structures of diepoxins γ (**11**), δ (**12**), ϕ (**13**), ι (**14**), and κ (**15**) and of the 4-*N*,*N*-dimethylaminobenzoates **8a**, **14a**, and **15a**

References, pp. 46–49

function in ring B, was also presented in this publication. It was isolated from the culture broth but also formed as an oxidative byproduct in the base-catalyzed addition of methanol to diepoxin ζ (**9**) in the presence of air. It is identical with palmarumycin C_{10} (*16*). This compound was also prepared by manganese dioxide oxidation of cladospirone bisepoxide (**9**) (*26*). Confirmation of the all *trans* relationship of the hydroxyl and methoxy groups in diepoxin κ (**15**) was obtained unambiguously by an X-ray structure analysis. This way, the *cis* orientation of the hydroxyl group at C-4 and the neighboring epoxy ring at C-2 and C-3 in diepoxin ζ (**9**) was also confirmed.

The first compound isolated from the culture broth of the fungus *Nattrassia mangiferae* (collected in Guatemala and provided by MYCOsearch) by Chu *et al.* at Schering Plough Research Institute was named Sch 49209 (*27*). The compound was found to be identical with diepoxin σ (**10**) (*24*), in particular in view of the similar specific optical rotation (see Table 1). Structure elucidation by extensive NMR experiments was mainly done on the corresponding acetate **10a**. Natural product **10** showed antitumor activity in the invasion chamber assay and was able to reduce the size of primary tumors and the number of metastases (*27, 34*). The Schering researchers were able to obtain crystals suitable for X-ray analysis of the triepoxy acetate Sch 50674 (**16**) prepared in two steps (epoxidation and acetylation) from the parent compound Sch 49209 (**10**), thus confirming the relative stereochemistry with *trans*-orientation of the epoxy and hydroxyl groups (Scheme 2).

The Schering group described the isolation of other related compounds in four subsequent papers. The data of the spirobisnaphthalene named Sch 53516 (*25*) show good agreement with those of diepoxin η (**8**) (*24*) and palmarumycin C_{14} (*16*). On the other hand, the data for Sch 53514 corresponded to those of diepoxin ζ (**9**) (*24*), palmarumycin C_{13}

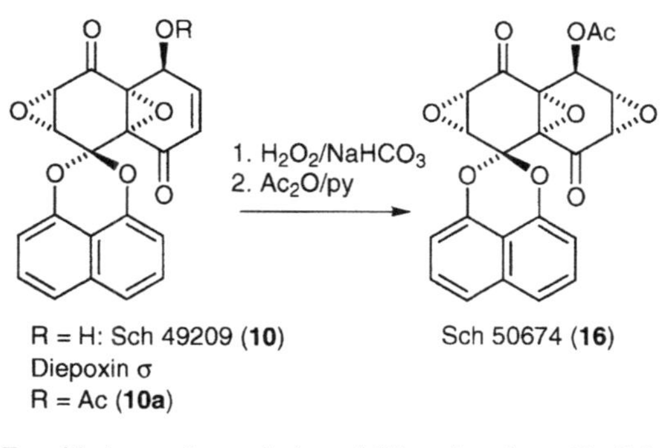

Scheme 2. Epoxidation and acetylation of **10** to the triepoxide Sch 50674 (**16**)

(*16*) and cladospirone bisepoxide (*26*). Their relative stereochemistry was supported by data from extensive NOESY experiments which independently arrived at the same result as the other authors for their respective compounds. Effective separation of the chromatographically very similar (saturated or unsaturated) derivatives **8** and **9** was accomplished using polyvinyl alcohol coated silica gel with methanol/1-chlorobutane as the solvent system.

The bisepoxydiol, Sch 49210 (**17**) (*25*) (Chart 4), a reduction product of the ketone **8**, was assigned the 5,8-*cis* configuration on the basis of a weak correlation of the diaxial 5,8-protons based on the NOESY NMR data. Palmarumycin C_{16} was assigned the same 5,8-diol structure without assignment of the configuration of 8-OH (*16*). The published spectroscopic data of the two compounds are similar, but the melting points of Sch 49210 (140–143°C) and palmarumycin C_{16} (187–188°C) deviate (for other data see Table 1). At the present time, the identity of the two compounds cannot be determined without direct comparison of the samples. It is interesting to note that the diacetates **8a** and **9b** also show antitumor activity in the invasion chamber assay. Particularly the diol **17** reduced primary and metastatic tumors. All compounds were inhibitors in the phospholipase D assay as well (*25*) (for a compilation of biological data see Chapter 3).

A structurally interesting hemiketal group was proposed on the basis of the spectroscopic data for Sch 50673 (**18**) (*23*). Signals for a methoxy group in the NMR spectra and resonances at $\delta = 94.10$ and 92.58 ppm indicated the presence of **two** ketal groups. The compound was crystallized from MeOH-CH_2Cl_2 (1:1) and is probably formed during crystallization by addition of methanol to the carbonyl group of palmarumycin C_9 (**52**). Similar observations were made in our laboratory and by Schlingmann (*36*) in that the water is added to the molecule to form the corresponding hydrate, first observed during NMR measurements of palmarumycin C_9 (see below). The normally unstable hemiketal in **18** may be stabilized by the neighboring carbonyl and epoxy groups.

Another compound, named Sch 50676 (m.p. 235–240°C), was also isolated from *N. mangiferae*. Structure **19** was proposed for this phenol. However, the excellent agreement of all physical and spectroscopic data, notably m.p. and specific optical rotation (see Table 1) with palmarumycin C_{11} (**5**) (m.p. 237–238°C), suggests that these compounds are identical. Palmarumycin C_{11} (**5**) was oxidized to palmarumycin C_2 (**6**) (see Scheme 1) and the structure of this metabolite was determined by X-ray single crystal analysis (*16*), proving the location of the phenol group in palmarumycin C_{11} (**5**). In addition, the structure of palmarumycin C_{11} (**5**) has been confirmed by total synthesis (*20*). Thus, the structure of Sch 50676 has to be revised and the phenolic group replaced from C-8 in **19** to C-5 as shown in **5** (Chart 4).

Further studies on the culture broth of *N. mangiferae* fermentations exposed two further spirobisnaphthalene derivatives, named Sch 49211 (**20**) and Sch 49212 (**21**) (*28*). Their structures are closely related to the diol **17** with rearrangement of the external epoxide to an enol structure. The regiochemistry of the enol was determined by SINEPT experiments and the *cis* stereochemisty of the diols was proven by a NOESY correlation. Both compounds were shown to be potent phospholipase D inhibitors in the fMLP-stimulated PLD assay and showed antitumor activity in the tumor cell invasion assay (IC50 were 2.8 and 6.1 M, respectively) (*23*).

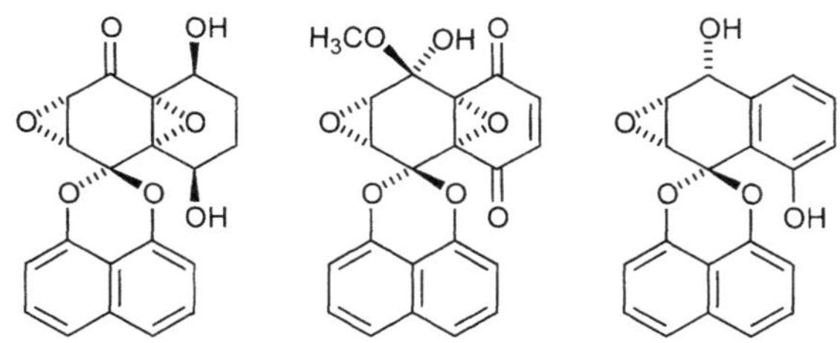

Sch 49210 (**17**) Palmarumycin C_{16} ? Sch 50673 (**18**) Sch 50676 (**19**) ?

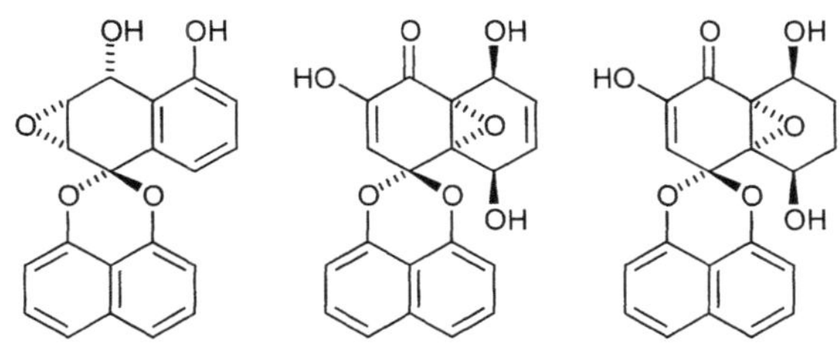

Palmarumycin C_{11} (**5**) Sch 50676 Sch 49211 (**20**) Sch 49212 (**21**)

Chart 4. Structures of Sch 49210 (**17**), Sch 50673 (**18**), Sch 50676 (**19**), Sch 49211 (**20**), and Sch 49212 (**21**)

Chemists at Pfizer Inc. discovered a DNA gyrase inhibitory effect for two new spirobisnaphthalenes, which they named CJ-12,371 (**22**) and CJ-12,372 (**23**) (*29*) (Chart 5). These compounds were isolated from the culture broth of an unidentified fungus provided by MYCOsearch Laboratory (USA). Their structures are closely related to palmarumycins

C_{11} and C_{12}, missing the epoxy oxygen (*vide infra*). Their 4(*S*) absolute configuration (see Chart 5) was determined by applying the CD exiton chirality method utilizing the corresponding bis-4-*N,N*-dimethylamino benzoates, similarly to that described by Schlingmann *et al.* (*11*). To improve on the gyrase inhibition, a number of derivatives (ethyl and methyl ethers, acetates, 4-methoxybenzoates) were prepared, but the selectivity towards the inhibition of topoisomerase II-mediated relaxation could not be increased.

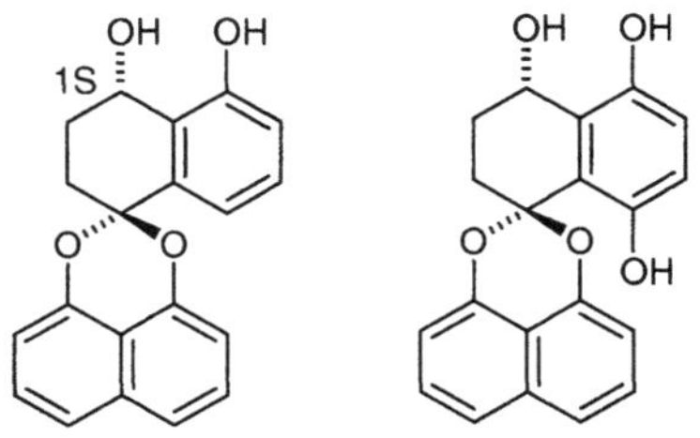

CJ-12,371 (**22**) CJ-12,372 (**23**)

Chart 5. Structures of CJ-12,371 (**22**) and CJ-12,372 (**23**)

Chemists at Ciba Geigy investigated the biologically active constituents from a saprophytic fungus isolated from a bed of needles collected in a forest of Montezuma pines about 200 km north of Mexico City. The fungus was initially classified as a *Cladosporium* strain but later reassigned to the *Sphaeropsidales* group (*31*). The spirobisnaphthalene isolated from the culture broth was named cladospirone bisepoxide and assigned structure **24** (*30*) (Scheme 3). The compound was oxidized with manganese dioxide to 4-oxo-cladospirone bisepoxide (**25**). Its structure was determined by X-ray single crystal analysis. This proved part of structure **24** but could not resolve the location of the original hydroxyl group. Later on, the authors performed an X-ray crystal analysis of the natural product and the structure was reassigned to **9**, identical with diepoxin ζ, Sch 49209 and palmarumycin C_{13}. Its oxidation product **25** was also isolated as a natural product by our group and named palmarumycin C_{10} (*16*) (see Table 1). Biological tests of cladospirone bisepoxide (**9**) revealed a wide spectrum of biological activities against bacteria, fungi, and yeasts. The compound also inhibited the germination of seeds.

The chemical reactivity of **9** was exploited by various authors in a number of chemical transformations (*26*) (Scheme 3). Thus, hydrogenation of the external double bond yielded compound **8**, identical with diepoxin η. Reaction with trimethylsilyl azide resulted in conjugate addition of azide to the double bond at C-7 to produce the azide **26**

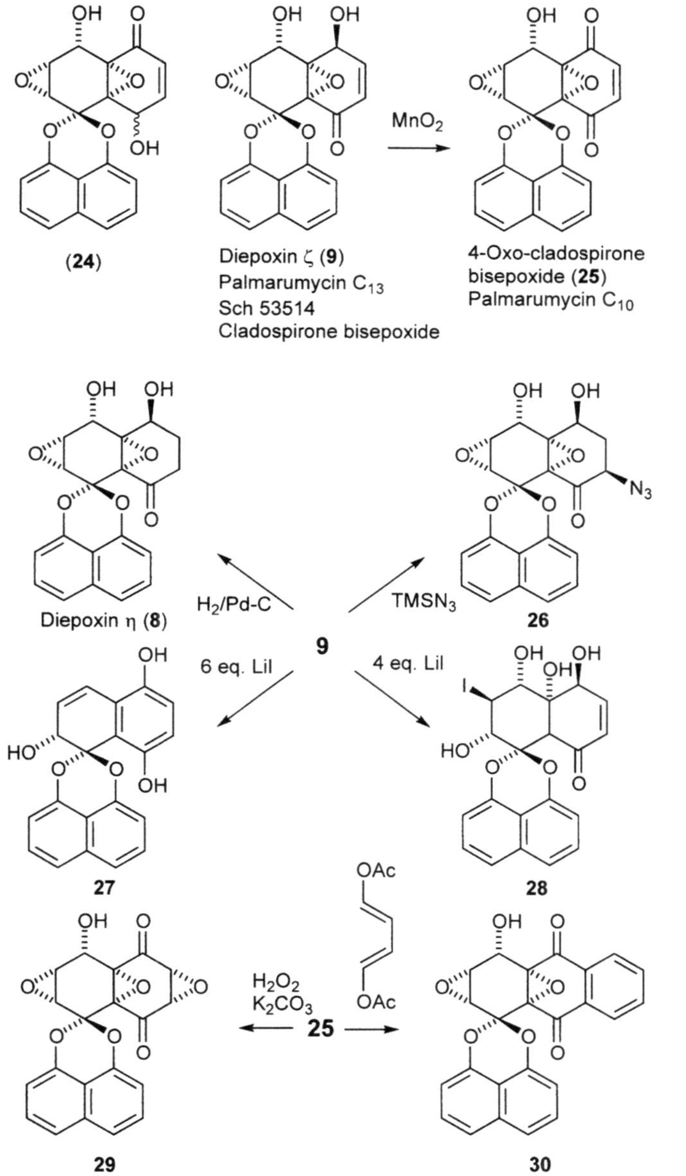

Scheme 3. Structure of cladospirone epoxide (**9**) and transformation to derivatives **25–30**

stereoselectively. Reaction with lithium iodide depended of the amount of reagent used to yield either the allylic alcohol **27** or the iodide **28**. The oxidation product, 4-oxo-cladospirone bisepoxide **25**, could be epoxidized under basic conditions at the electron-deficient enedione double bond to form the triepoxide **29**. Addition to this double bond was also used in a Diels-Alder reaction with 1,4-diacetoxy-butadiene to afford the anthraquinone derivative **30**.

With consideration of the interesting biological activities, Zeeck and coworkers reinvestigated the formation of secondary metabolites from the cladospirone bisepoxide producing fungus *Sphaeropsidales* sp. (strain F-24′707) under a variety of culture conditions. Following the OSMAC method (one strain/many compounds) (*37*), they used different culture media, vessels, and general culture conditions and were able to isolate fifteen compounds. Eight products were shown to be new and named cladospirones B to I (**31**–**38**) (Chart 6). The known compounds isolated were (in addition to cladospirone bisepoxide) diepoxin ζ (**9**), diepoxins η (**8**), σ (**10**) and δ (**12**) and palmarumycins C_2, C_3, and C_{12} (Chart 8). The eight new compounds showed great variation in their functionalities at rings A and B, resulting in compounds with opened rings, reductions, or rearrangement of the reactive epoxide groups. "The spirobisnaphthalenes may be considered the most diverse secondary metabolites, allowing for the small number of carbon atoms available to host theses variations" was stated by Zeeck *et al.* (*9*). The production of cladospirone bisepoxide predominates under oxygen-rich conditions with long incubation times, whereas palmarumycin C_3 is produced in high quantities under oxygen-limited conditions existing under static surface culture. Particularly the cladospirones C (**32**) and D (**33**) showed good antibacterial activity in the agar diffusion assay against *Escherichia coli*, *Staphylococcus aureus*, and *Bacillus subtilis*. They are not active against certain fungi and yeasts such as *Candida albicans* or *Mucor hiemalis*, but **33** shows activity against *Chlorella vulgaris* and *Chlorella sorokiniana*. An interesting rearrangement with opening of the epoxide ring of cladospirone G (**36**) to the bridged derivative **39** occurred upon treatment with acetic acid anhydride.

Coniothyrium species also proved to be a particularly rich source of structurally diverse spirobisnaphthalenes. Six of these interesting compounds were isolated from *Coniothyrium palmarum*, an endophytic fungus isolated from *Lamnium purpureum* (*12*, *33*). The genus *Coniothyrium* belongs to the Deuteromycota (Fungi imperfecti), class Coelomycetes, order Sphaeropsidales, family Sphaeropsidaceae; synonyms for these names are *Clipsosporium* and *Microsphaeropsis*, respectively. According to their origin, they were named palmarumycins and their observed

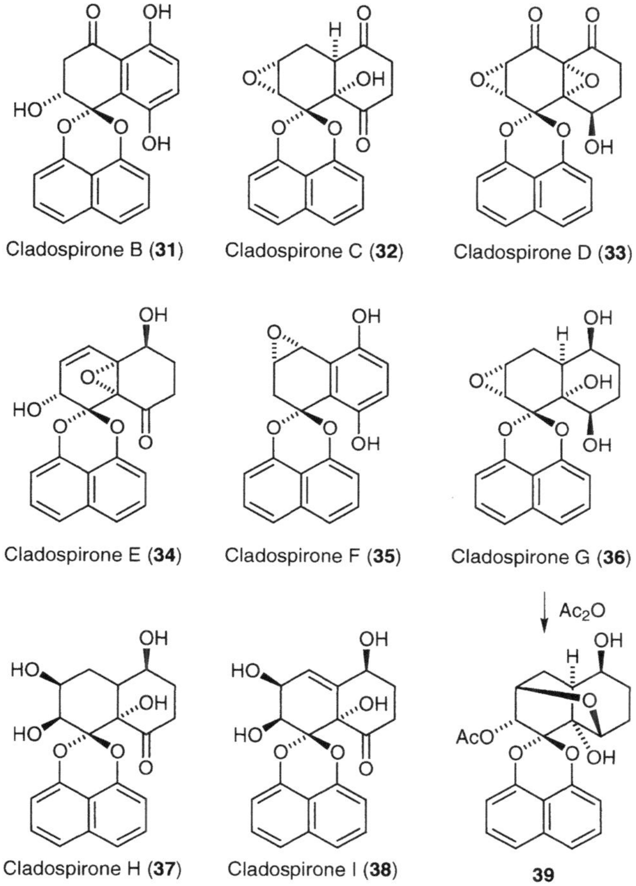

Chart 6. Structures of cladospirones B to I (**31–38**) (*37*)

polarity determined the index, e.g. palmarumycin CP_1 (**2**) being the least polar, and CP_5 (**44**) the most polar compound (see Table 1). The structure elucidation relied on a combination of spectroscopic methods and – in the case of CP_3, CP_{4a}, and CP_5 – also on X-ray structure analysis. Palmarumycin CP_1 (**2**) (Chart 1) is probably the biosynthetic parent of all the spirobisnaphthalene antibiotics (*vide infra*) including CP_2 (**40**) (Chart 7), the corresponding hydrogenation product. Palmarumycin CP_3 (**41**) (Chart 7) has an interesting additional oxygen bridge, connecting C-2 with C-8. CP_4 (**42**) can be regarded as an open chain precursor of **41**. Noteworthy is that in contrast to most of the other spirobisnaphthalenes, none of the isolated compounds from cultures of

Coniothyrium palmarum carried epoxide functions. The absolute configurations of **2**, **41**, **43**, and **44** were determined by quantum chemical calculations of the CD spectra (*vide infra*) (*22*, *33*).

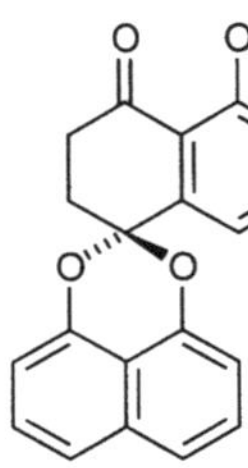

Palmarumycin CP_2 (**40**)
Deoxypreussomerin B

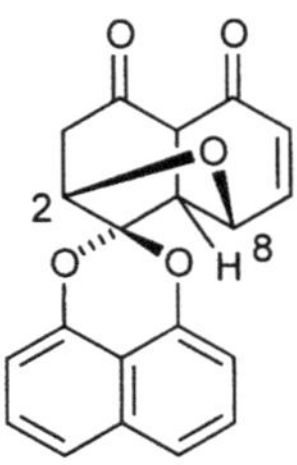

Palmarumycin CP_3 (**41**)

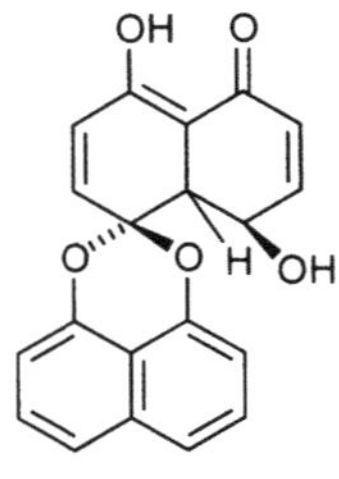

Palmarumycin CP_4 (**42**)

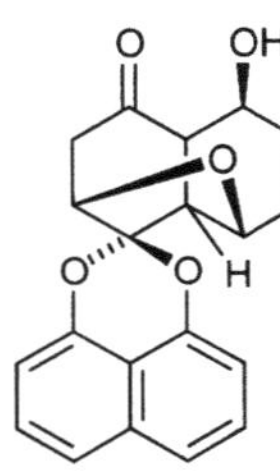

Palmarumycin CP_{4a} (**43**)

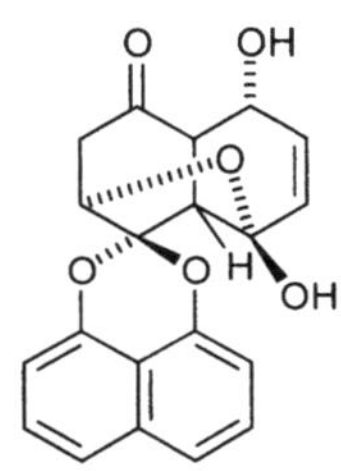

Palmarumycin CP_5 (**44**)

Chart 7. Structures of palmarumycins CP from *Coniothyrium palmarum* (*12*, *33*)

An unidentified *Coniothyrium* species from soil from a forest on West Borneo produced a large variety of secondary metabolites in submerged fermentation culture in contrast to *Coniothyrium palmarum* that only produced active metabolites when grown on semisolid agar (*16*). Sixteen compounds, named palmarumycins C_1–C_{16}, (**5**, **6**, **8**, **9**, **13**, and **45–54**) were structurally identified from the culture broth of this fungus (Chart 8). The compounds show antibacterial, antifungal and herbicidal activity at concentrations of 10^{-6}–10^{-4} mol/l. The structures of palmarumycin C_2 (**6**) (Scheme 1), palmarumycin C_3 (**46**), and palmarumycin C_5 (**48**) (Chart 8) were confirmed by X-ray structure analysis.

As found for Sch 53825 (**5b**), isolated by Chu *et al.* (*21*), four products, the palmarumycins C_1 (**45**), C_4 (**47**), C_7 (**50**), and C_8 (**51**) contained a chlorine atom. The mechanism by which these compounds are formed is most likely due to a nucleophilic attack of a chloride on the external epoxide of the respective precursor. This hypothesis was tested by a chemical experiment in which palmarumycin C_9 (**52**) was converted to palmarumycin C_4 (**47**) by treatment with hydrochloric acid, resulting

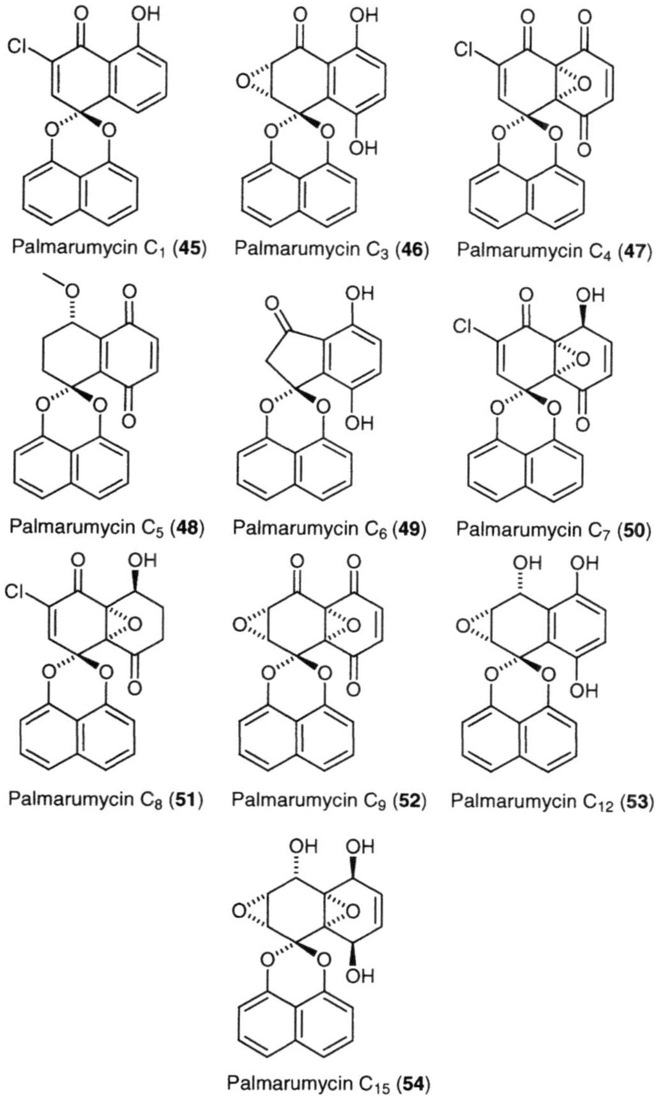

Chart 8. Palmarumycins C_1–C_{16}, (**45–54**) (*16*)

in the addition of chlorine and elimination of water. When palmarumycin C_2 (**6**) was treated with hydrochloric acid, the intermediate **55** could be isolated, but in chloroform solution slowly decomposed to form palmarumycin C_1 (**45**) (Scheme 4) (*16*).

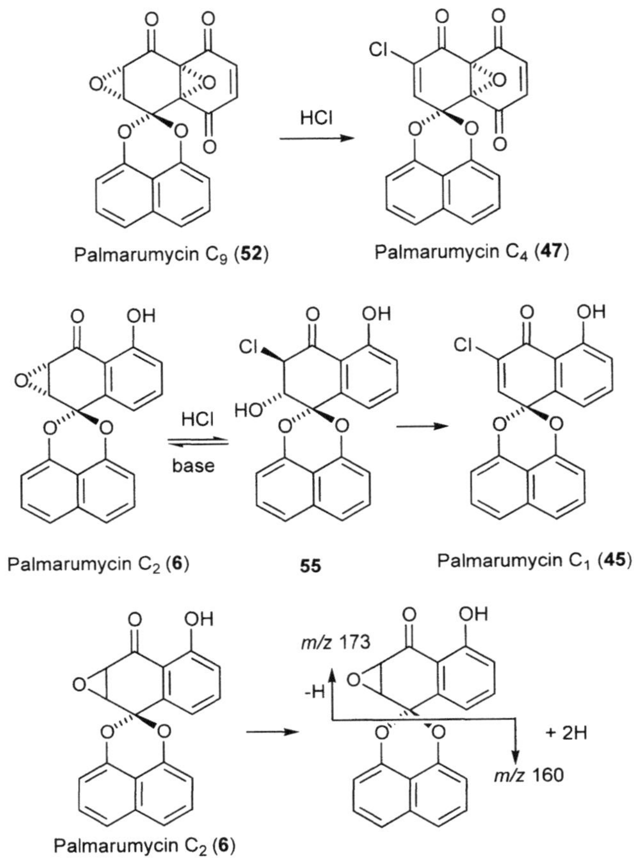

Scheme 4. Reaction of palmarumycins C_2 (**6**) and C_9 (**52**) with HCl and noted fragments in the mass spectra

Singh *et al.* isolated a number of new preussomerins (see below) as well as the palmarumycins C_2 (**6**) and CP_2 (**40**), named deoxypreussomerin A and B, from a culture of a dung inhabiting coelomycetous fungus isolated at Chaco Province, Argentina (*32*). The production of both types of spirobisnaphthalenes by the same fungus is an indication for the close biosynthetic relationship of these compounds. The author drew special attention to a common fragmentation pattern of all spirobisnaphthalenes as exemplified by compound **6**, where typical fragments in the mass spectrum were observed at $m/z = 160$ and 173 (Scheme 4).

The structures of the other palmarumycins (see Chart 8) reflected the previously observed oxidative transformation of palmarumycin C_1

(monophenol **45**) to palmarumycin C_3 (*p*-bisphenol **46**), and further oxidation (or epoxidation) of the hydroquinone to the quinone found in the palmarumycins C_4 (**47**), C_5 (**48**), and C_9 (**52**). If naphthoquinones that still contain a double bond, as is the case with palmarumycin C_9, are epoxidized, reduction of either of the carbonyl groups and/or the external double bond can occur, as exemplified by the palmarumycins C_7 (**50**), C_8 (**51**), C_{13} (**9**), C_{14} (**8**), and C_{15} (**54**). Interestingly, the hydroxyl group in ring B is orientated *trans* to the remaining epoxide (at positions 4a/8a, numbering see **2**, Chart 1). This was confirmed by X-ray structure analysis of diepoxin κ (**15**) and couplings observed in NMR spectra. In many cases, as in the palmarumycins C_5, C_{10}–C_{16}, and in most of the diepoxins, the carbonyl group at C-4 is stereoselectively reduced *cis* to the remaining epoxide rings. In palmarumycin C_5 (**48**) a methoxy group replaced the usual hydroxy group which was confirmed by X-ray structure analysis (*16*). An unusual feature is the ring contraction of the six-membered ring A to a five-membered ring as found in palmarumycin C_6 (**49**).

2.2. Spirobisnaphthalenes with Three Oxygen Bridges (Preussomerins)

Preussomerins A–F (**3**, **56**–**61**) were first isolated by Gloer *et al.* from cultures of the coprophilous (dung-colonizing) fungus *Preussia isomera* Cain. (*13*, *38*). The isolation of these metabolites was guided by bioassays for antifungal activity toward the early successional coprophilous fungus *Sordaria fimicola*. The six isolated preussomerins (A-F) showed pronounced antifungal activity toward early-successional coprophilous fungi, thus supporting the theory of an interspecies competition among coprophilous fungi. Preussomerin D (**59**) was subsequently also isolated from the endophyte *Hormonema dematioides* (*39*).

The structures of the preussomerins (see Chart 9 and Table 2) differ from those of the previously described spirobisnaphthalenes in that they carry an additional oxygen atom (**three** instead of **two**) in the bridge that links the two naphthalene units. Furthermore, the two naphthalene units are both partially hydrogenated and the "bottom right" ring bears an additional oxygen. Similar to the palmarumycins, the double bonds are here often epoxidized, the epoxides then reductively opened and/or the carbonyl groups reduced to hydroxyl groups. The relative stereochemistry of preussomerin A (**3**) is based on single crystal X-ray analysis (*38*) and its absolute stereochemistry deduced from acid catalyzed degradation to afford (–)-regiolone.

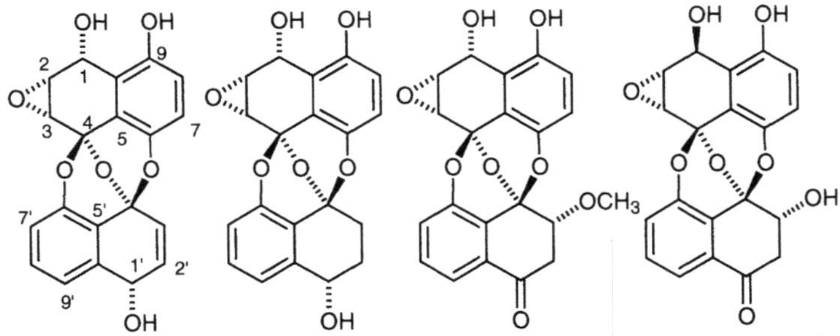

Preussomerin A (**3**) Preussomerin B (**56**) Preussomerin C (**57**) 3'-*O*-Desmethyl-1-epi-preussomerin C (**58**)

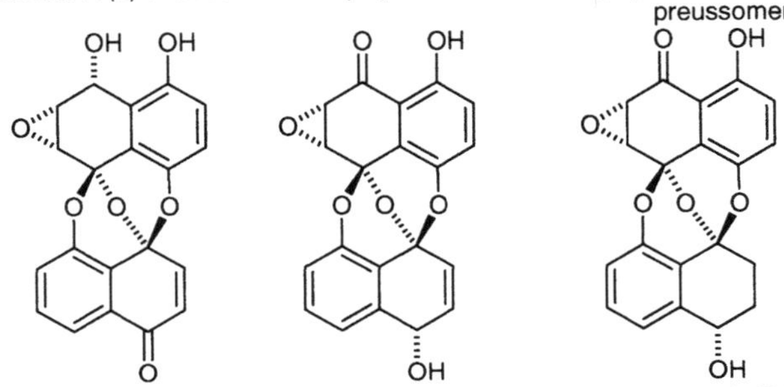

Preussomerin D (**59**) Preussomerin E (**60**) Preussomerin F (**61**)

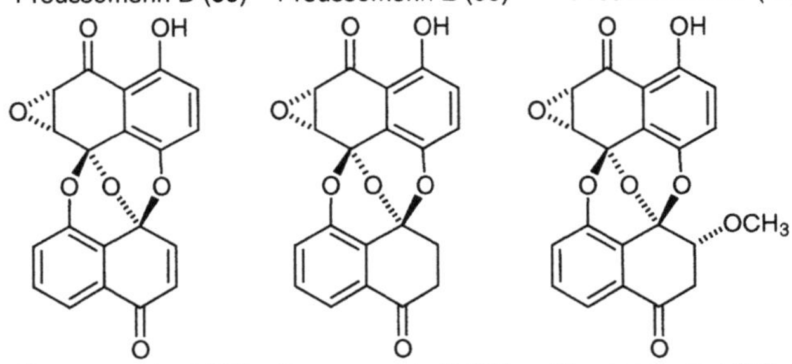

Preussomerin G (**62**) Preussomerin H (**63**) Preussomerin I (**64**)

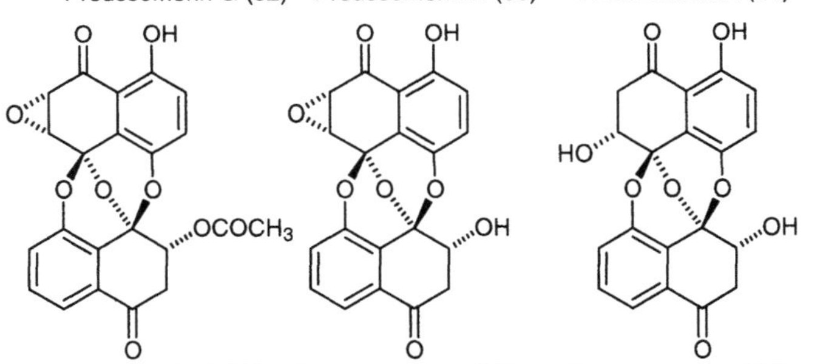

Preussomerin J (**65**) Preussomerin K (**66**) Preussomerin L (**67**)

Chart 9. Structures of preussomerins A–L (**3** and **56–67**)

Table 2. List of Preussomerins (Chart 9). Source (organism), melting point, specific optical rotation, absolute configuration, structural confirmation by X-ray analysis, and biological activity

Name	Ref.	Source	*M.p.* °C	$[\alpha]_D$	X-ray	CD/abs. conf.	Bioactivity
Preussomerin A (**3**)	*(13, 38)*	*Preussia isomera* Cain	235–240	−212	+	+	antifungal
Preussomerin A acetate	*(38)*	semisynthetic	–	–			
Preussomerin B (**56**)	*(38)*	*P. isomera* Cain	–	−242	–	–	antifungal
Preussomerin C (**57**)	*(38)*	*P. isomera* Cain	–	−155	–	–	antifungal
3′-*O*-Desmethyl-1-epipreussomerin C (**58**)	*(38)*	*P. isomera* Cain	–	−155	–	–	
Preussomerin D (**59**)	*(38)*	*P. isomera* Cain,	–	−144	–	–	antifungal
Preussomerin D (**59**)	*(32)*	coelomycetous f.					
Preussomerin E (**60**)	*(38)*	*P. isomera* Cain	–	−244	–	–	antifungal
Preussomerin F (**61**)	*(38)*	*P. isomera* Cain	–	−240	–	–	antifungal
Preussomerin G (**62**)	*(32)*	coelomycetous f.	222–225	−688	+	–	
Preussomerin G (**62**)	*(41)*	*Mycelia sterila*	222–225			+	
Preussomerin H (**63**)	*(32)*	coelomycetous f.	–	−371	–	+	
Preussomerin H (**63**)	*(41)*	*Mycelia sterila*				+	
Preussomerin I (**64**)	*(41)*	*Mycelia sterila*	130–132			+	
Preussomerin J (**65**)	*(41)*	*Mycelia sterila*	oil	−218	–	+	
Preussomerin K (**66**)	*(41)*	*Mycelia sterila*	oil	−150	–	+	
Preussomerin L (**67**)	*(41)*	*Mycelia sterila*	171–173	−557	+	+	

The stereochemistry of the other members of the preussomerin family were assigned accordingly as shown in Chart 9, with respect to their NMR spectral data and negative specific optical rotation. The putative intermediate **3a** resulting from the degradation of **3** shows chemical and stereochemical congruency to the palmarumycins, as exemplified by palmarumycin C_{10} (**13**) (Scheme 5). However, there

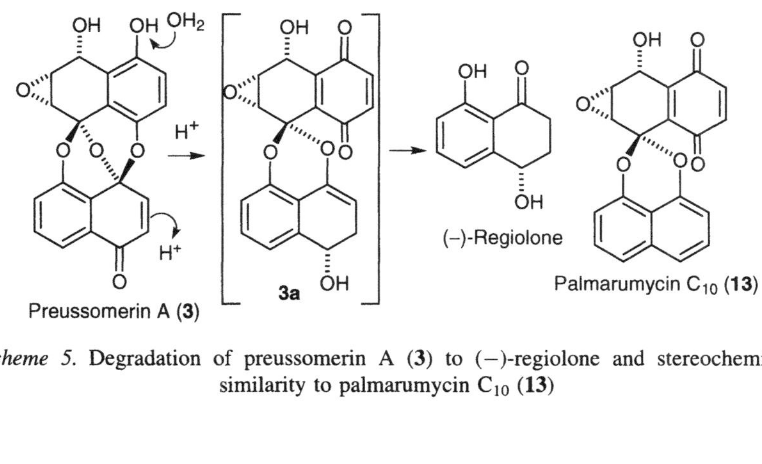

Scheme 5. Degradation of preussomerin A (**3**) to (−)-regiolone and stereochemical similarity to palmarumycin C_{10} (**13**)

appear to be exceptions to this rule, as a compound isolated from the coprophilous fungus *Sporomiella vexans* was assigned the C-1-epimeric structure 3′-*O*-desmethyl-1-epipreussomerin C (**58**) (*40*).

The biosynthetic bridge between the palmarumycins and the preussomerins was corroborated by the isolation of the preussomerins D (**59**) and G–I (**62–64**) and the palmarumycins C_2 (**6**) (deoxypreussomerin A) and CP_2 (**40**) (deoxypreussomerin B) from one culture. Singh *et al.* at Merck Laboratories isolated these compounds from the culture broth of a dung-inhabiting coelomycetous fungus (*32*). This group also made the very important discovery that preussomerin G (**62**) inhibits farnesyl-protein transferase (FPTase). FPTase inhibitors are potential agents against cancers with mutated *ras* gene such as colon or pancreatic carcinomas. The relative stereochemistry of preussomerin G (**62**) was confirmed by X-ray structure analysis (*32*).

Very recently, three of the known preussomerins, namely G (**62**), H (**63**), and I (**64**) and three new representatives, J (**65**), K (**66**), and L (**67**) were isolated by our group from an unknown endophytic fungus that was isolated following surface sterilization from the root of *Atropa belladonna* (*41*). Since the fungus grew only as a *mycelium sterilum*, it could not be characterized taxonomically. The culture extract of the fungus was found to have antibacterial and antifungal properties due to the presence of preussomerins of which preussomerin G (**62**) is the most active antifungal. The relative configuration of the most polar metabolite in this series, preussomerin L (**67**), was elucidated by X-ray analysis and its absolute configuration was determined by comparison of calculated and experimental CD spectra (*vide infra*) (*41*). Preussomerin L (**67**) is presumed to be derived from a putative bisepoxide via reduction, that itself, however, has not yet been isolated as a natural product.

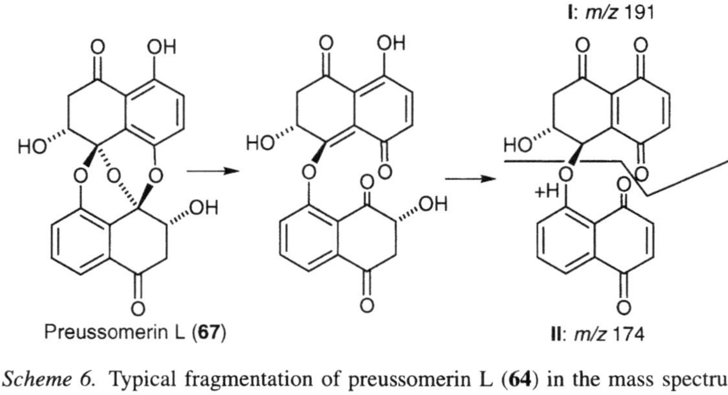

Scheme 6. Typical fragmentation of preussomerin L (**64**) in the mass spectrum

Similar to the spirobisnaphthalenes with two oxygen bridges, the preussomerins also exhibit a typical fragmentation pattern as shown in Scheme 6 for preussomerin L (**67**) the mass spectra thus give an additional indication for the substituents on the top or bottom fragments (*41*).

2.3. Spirobisnaphthalenes with Two Oxygen Bridges and One C-C Bridge (Spiroxins)

Also very recently, a new series of natural products with bisnaphthospiroketal structure was isolated from a fungal strain, found in a soft coral collected from the waters of Dixon Bay, Vancouver Island (*10*). The culture was not sporulating and, therefore, the strain could not be classified taxonomically. The compounds were named spiroxins A–E (**68**)–(**72**) (Chart 10, Table 3) and contain two oxygen and one carbon-carbon linkage in the bridge region between the two naphthalene units. Similar to the diepoxins, the double bonds are all epoxidized. The phenolic aromatic rings are in part chlorinated and the carbonyl groups in part reduced. In the short communication (*10*) no melting points were

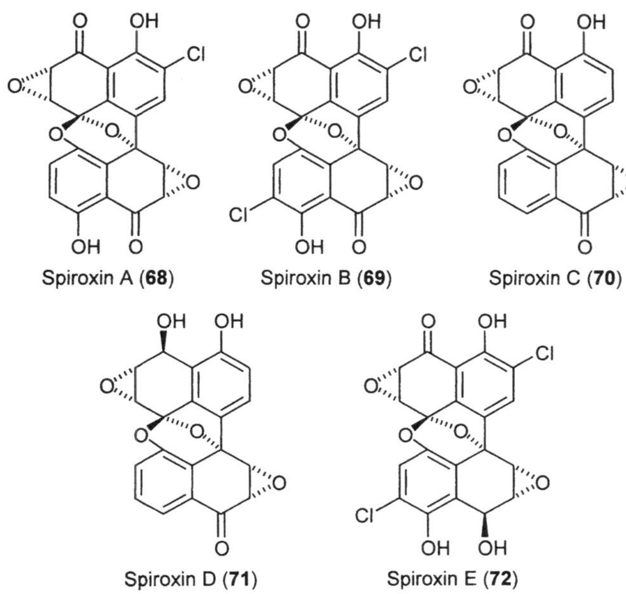

Chart 10. Structures of spiroxins A–E (**68**)–(**72**) (*10*)

Table 3. List of Spiroxins (**68–72**) (Chart 10). Source, specific optical rotation, and biological activity

Name	Ref.	Source	$[\alpha]_D$	Bioactivity
Spiroxin A (**68**)	(*10*)	marine endophyte	−644	antibacterial, DNA cleaving
Spiroxin B (**69**)	(*10*)	marine endophyte	−475	antibacterial, DNA cleaving
Spiroxin C (**70**)	(*10*)	marine endophyte	−706	antibacterial, DNA cleaving
Spiroxin D (**71**)	(*10*)	marine endophyte		antibacterial, DNA cleaving
Spiroxin E (**72**)	(*10*)	marine endophyte		antibacterial, DNA cleaving

reported and also only one HRMS for spiroxin A (**68**) (Table 3). Based on weak NOE correlations between H-2′ and H-3′ with H-7, the authors suggested the location of the hydroxyl groups on the upper side and the epoxides on the lower ring assembly as shown in Chart 10 in **71** and **72**. The spiroxins showed marked cytotoxic effects through a concentration dependent single-stranded DNA cleavage.

2.4. Determination of Relative and Absolute Stereochemistry

NMR data, notably the coupling constants of relevant proton-proton interactions, as well as NOESY or ROESY correlations, are useful tools for the determination of relative stereochemistry. To solve stereochemical questions encountered in the series of spirobisnaphthalene antibiotics, NMR considerations have found only limited application due to the lack of connectivities. In particular, the relative stereochemistry of the epoxide groups at C2–C3, C4a–C8a, and C6–C7 with respect to hydroxyl groups at C-4 or C-5 are not readily determined from coupling constants. Very often, NOESY correlations are also not unambiguous (*38*). However, X-ray analysis (see Tables 1–3) of quite a number of the spirobisnaphthalene antibiotics served as "relays" to determine the relative stereochemistry also for most of the other compounds by comparison of the relevant NMR data. In some cases, chemical correlations helped to bridge the gap to compounds for which the X-ray structure had been determined. This could not be done in all cases, and the structure of Sch 50676 (**19**) (Chart 4) and Sch 53823 (**5a**) (Scheme 1) have to be revised (see above).

Some examples may illustrate here how correlating X-ray structure determination with NMR observations can be used to elucidate relative and absolute stereochemistry. X-ray structure analysis of Sch 50674 (**16**) (Scheme 2), a triepoxy-acetate derived from Sch 49209 (**10**), revealed the *cis*-orientation of all three epoxy groups and thus also established the *cis* orientation of epoxy groups in the natural product **10** (Sch 49209 and diepoxin σ). With respect to the very similar patterns in the NMR spectra, this same configuration was therefore indicated for the other diepoxides as well. Similar results were obtained from the X-ray structure determination of cladospirone bisepoxide (**9**) (*31*) and the corresponding 4-oxo-compound (**25**) (*30*).

The relative configuration of preussomerin A (**3**) (see Chart 1) was known from X-ray analysis and the absolute configuration was correlated by degradation to (–)-regiolone as shown in Scheme 5. The absolute configurations of preussomerins B–F (**56–61**) were assumed to be analogous to that of (**3**) on the basis of the NMR, structural similarities and similar CD spectra, and PCC oxidation of preussomerins A (**3**) and D (**59**) to the dione (**62**). This dione was later isolated as a natural product and named preussomerin G (**62**) (*38*). The preussomerins were stereochemically correlated with the palmarumycins by degrading **3** to (–)-regiolone via the putative intermediate **3a** that ought to have the same stereochemistry at C3 and C4 as palmarumycin C_{10} (**3**) (Scheme 5) (*16*).

In two cases, the absolute configuration was determined directly from X-ray analysis. The first case was palmarumycin CP_3 where the refined data allowed the assignment of the absolute configuration shown in **41** (Chart 7), with the oxygen bridging the rings A and B at the front of the molecule (*12*). It is noteworthy that the oxygen bridge in palmarumycin CP_5 (**44**) is located on the back side of the molecule. The second case was the iodinated compound **28** (Scheme 3) derived from cladospirone bisepoxide (**9**) (*26*). Since this compound (**28**) has nearly an identical specific optical rotation (+67° versus +75°) as that recorded for diepoxin ζ, it is reasonable to assume that these compounds, although isolated from two different sources, have the same absolute configuration as shown in structure **9** (Chart 2). The absolute configuration of all diepoxins was confirmed independently by recording CD spectra of the bis-dimethylaminobenzoate derivatives of the diepoxins η (**8**), ι (**14**), and κ (**15**) (*11*). In all cases, the CD spectra revealed a positive twist between the two hydroxyl groups at C-4 and C-5 and thus inferred the (*S*) configuration at both of these stereogenic centers. The absolute configuration of CJ-12,371 (**22**) and CJ-12,372 (**23**) (Chart 5) was established in the same manner (*29*).

Quantum chemical calculations of CD spectra, in collaboration with G. Bringmann (Würzburg), served to determine the absolute configurations of the palmarumycins C_2 (**6**), CP_3 (**41**) (*22*), CP_{4a} (**43**), CP_5 (**44**) (*33*), the semisynthetic C_{17} (**76**) (*18*), and the preussomerins G–L (**62–67**) (*41*). The procedure is briefly explained using the example of preussomerin L (**67**) (Chart 9). The relative configuration of this metabolite was established by X-ray crystal analysis. Next, the con-

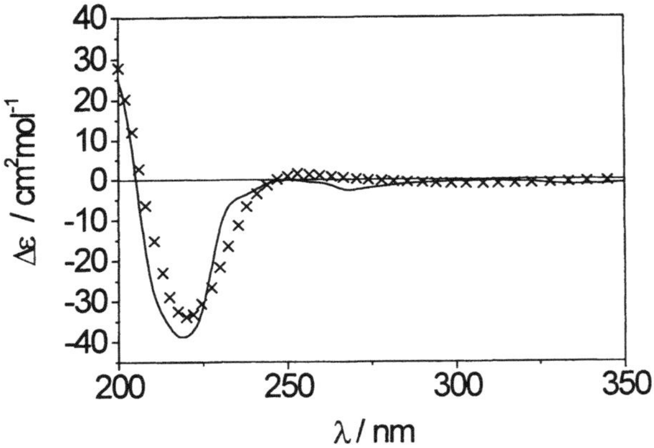

Fig. 1. Experimental CD spectrum (CH_3OH) and calculated spectrum (crosses) of structure **67** of preussomerin L

formation of the molecule was calculated using semiempirical methods. Not surprisingly, the calculated minimum energy conformation matched the X-ray structure very well because the bridged and rigid preussomerins have very little conformational freedom. The existence of a limited number of relatively well defined conformations is in fact a prerequisite to make use of quantum chemical calculations of CD spectra. In the next step, the CD spectrum is calculated (see Figure 1) on the basis of the computed conformation using the program package BDZDO/MCDPPD (*42*). Finally, the experimental CD spectrum is compared with the calculated one and the absolute configuration is

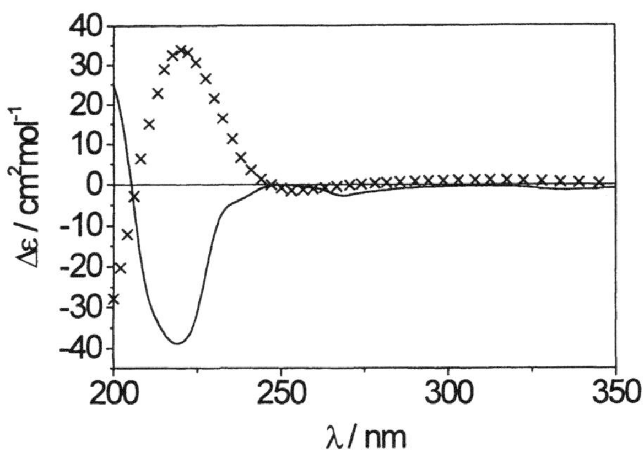

Fig. 2. Experimental CD spectrum (CH_3OH) and calculated spectrum (crosses) of structure *ent*-**67**

assigned to that enantiomer with the matching spectrum (see Figure 1). The CD spectra of the two enantiomers are mirror images (Figure 2) and an unambiguous decision can usually be made to determine the absolute configuration. In the case of the palmarumycins, the conformational flexibility was greater and the Boltzmann-weighted addition of the respective calculated CD spectra have to be taken for the comparison with the experimental spectra (*33*). In the meantime, additional quantum chemical calculations of CD spectra of the palmarumycins C_9 (**52**), C_{10} (**13**), and C_{12} (**53**) and comparison with their experimental spectra were performed confirming the absolute configurations depicted in Charts 7 and 8 (*43*).

3. Biological Activity

The biological activities have been mentioned together with the discussions of the compounds and are also listed in Tables 1–3. This chapter gives a brief summary. MK 3018, the first spirobisnaphthalene mentioned in the patent literature (*14*), showed a broad spectrum of antibacterial activity. In general, antibacterial and antifungal activity as well as germination inhibition of *Lepidum sativum* have been observed for most of the palmarumycins (*12*, *16*) (Chart 7), cladospirone epoxide (**9**) (*31*), and the diepoxins (*24*). Interestingly, this activity was completely missing for palmarumycins CP_2 and C_6, as well as the cladospirones B (**31**), and E–I (**34–38**) having neither an epoxide ring nor an α,β-unsaturated ketone as potential electrophilic reaction sites. It is also noteworthy that the other three compounds with reduced antibacterial and antifungal activity, the palmarumycins C_{11}, CP_{15}, and CP_{16} have epoxide rings, but lack the carbonyl groups. Similar results were reported for diepoxin η (**8**) (*24*). It is thus conceivable that the general antibacterial (and to a great extent also antifungal) activity is due to the presence of electrophilic groups such as Michael acceptors and to a lesser extent to oxirane rings in the molecule.

Chu *et al.* at Schering-Plough first observed the antitumor activity in the tumor cell invasion assay of their spirobisnaphthalenes isolated from *Nattrassia mangiferae* (*23*). Part of this effect may be attributed to the phospholipase D inhibitory activity (*21*, *25*, *28*). Antitumor activity was also observed in the epoxide group bearing spiroxins (Chart 10), as seen in a mouse xenograph model against human ovarian carcinoma. This activity was attributed in part to their effect on DNA.

Two closely related compounds, CJ-12,371 (**22**) and CJ-12,372 (**23**), inhibited DNA supercoiling and relaxation mediated by *E. coli* DNA gyrase (*29*). However, the compounds did not inhibit DNA gyrase selectively, as they also inhibited eukaryotic topoisomerase II-mediated relaxation. Interestingly, these compounds do not have any highly electrophilic functional groups in their molecules.

Potent antifungal activity and inhibition of Gram-positive bacteria was most pronounced in the preussomerins A–F (Cart 9) as investigated by Gloer *et al.* (*38*) in assays against the coprophilous fungal competitors *S. fimicola* and *Ascobolus furfuraceus*. Interestingly, none of these preussomerins was active against *Candida albicans* and preussomerin A (**3**) did show no activity against assays against herpes virus and was only mildly cytotoxic toward a mammalian cell line.

Great attention attracted a report of Singh *et al.* (*32*) on the inhibition of Ras farnesyl-protein transferase because this effect is correlated to a modern concept of antitumor activity. Preussomerins D (**59**) and G (**62**) exhibited an IC_{50} at 1.4 and 1.2 μM, respectively. Interestingly, the most active compounds in this test contain conjugated ketone functions as well as epoxides. These groups may serve as Michael acceptors for the Ras-CVLS nucleophile, but the exact mechanism of inhibition is still unclear (*32*).

4. Biosynthesis

The dimeric nature of the spirobisnaphthalene structures suggested that these molecules are formed by oxidative coupling of the monomers. We previously proposed a plausible mechanism for the dimerisation reaction as outlined in Scheme 7 (*16*). According to our proposal, the radical reaction of 1,8-dihydroxy-naphthalene (DHN) (**73**) could occur via the formation of intermediates such as the dinaphthyl ether **74** and then be converted to the parent of all spirobisnaphthalene compounds, palmarumycin CP_1 (**2**), in a second oxidation step. This idea is supported by the finding of K. Beckmann in our group, who succeeded in isolating compound **75**, the first naturally occurring dinaphthyl ether reported. Chemical oxidation of **75** with silver oxide yielded palmarumycin C_{17} (**76**), a compound that has not yet been isolated from fungal cultures.

Zeeck *et al.* studied the biosynthesis of cladospirone bisepoxide (**9**) by feeding experiments with labeled acetate and established that the naphthyl units (e.g. DHN) were formed from pentaketide-derived precursors such as **78–81** (*17*). The likelihood that DHN is the probable precursor of the spirobisnaphthalenes can also be concluced from the fact that, very characteristically, production of these compounds is accompanied by brown, DHN-derived melanin components. However, labeling experiments indicated that DHN (or a similar compound) is not

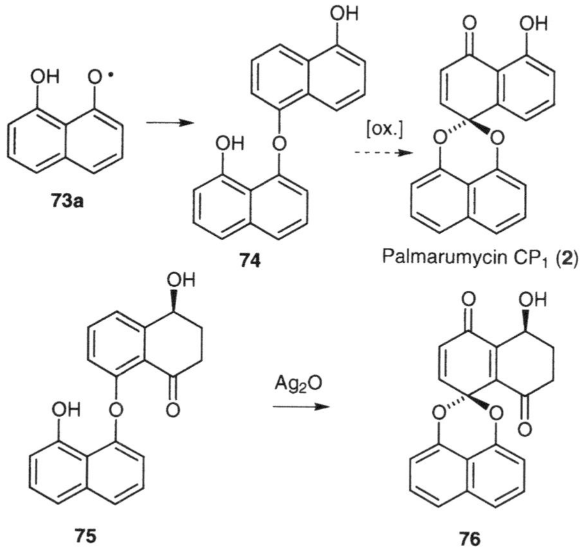

Scheme 7. Proposed dimerization of 1,8-dihydroxynaphthalene (**73**) via the dinaphthyl ether **74** to **2**, as well as chemical oxidation of **75** to palmarumyin C_{17} (**76**)

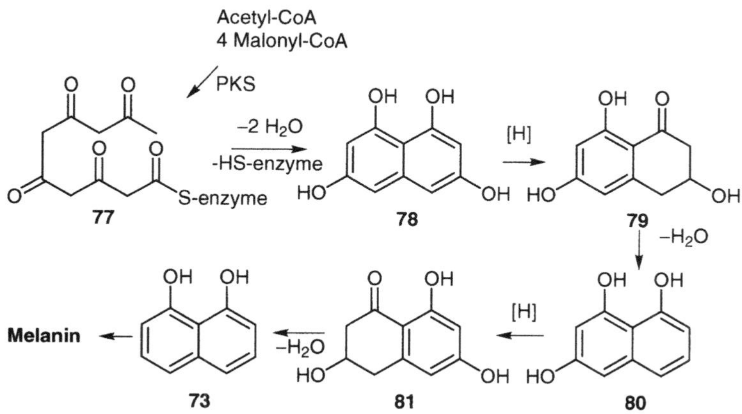

Scheme 8. Formation of DHN (**3**) from pentaketide **77** via **78–81**

incorporated in a straightforward manner since scrambling of the label into the spirobisnaphthalenes occurred (*36*). Therefore, it is possible that the biological dimerisation process (especially if this proceeds via a radical mechanism) is much less controlled than previously assumed.

UV mutagenesis and enzyme inhibitors were used as tools to elucidate the later steps of the spirobisnaphthalene biosynthesis (*8*). The

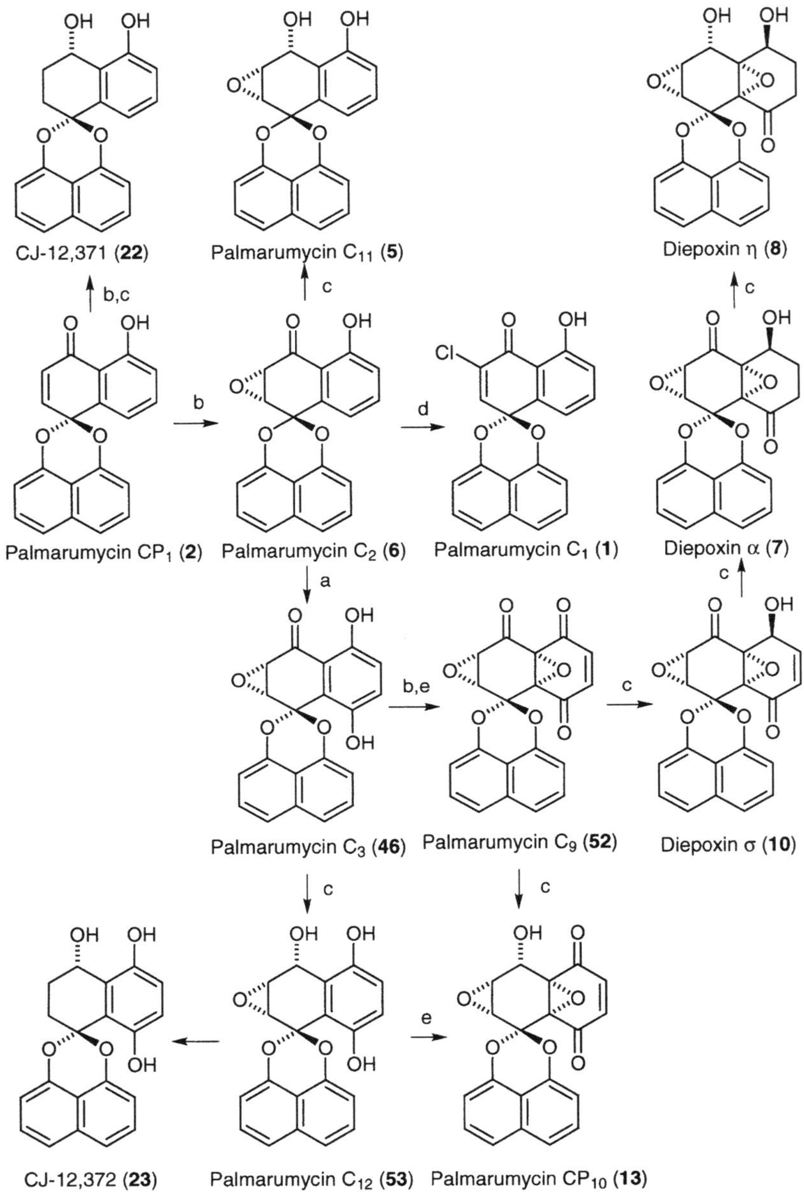

Scheme 9. Chemical diversity using simple transformations starting from palmarumycins CP_1 (**2**) and C_3 (**46**)

summary presented in Scheme 9 is only brief, but shows how variable products are that are formed from the same basic molecule by a series of hydroxylation (a), epoxidation (b), reduction (c), chlorination (d), or oxidation (e) steps. Thus, two families can be generated from the "parent" spirobisnaphthalene, palmarumycin CP_1 (**2**), by either epoxidation to palmarumycin C_2 (**6**) and further steps, or indirectly, from the hydroxylated product of palmarumycin C_2, palmarumycin C_3 (**46**). Further epoxidation and oxidation then yields palmarumycin C_9 (**52**) that in turn undergoes a cascade of reactions that leads to the generation of the compounds shown at the bottom and left hand part of Scheme 9. Some other interesting chemical transformations in the cladospirone family, shown in Chart 6, are worth mentioning here. Of course the picture of these transformations follows chemical "logic" as the actual sequence of the transformation events may perhaps vary from one microorganism to another.

It has already been mentioned that the stereochemical link between the spirobisnaphthalenes listed in Table 1 and the preussomerins was established with the degradation experiment leading over the putative intermediate **3a** to (−)-regiolone (Scheme 5). It is conceivable that preussomerin G (**62**) is generated from palmarumycin C_3 (**46**) via the diradical **46a**, which could then form another bridging bond as suggested by Zeeck and Bode (*8*) (Scheme 10). Likewise, the postulated diradical

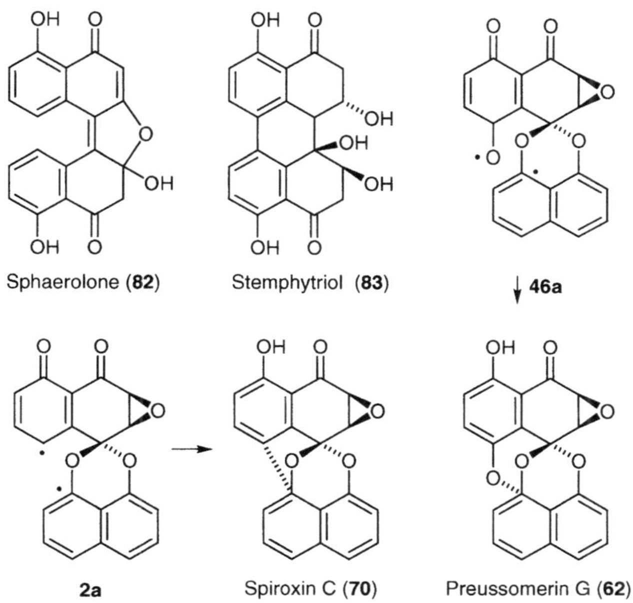

Scheme 10. Late biosynthesis of perylenes, preussomerins, and spiroxins

2a, generated from palmarumycin CP_1 (**2**) could lead to the formation of the spiroxin series of spirobisnaphthalenes with two oxygen atoms and one carbon atom in the bridge region (shown here as spiroxin C (**70**)). Direct dimerization of DHN (**73**) or DHN type precursors would generate compounds such as sphaerolone (**82**) or the perylenes exemplified here by stemphytriol (**83**) (*44*). In fact, Bode and Zeeck were able to isolate sphaerolone (**82**) from a fermentation that was incubated with tricylazole (*2*). Thus, the biosynthesis of the carbon-bridged perylenes or and the oxygen bridged spirobisnaphthalenes are probably formed from dihydroxynaphthalene or a similar precursor by (presumably radical) dimerization of dihydroxynaphthalene, albeit in different manners.

5. Synthesis

The interesting structure and biological properties of the spirobisnaphthalenes have encouraged several synthetic chemistry groups worldwide to study their chemical synthesis. Two major strategies have been used:

1. Biomimetic type oxidative cyclization of binaphthyl ethers;
2. Ketal formation of substituted tetralones with 1,8-dihydroxynaphthalene (1,8-DHN).

5.1. Biomimetic Type Approach

The feasibility of the first approach was demonstrated in our early paper (*18*) by the silver oxide mediated cyclization of the naturally occurring biaryl precursor **75** to the semisynthetic palmarumycin C_{17} (**76**) (Scheme 7). Based on this concept, Wipf *et al.* have developed elegant syntheses of palmarumycin CP_1 (**2**), racemic palmarumycin C_2 (*rac*-**6**) (deoxypreussomerin A) (*45*), racemic diepoxin σ (*rac*-**10**) (*46*), and a formal synthesis of (+)-diepoxin σ (**10**) (*47*). A prerequisite for the success of this approach was the good availability of the required dinaphthyl ethers. However, there are limitations in the Ullmann biaryl ether synthesis concerning the coupling of naphthalenes. For example, it was not possible to couple the iodide **84** with the tetralone **87** (Scheme 11) (*45*). This may be caused by the poor nucleophilicity of the chelated phenolic hydroxyl group in **87** in addition to steric hindrance. Steric hindrance is probably responsible for the poor yield in the coupling of **84** with the protected naphthol **85** as well, as the yield was only 7% of the

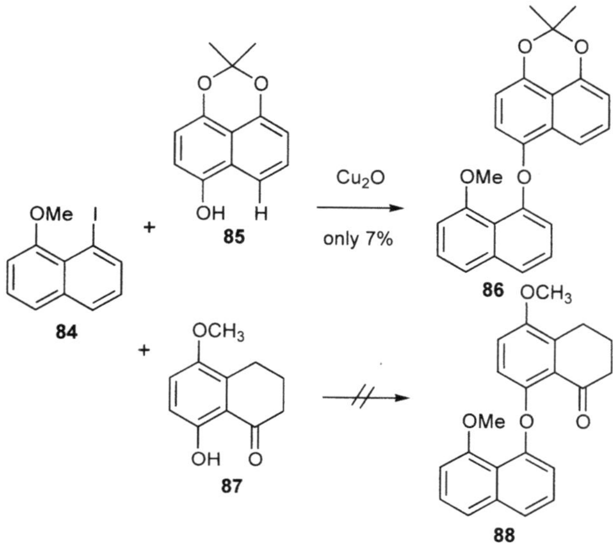

Scheme 11. Limitations in the Ullmann binaphthyl ether synthesis

required bisnaphthol **86** (*48*). On the other hand, it was possible to couple the corresponding alcohol version of ketone **87** (*47*).

In the synthesis of palmarumycin CP_1 (**2**) and racemic palmarumycin C_2 (*rac*-**6**) (deoxypreussomerin A) (*45*), the coupling of the iodide **84** with the naphthol **89** succeeded in 78% yield under standard Ullmann conditions (Cu_2O, K_2CO_3, pyridine, reflux) to give the biaryl ether **90**. The tetralone corresponding to the ketal **89** did **not** react. Simple acetal and methyl ether cleavage, followed by LAH reduction of the ketone to the alcohol, afforded an intermediate binaphthyl ether, that was cyclized by treatment with phenyliodonium diacetate [$PhI(OAc)_2$] to yield 87% of the 1,8-DHN ketal **91** (Scheme 12). Oxidation of the benzylic alcohol **91** to the ketone **92** was achieved using the Dess-Martin periodinane and dehydrogenation (aromatization) reaction. Using activated manganese dioxide gave palmarumycin CP_1 (**2**), the achiral parent compound of the spirobisnaphthalene antibiotic family. Alternatively, epoxidation of the electron-deficient double bond in the enone **91**, using cumene hydroperoxide/NaH, gave the epoxide **93** in 47% yield. Aromatization of **93** was then performed using the same procedure as described above to afford palmarumycin C_2 (*rac*-**6**) in 55% yield.

Protection of the external double bond by Diels-Alder reaction with the option of reconstruction by retro-Diels-Alder reaction, was a key feature in the synthesis of diepoxin σ (*rac*-**10**) (*46*) and a formal synthesis of

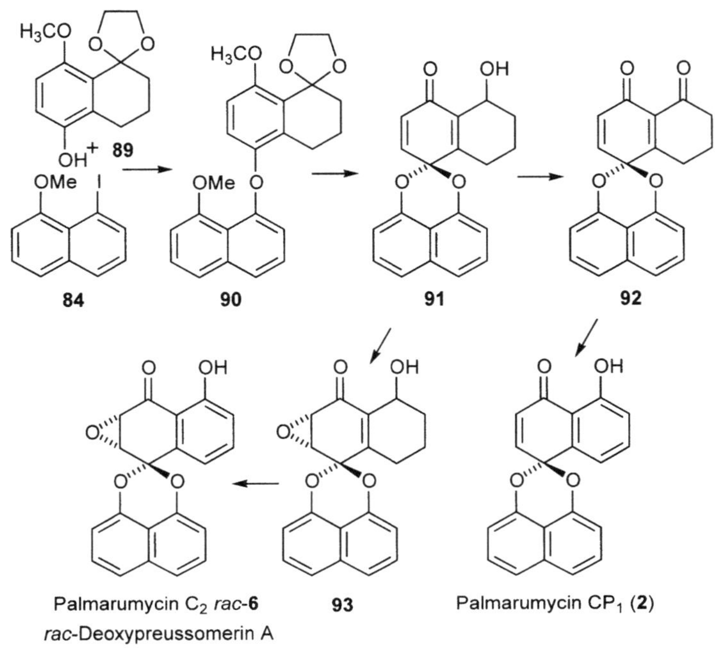

Scheme 12. Procedure of Wipf and Jung (*45*) for the synthesis of palmarumycin CP_1 (**2**) and racemic palmarumycin C_2 (*rac*-**6**)

(+)-diepoxin σ (**10**) (*47*). Again, the coupling of the iodide **84** with the phenol **94** under standard Ullmann conditions presented no problem and the ether **95** was isolated in 70% yield (Scheme 13). Demethylation of **95** proved to be a challenging reaction due to the acid/base lability of the two benzylic hydroxyls. After considerable experimentation, lithiated diphenylphosphine proved to cleave the aryl methyl ether in 95% to the corresponding bisphenol, which was cyclized using $PhI(OAc)_2$ in the presence of hexafluoro-2-propanol and 4 Å molecular sieves to afford the spiroketal **96** in 61% yield. Selective protection of the less hindered hydroxyl group as the TBS-ether (91%) and PDC oxidation (72%) afforded the ketone **97**, which was easily epoxidized to the bisepoxide **98** (88%) using hydrogen peroxide and potassium carbonate. This bisepoxide proved to be surprisingly stable under thermal conditions of the retro-Diels-Alder reaction (PhOPh, reflux). Fluoride-assisted (HF in acetonitrile) cleavage of the silyl ether terminated the synthesis of racemic diepoxin σ (**10**) (73%, two steps).

In the full paper (*47*), describing this synthesis and also many details of related reactions, Wipf and Jung proposed a synthesis of the highly

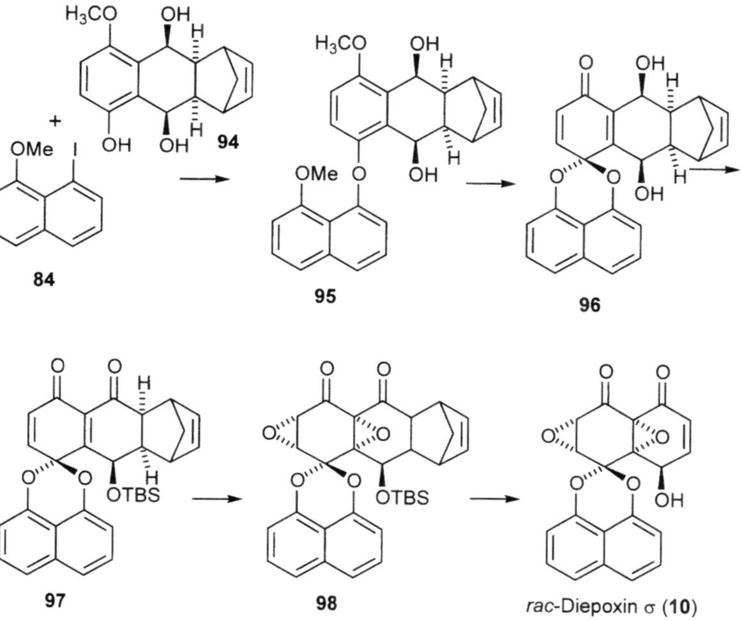

Scheme 13. Synthesis of racemic diepoxin σ (**10**) (*46, 47*)

enantiomerically enriched starting material **102**. Reaction of naphthazarine monomethyl ether (**99**) with cyclobutadiene **101** gave **102** in 94% ee in the presence of the chiral catalyst **100** and BF_3-THF (Scheme 14). In this reaction, the free hydroxyl group in the naphthoquinone forms a chelate boron complex with the chiral bisnaphthol **100**, thus ensuring chiral modification and catalysis at the same time. The principle of this chiral catalysis was investigated by Kelly *et al.* (1985) (*49*) and Yamamoto *et al.* (1986) (*50*) using juglone as the substrate.

In our group, we initially wanted to couple two unsaturated naphthalene building blocks such as **84** and **85** (*48*). Soon we had to realize the limitation of this Ullmann coupling that gave only very poor yields in the coupling reaction (see Scheme 11). By contrast, the coupling of **84** with hydroquinone monomethyl ether **104** proceed in excellent yields under standard Ullmann conditions to yield **105** in 89% yield (Scheme 15). Nearly quantitative cleavage of the methyl ethers was effected by boron tribromide to afford the phenol **106** (98%), which was cyclized to the ketal **107** by using either silver oxide (50%) or $PhI(OAc)_2$ (64%) (Scheme 15). Diels-Alder reaction of the dienone **107** was sluggish and the relatively unreactive dienophile had to be refluxed with excess of

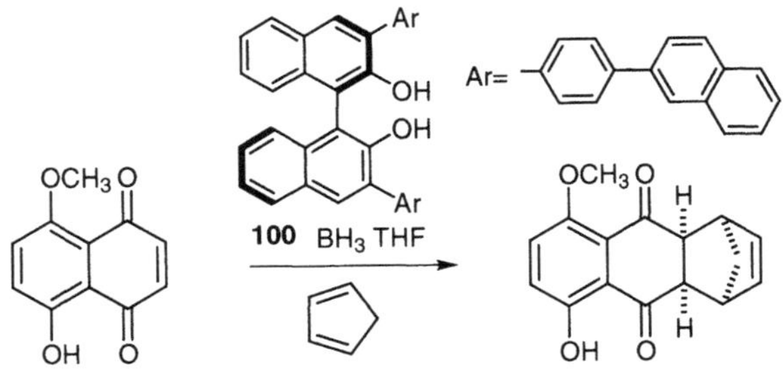

99 101 102 (max. 94% ee)

Scheme 14. Synthesis of enantioselective starting material **102** by Diels-Alder reaction (*47*)

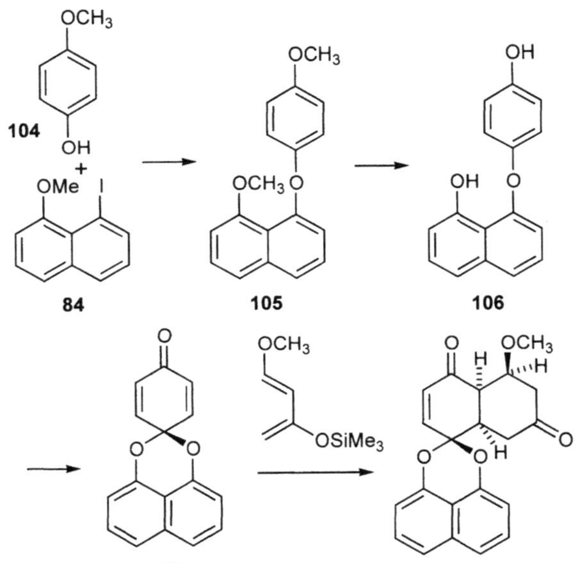

Scheme 15. Thermal Diels-Alder reaction of enedione **107** with Danishefsky's diene (*48*)

Danishefsky's diene in toluene to afford the desired adduct **108** in ca. 50% yield, introducing an oxygen function at C-7 that is not frequently observed in the natural products. This method combines the advantages of the biomimetic type cyclization and the very flexible Diels-Alder reaction. Many dienes can potentially be used to yield naturally and not naturally occurring derivatives of the spirobisnaphthalene antibiotics.

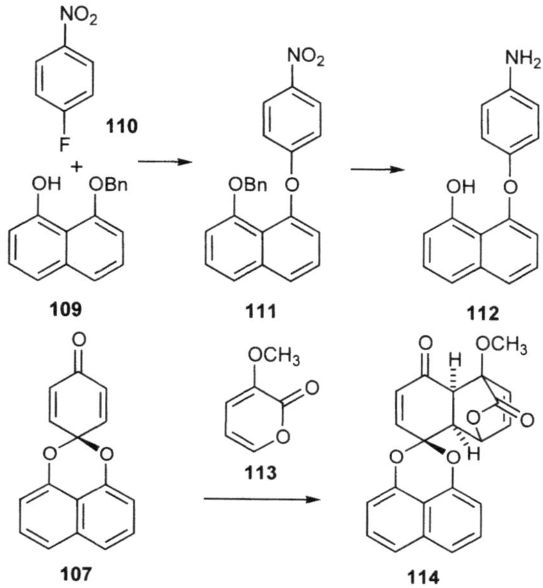

Scheme 16. Nucleophilic aromatic displacement and high pressure Diels-Alder reactions used in spirobisnaphthalene synthesis (*51*)

The work of Coutts *et al.* solved two problems encountered with the above method (*51*). First, the naphthalene iodide **84** was not easily available in larger quantities. The authors used a nucleophilic displacement reaction instead of the Ullmann coupling, employing the monobenzyl ether **109** of 1,8-DHN as the nucleophile and 1-fluoro-4-nitrobenzene (**110**) as the electron-deficient electrophile. The coupling to **111** proceeded in good yield and the nitro compound was hydrogenated in one step to the aminophenol **112**. Active manganese dioxide was used in the subsequent cyclization step to afford the requisite spiroketal **107** in 86% yield. Secondly, the authors used high pressure (12–15 kbar) for the following Diels-Alder reaction. When pyrone **113** was used as the diene, the adduct **114** was isolated almost quantitatively. Decarboxylation proved to be difficult but other dienes (including open chain ones) can potentially be used in this reaction.

5.2. Spiroketalization Approach

A very convincing synthetic approach to the spirobisnaphthalene antibiotics would be the simple ketalization of 1,8-DHN (**73**) with the

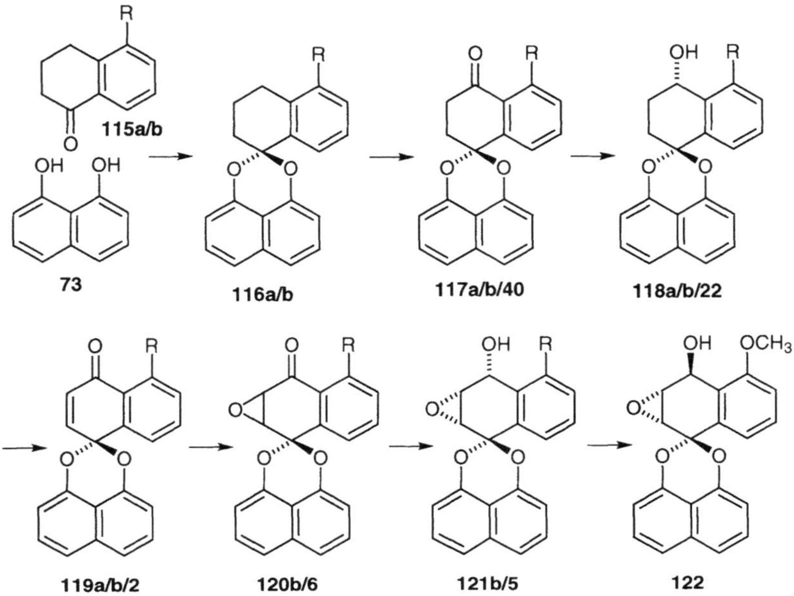

a: R = H
b: R = OCH_3
c: OH

Scheme 17. Synthesis of Barrett *et al.* (*52*) and Taylor *et al.* (*53, 20*) employing ketalization

tetralones **115** (**a** = H, **b** = OCH_3). However, the nucleophilicity of the strongly chelated phenol groups in **73** is very poor, and in our hands we obtained only low yields in the ketalization reaction. More vigorous conditions were required, and the groups of Barrett *et al.* (*52*) and Taylor *et al.* (*20, 53*) succeeded in converting **115a/b** and **73** to the spiroketals **116a/b** in reasonable yields (86% and 74%) using a Dean-Stark trap and *p*-toluene sulfonic acid or trifluoro acetic acid in boiling benzene (Scheme 17). Oxidation of the benzylic position in **116** to the ketones **117a/b** was performed by both groups in a radical oxidation, employing *tert*-butyl hydroperoxide in the presence of chromium (VI) (*54*) (61% and 64% yields).

The first natural product, palmarumycin CP_2 (**40**), was obtained by cleavage of the aryl-methyl ether using freshly prepared magnesium iodide in benzene (84% (*52, 55*)) or boron tribromide (58% (*20, 53*)). Not naturally occurring derivatives (**117a** R = H, **117b** R = OCH_3) were also available this way (*20*). Enantioselective reduction of synthetic **40** (R = OH) employing Brown's (+)-*B*-chlorodiisopinocamphenyl borane (*56*) gave CJ-12,371 (**22**), the DNA gyrase inhibitor isolated by Sakemi *et al.* (*29*).

References, pp. 46–49

The spectroscopic data including specific optical rotation were in agreement with those published for the natural product. Interestingly, the usual stereochemistry observed with the borane reducing reagent was reversed due to the intramolecular nature of the hydride transfer. Desaturation of the cyclohexanone derivatives **117a/b** and **40** were performed either by treatment with DDQ (**2**: 65% (*52*)) or benzeneseleninic anhydride (**119b**: 64%, (*53* (*20*)). The methyl ether in **119b** could be cleaved using boron tribromide to palmarumycin CP_1 (**2**), the parent compound of the entire spirobisnaphthalene family. The structures of the natural products **2** and **40** were confirmed by direct comparison with authentic samples (*52*).

Both the methyl ether and the phenol were epoxidized to **120b** and racemic palmarumycin C_2 (deoxypreussomerin A) (**6**), using *tert*-butyl hydroperoxide/1,5,7-triazabicyclo[4.4.0]dec-5-ene (53% (*20, 53*)). Taylor *et al.* also addressed the question of the stereochemistry of palmarumycin C_{11} (**5**) and of bipendensin and Sch 53823 (**5a**) (Scheme 1). As already mentioned, the NMR spectra of bipendensin and Sch 53823 (**5a**) are in good agreement but differ from those of palmarumycin C_{11} (**5**), notably for the diagnostic chemical shift for C-4 in the ^{13}C NMR spectrum. Reduction experiments with **120b** ($LiEt_3BH$: **121b**:**122** = 3.6:1); DIBAL-H: 5:95) supported the *syn*-configuration **5** for palmarumycin C_{11} and the *anti* configuration (**5a**) for bipendensin and Sch 53823. The NMR data for the synthetic *syn* product **5** were in agreement with those reported for of palmarumycin C_{11} (*16*).

Another noteworthy observation was made on the attempted arylmethyl ether cleavage of racemic **118b**: only binaphthyl ether **74** (Scheme 7) and its monomethyl ether were formed upon boron tribromide treatment of **118b** (*20*).

A simple, possibly biomimetic construction to the unusual ring system of the preussomerins including the first total synthesis of preussomerins G (**62**) and I (**64**) (Chart 9) was published by Chi and Heathcock (*57*) (Scheme 18). Acid-catalyzed condensation of 4-acetoxy-1,8-naphthalenediol (**124**) with the methyl enol ether (**123**) gave the requisite ketal **125** (62%). To prevent the "bottom" naphthalene ring from oxidation, the acetate was converted to the more electron-withdrawing trichloroacetate and ceric ammonium nitrate oxidation afforded the quinone **126** (87%). In the key step of the synthesis, the treatment of quinone **126** with lithium hydroxide in THF-water induced the rearrangement to the hexacyclic preussomerin ring system **127**. The resonance energy gained from the formation of two isolated benzene rings in **127** provides a strong driving force for that cyclization, as confirmed by ab initio calculations. In the remaining steps, functional

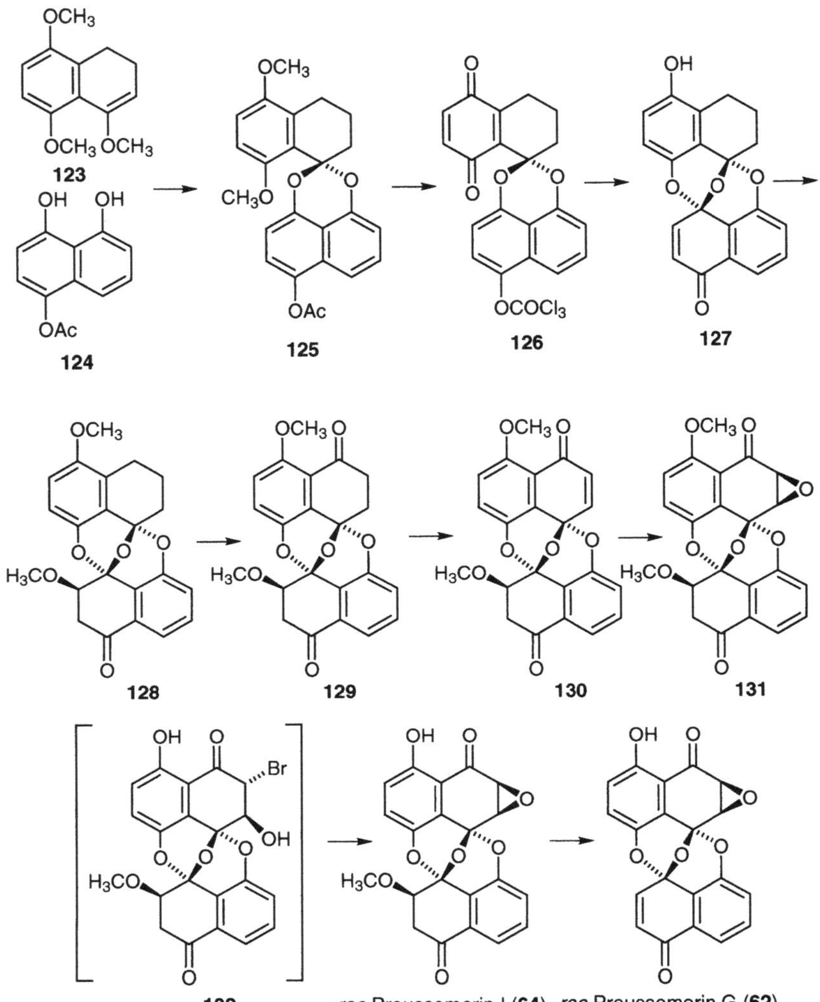

Scheme 18. Synthesis of raemic preussomerins G (**62**) and I (**64**) by Chi and Heathcock (*57*)

group transformations converted **127** to the racemic preussomerins G (**62**) and I (**64**). Conjugate addition of methoxide (LiOMe) and methylation of the phenol with diazomethane gave the dimethyl ether **128** which was oxidized in a three step sequence by bromination, hydrolysis, and Dess-Martin periodinane oxidation to the ketone **129** in good overall yield (56%). Desaturation to the α,β-unsaturated ketone

129 was effected by treatment of the corresponding silyl enol ether with $Pd(OAc)_2$ using the Saegusa conditions (65% (*58*)). Epoxidation with H_2O_2 under basic conditions ($NaHCO_3$) afforded the epoxide **131**. Boron tribromide-mediated demethylation resulted in the opening of the epoxide to **132** and aryl-methyl ether cleavage. Fortunately, the epoxide was easily reclosed under mild basic condition ($LiOCH_3$) to form racemic preussomerin I (**64**) (60%). β-Elimination to form *rac*-preussomerin G (**62**) was induced under Lewis acidic conditions (TMSOTf, TEA).

An interesting alternative approach of Taylor *et al.* (*59*) allowed access to preussomerin analogues with the full carbon skeleton of the natural products. The idea relied on the dimerization of 2-hydroxybenzaldehydes such as **133** to the spiroacetal **134**, a reaction known for more than a century, realized in 96% yield using pivalic anhydride/sulfuric acid as the dehydrating agent (Scheme 19). Monodeprotonation of **134** proved to be easy, but the diallylation had to be done in two successive steps (*n*-Bu-Li, allyl bromide, then *sec*-BuLi, allyl bromide, 52%). Double

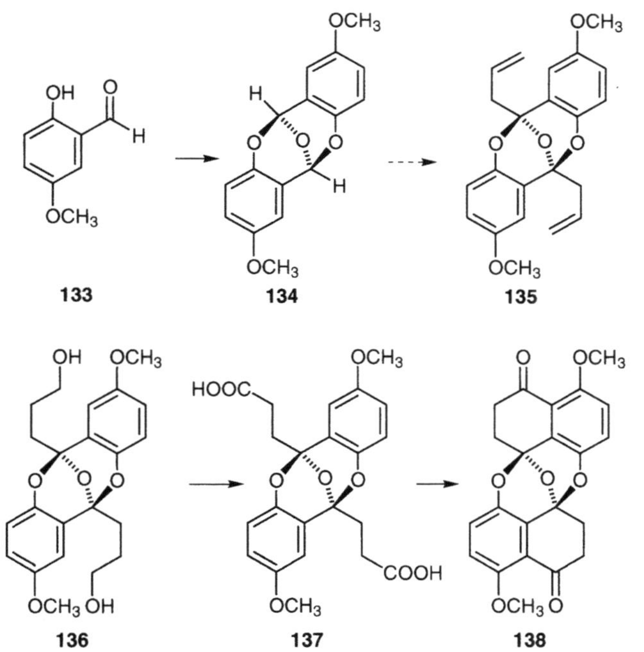

Scheme 19. Conversion of the salicylic aldehyde anhydride **134** to the preussomerin analogue **138** (*59*)

hydroboration of the diolefin **135** using 9-BBN followed by oxidative workup with alkaline hydrogen peroxide to the diol **136** (32%) was followed by PDC oxidation in DMF to the diacid **137** (66%). The diacid **137** was then converted to the corresponding diacid dichloride and Friedel-Crafts cyclization afforded the diketone **138** with the preussomerin skeleton in 73% yield over the two steps.

Based on experience with their earlier racemic approach (*52*), Barrett *et al.* recently published the enantioselective synthesis of several palmarumycins and preussomerin G (**62**). (*60*) The key step was the enantioselective epoxidation of the prochiral synthetic palmarumycin CP_1 (**2**) with *tert*-butyl hydroperoxide/NaOH and *N*-benzylcinchoninium chloride as the chiral catalyst to afford **6** in 81% yield and 95% ee (Scheme 20) (*60*). Sodium borohydride reduction of the carbonyl group in **6** gave palmarumycin C_{11} (**5**). X-ray crystal analysis of the ester of **N**-phenylsulfonyl proline ester of **6** confirmed the absolute configurations of **6** and indirectly also that of **5**. The assignments were in agreement with results of quantum-mechanical CD calculations in our group (*22*, *43*). This result also confirms that Sch 53823, assigned structure **5** by Chu *et al.* (*21*) but deviating in the NMR data from palmarumycin C_{11} (**5**), is most probably the epimeric secondary alcohol **5a**.

Hydroxylation of the phenol **6** was achieved using the hypervalent iodine reagent $PhI(OCOCF_3)_2$ to yield palmarumycin C_{12} (**53**) after dithionite reduction. The same reagent (followed by oxidation with active manganese dioxide and reduction) also converted **5** to palmarumycin C_3 (**46**) (Scheme 20). The specific optical rotations of the synthetic products were again in good agreement with those of the natural products and established the absolute configurations of palmarumycin C_3 (**46**) as 2(*R*), 3(*R*) and that of palmarumycin C_{12} (**53**) as 2(*R*), 3(*R*), 4(*R*) in accord with quantum-mechanical CD calculations (*43*).

In this unified route to spirobisnaphthalenes, the synthetic palmarumycin C_3 (**46**) served as the precursor for preussomerin G (**62**). For that purpose, hydroxylation of the bottom 1,8-dioxonaphthalene of **46** was required. The phenolic groups in **46** were protected as the methoxy acetic esters and this diester was acetoxylated at the aromatic bottom naphthalene ring using lead tetraacetate to afford the acetic ester **139** as a mixture of diastereoisomers in 34% yield. The cyclization was performed in analogy to the procedure of Heathcock *et al.* (*57*) (Scheme 18) but employing lead tetraacetate treatment of the phenol obtained by LiOH saponification of **139**. The two isomeric cyclization products formed in a 1:1 ratio in 55% yield could be separated. One isomer was identical in all respects to the important Ras farnesyl transferase inhibitor preussomerin G (**62**).

References, pp. 46–49

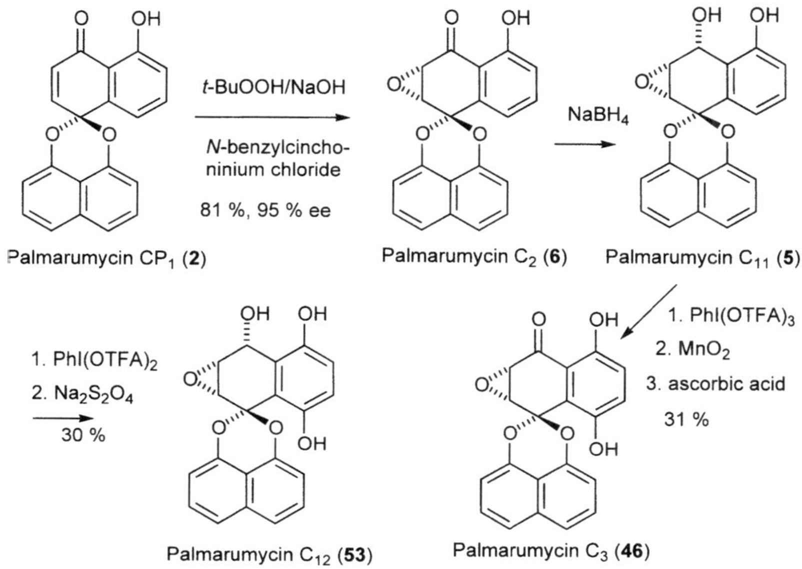

Scheme 20

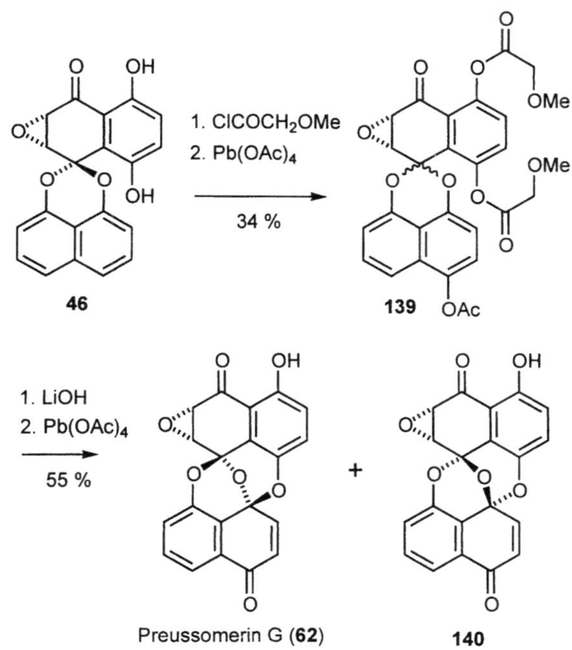

Scheme 21

Acknowledgements

I thank the BMBF and the BASF AG for financial support and T. van Ree and G. Schlingmann for reviewing the manuscript.

References

1. Tsuboi M, Minami M, Nonaka GI, Nishioka I (1977) Studies on Rhubarb (Rhei Rhizoma) IV. Naphthalene Glycosides. Chem Pharm Bull (Jn) **25**: 2708
2. Bode HB, Zeeck A (2000) Sphaerolone and Dihydrosphaerolone, Two New Bisnaphthyl Pigments from the Fungus Sphaeropsidales sp. F24′707. Phytochem **54**: 597
3. Arnone A, Nasini G, Merlini L, Assante G (1986) Secondary Mould Metabolites. Part 16. Stemphyltoxins, New Reduced Perylenequinone Metabolites from *Stemphylium botryosum* var. *Lactucum*. J Chem Soc Perkin Trans 1, 525
4. Diwu Z, Lown JW (1992) A Simple High-Yielding Approach to Perylenequinones from the Novel One-Step Double Coupling Reaction of 1,2-Naphthoquinone. Tetrahedron **48**: 45
5. Lown WS, Diwu Z, Lown JW (1996) 1996 Hoffman-LaRoche Award Lecture. Photochemistry and Photobiology of Perylenequinones. Can J Chem **75**: 99
6. Arnone A, Merlini L, Mondelli R, Nasini G, Ragg E, Scaglioni L (1993) Structure, Conformational Analysis and Absolute Configuration of the Perylenequinone Pigments Elsinochromes B1, B2, C1 and C2. Gazz Chim Ital **123**: 131
7. Gill M, Steglich W (1987) Pigments from Fungi. In: Herz W, Grisebach H, Kirby GW, Tamm Ch (eds) Prog Chem Org Nat Prod, vol 51. Springer, Wien New York, p 118ff
8. Bode HB, Zeeck A (2000) UV Mutagenesis and Enzyme Inhibitors as Tools to Elucidate the Late Biosynthesis of the Spirobisnaphthalenes. Phytochem **55**: 311
9. Bode HB, Walker M, Zeeck A (2000) Secondary Metabolites by Chemical Screening. 42. Cladospirones B to I from *Sphaeropsidales* sp. F-24′707 by Variation of the Culture Conditions. Eur J Org Chem 3185
10. McDonald LA, Abbanat DR, Barbieri LR, Bernan VS, Discafani CM, Greenstein M, Janota K, Korshalla JD, Lassota P, Tischler M, Carter GT (1999) Spiroxins, DNA Cleaving Antitumor Antibiotics from a Marine-Derived Fungus. Tetrahedron Lett **40**: 2489
11. Schlingmann G, Matile S, Berova N, Nakanishi K, Carter GT (1995) Absolute Stereochemistry of the Diepoxins. Tetrahedron **52**: 435
12. Krohn K, Michel A, Flörke U, Aust H-J, Draeger S, Schulz B (1994) Biologically Active Metabolites from Fungi, 4. Palmarumycins CP_1 to CP_4 from *Coniothyrium palmarum*: Isolation, Structure Elucidation and Biological Activity. Liebigs Ann Chem 1093
13. Weber HA, Baenziger NC, Gloer JB (1990) Structure of Preussomerin A: An Unusual New Antifungal Metabolite from the Coprophilous Fungus *Preussia isomera*. J Am Chem Soc **112**: 6718
14. Ohkishi H, Chiba N, Mikawa T, Sasaki T, Miyaji S, Sezaki M (1989) Mitsubishi Kasei Corp, JP, 01294686 (November 28, 1989); (1990) Chem Abst **113**: 38906q
15. Connolly JD (1990) Structural Elucidation of Some Natural Products. In: Atta-ur-Rahman (ed) Frontiers in Natural Product Chemistry, Elsevier, Amsterdam Oxford New York Tokyo, pp 524–539

16. Krohn K, Michel A, Flörke U, Aust H-J, Draeger S, Schulz B (1994) Biologically Active Metabolites from Fungi, 5. Palmarumycins C_1 to C_{16} from *Coniothyrium* sp.: Isolation, Structure Elucidation, and Biological Activity. Liebigs Ann Chem 1099
17. Bode HB, Wegner B, Zeeck A (2000) Biosynthesis of Cladospirone Bisepoxide, A Member of the Spirobisnaphthalene Family. J Antibiot **53**: 153
18. Krohn K, Beckmann K, Aust H-J, Draeger S, Schulz B, Busemann S, Bringmann G (1997) Biologically Active Metabolites from Fungi, 10. Generation of the Palmarumycin Spiroacetal Framework by Oxidative Cyclization of an Open Chain Metabolite from *Coniothyrium palmarum*. Liebigs Ann Chem 2531
19. Kouam TNM, Lavaud C, Massiot G, Nuzillard J-M, Connolly JD, Rycroft D (1993) Bipendensin, an Unusual Phenolic Acetal from *Afzelia Bipendensis*. Nat Prod Lett **3**: 299
20. Ragot JP, Steeneck C, Alcaraz M-L, Taylor RJK (1999) The Synthesis of 1,8-Dihydroxynaphthalene-Derived Natural Products: Palmarumycin CP_1 and CP_2, CJ-12,371, Deoxypreussomerin A and Novel Analogues. J Chem Soc Perkin Trans 1, 1073
21. Chu M, Patel MG, Pai MG, Das J-K, Puar MS (1996) Sch 55823 and Sch 53825, Novel Fungal Metabolites with Phospholipase D Inhibitory Activity. Bioorg Med Chem Lett **6**: 579
22. Bringmann G, Busemann S, Krohn K, Beckmann K (1997) Quantum Mechanical Calculation of CD Spectra: The Absolute Configuration of Palmarumycins CP_3 and C_2. Tetrahedron **53**: 1655
23. Chu M, Truumees I, Patel MG, Blood C, Das PR, Puar MS (1995) Sch 50673 and Sch 50676, Two Novel Antitumor Fungal Metabolites. J Antibiot **48**: 329
24. Schlingmann G, West RR, Milne L, Pearce CJ, Carter GT (1993) Diepoxins, Novel Fungal Metabolites with Antibiotic Activity. Tetrahedron Lett **34**: 7225
25. Chu M, Truumees I, Patel MG, Blood C, Das PR, Puar MS (1994) A Novel Class of Antitumor Metabolites from the Fungus *Nattrassia mangiferae*. Tetrahedron Lett **35**: 1343
26. Thiergardt R, Hug P, Rihs G, Peter HH (1995) Cladospirone Bisepoxide: Definite Structure Assignment Including Absolute Configuration and Selective Chemical Transformations. Tetrahedron **51**: 733
27. Chu M, Truumees I, Patel MG, Gullo VP, Puar M (1994) Structure of Sch 49209: A Novel Antitumor Agent from the Fungus *Nattrassia mangiferae*. J Org Chem **59**: 1222
28. Chu M, Truumers I, Patel MG, Gullo VP, Pai I-K, Das PR, Puar MS (1994) Two New Phospholipase D Inhibitors, Sch 49211 and Sch 49212, Produced by the Fungus *N. mangiferae*. Bio Med Chem Lett **4**: 1539
29. Sakemi S, Inagaki T, Kaneda K, Hirai H, Iwata E, Sakakibara T, Yamauchi Y, Norcia M, Wondrack LM, Sutcliffe JA, Kojima N (1995) CJ-12,371 and CJ-12,372, Two Novel DNA Gyrase Inhibitors. Fermentation, Isolation, Structural Elucidation, and Biological Activities. J Antibiot **27**: 134
30. Thiergardt R, Hug P, Rihs G, Peter HH (1994) Cladospirone Bisepoxide – A Novel Fungal Metabolite Structure Determination. Tetrahedron Lett **35**: 1043
31. Petersen F, Moerker T, Vanzanella F, Peter HH (1994) Production of Cladospirone Bisepoxide, A New Fungal Metabolite. J Antibiot **47**: 1098
32. Singh SB, Zink DL, Liesch JM, Ball RG, Goetz MA, Bolessa EA, Giacobbe RA, Silverman KC, Bills GF, Pelaez F, Cascales C, Gibbs JB, Lingman RB (1994) Preussomerins and Deoxypreussomerins: Novel Inhibitors of Ras Farnesyl-Protein Transferase. J Org Chem **59**: 6296

33. Krohn K, Beckmann K, Flörke U, Aust H-J, Draeger S, Schulz B, Busemann S, Bringmann G (1997) Biologically Active Metabolites from Fungi, 9. New Palmarumycins CP_{4a} and CP_5 from *Coniothyrium palmarum*: Structure Elucidation, Crystal Structure Analysis and Determination of the Absolute Configuration by CD-Calculations. Tetrahedron **53**: 3101
34. Chu M, Gullo VP, Horan AC, Patel M (1994) Schering Corp. WO, 9320081 (October 14, 1993); Chem Abstr **120**: 572
35. Schlingmann GR, Milne L, Borders DB, Greenstein M, Carter GT (1995) Isolation, Characterization and Structure of a New Allenic Polyyne Antibiotic Produced by Fungus LL-07F275. J Antibiot **48**: 375
36. G. Schlingmann (private communication)
37. Höfs R, Walker M, Zeeck A (2000) Hexacyclinsäure, ein Polyketid mit neuartigem Grundgerüst aus Streptomyces. Angew Chem **112**: 3400
38. Weber HA, Gloer JB (1991) The Preussomerins: Novel Antifungal Metabolites from the Coprophilous Fungus *Preussia isomera* Cain. J Org Chem **56**: 4355
39. Polishook JD, Dombrowski AW, Tsou NN, Salituro GM, Curoto JE (1993) Preussomerin D from the Endophyte *Hormonema dematioides*. Mycologia **85**: 62
40. Soman AG, Gloer JB, Koster B, Malloch D (1999) Sporovexins A-C and a New Preussomerin Analog: Antibacterial and Antifungal Metabolites from the Coprophilous Fungus *Sporomiella vexans*. J Nat Prod **62**: 659
41. Krohn K, Flörke U, John M, Root N, Steingröver K, Aust H-J, Draeger S, Schulz B, Antus S, Simonyi M, Zsila F (2001) Biologically Active Metabolites from Fungi, 16. New Preussomerins J, K and L from an Endophytic Fungus: Structure Elucidation, Crystal Structure Analysis and Determination of Absolute Configuration by CD Calculations. Tetrahedron **57**: 4343
42. Browning JW BDZDO/MCDPPD. Program package BDZDO/MCDPPD, Department of Chemistry and Biochemistry, University of Colorado, Boulder, USA. Modified by J Fleischauer J, Schlecker W, and Kramer B, University of Aachen, Germany
43. Krohn K, Steingröver K, Zsila F (2001) The Absolute Configuration of the Palmarumycins C_9, C_{10}, and C_{12} by Quantum-Mechanical Calculations of CD Spectra. Tetrahedron Asymm **12**: 1961
44. Krohn K, John M, Aust H-J, Draeger S, Schulz B (1999) Biologically Active Secondary Metabolites from Fungi, 13. Stemphytriol, a New Perylene Derivative from *Monodictys fluctuata*. Nat Prod Lett **14**: 31
45. Wipf P, Jung J-K (1998) Total Synthesis of Palmarumycin CP_1 and (±)-Deoxypreussomerin A. J Org Chem **63**: 3530
46. Wipf P, Jung J-K (1999) Total Synthesis of the Spiroketal Naphthoquinone (±)-Diepoxin σ. J Org Chem **64**: 1092
47. Wipf P, Jung J-K (2000) Formal Total Synthesis of (+)-Diepoxin σ. J Org Chem **65**: 6319
48. Krohn K, Schlummer S (unpublished results)
49. Kelly TR, Whiting A, Chandrakumar NS (1986) A rationally designed chiral Lewis acid for the asymmetric induction of some Diels-Alder reactions. J Am Chem Soc **108**: 3510
50. Maruoka K, Sakurai M, Fujiwara J, Yamamoto H (1986) Asymmetric Diels-Alder Reaction Directed Toward Chiral Anthracycline Intermediates. Tetrahedron Lett **27**: 4895
51. Coutts IGC, Allock RW, Scheeren HW (2000) Novel Synthetic Approaches to the Palmarumycin Skeleton. Tetrahedron Lett **41**: 9105
52. Barrett AGM, Hamprecht D, Meyer T (1998) Total Syntheses of Palmarumycins CP_1 and CP_2 and CJ-12,371. Novel Spiro-Ketal Fungal Metabolites. J Chem Soc Perkin Trans 1, 809

53. Ragot JP, Alcaraz M-L, Taylor RJK (1998) Synthesis of Palmarumycin CP_1 and CP_2, CJ-12,371 and Novel Analogues. Tetrahedron Lett **39**: 4921
54. Chidambaram N, Chandrsekaran S (1987) *tert*-Butylhydroperoxid-Pyridinium Dichromate: A Convenient Reagent System for Allylic and Benzylic Oxidations. J Org Chem **52**: 5048
55. Bycroft BW, Roberts JC (1963) Mycological Chemistry. XIV. Synthesis of Flavasperone. J Chem Soc 4868
56. Ramachandran PV, Gong B, Brown HC (1985) Chlorodiisopinocamphenyl Borane. Tetrahedron Lett **26**: 1415
57. Chi S, Heathcock CH (1999) Total Synthesis of (±)-Preussomerins. Org Lett **1**: 3
58. Saegusa T, Hirao T, Ito Y (1978) Dehydrierungen. J Org Chem **43**: 1011
59. Ragot JP, Prime ME, Archiblad SJ, Taylor RJK (2000) A Novel Route to Preussomerins via 2-Arylacetal Anions. Org Lett **2**: 1613
60. Barrett AGM, Blaney F, Campbell AD, Hamprecht D, Meyer T, White AJP, Witty D, Williams DJ (2002) A Unified Route to the Palmarumycin and Preussomerin Natural Products. An Enantioselective Synthesis (−)-Preussomerin G. J Org Chem **67** (2735)

Prokaryotic Glycoproteins

P. Messner and C. Schäffer*
Zentrum für Ultrastrukturforschung und Ludwig-Boltzmann-Institut für Molekulare Nanotechnologie, Universität für Bodenkultur Wien, Austria

Contents

1. Introduction

A large number of proteins and lipids in biological systems are glycosylated. Comparisons of well characterized protein sequence data base entries indicate that more than half of all proteins in nature will eventually be identified as glycoproteins. Only recently, the importance of carbohydrates in biology has been more fully appreciated. With the introduction of molecular genetics, simplification of expression

* Corresponding author; crs@edv1.boku.ac.at

methods, and improvement of analytical techniques, a rapid expansion of the field of glycobiology has taken place, particularly in the domain of eukaryotic organisms. For an in-depth review of the structure, function and molecular biology of eukaryotic glycoproteins, the reader is referred to a number of recent reviews and monographs (*1–9*).

For many years, the glycosylation of proteins was thought to be limited to eukaryotes. Presently, however, the existence of prokaryotic glycoproteins is firmly established. In the past two decades, a considerable body of knowledge has accumulated on structure, function, biosynthesis, and molecular biology of these glycoconjugates. The data provide us with first insights into the different glycosylation pathways which have evolved in nature over three billion years of prokaryotic evolution (for recent review see *10–20*). In the early days of prokaryotic glycoprotein research, the evidence for the occurrence of a glycoprotein was almost exclusively inferred from carbohydrate staining of protein bands after sodium dodecylsulfate polyacrylamide gel electrophoresis (SDS-PAGE) and/or lectin binding assays (for review see *16*, *17*). Although today more straightforward analytical techniques like mass spectrometry (MS) and nuclear magnetic resonance spectroscopy (NMR) gradually supersede these methods, they still continue to be applied in recent investigations (for details see Table 1). With the expansion of medical research, some of the prokaryotic organisms that synthesize glycoproteins were found to be significant pathogens, including *Neisseria*, *Mycobacteria*, and *Streptococci*. These findings have opened up a new aspect of microbial pathogenesis, based on the suggestion that glycosylation of prokaryotic proteins can be critical to pathogenicity. It remains to be determined how and where protein glycosylation takes place in or on prokaryotic cells and how it is regulated in response to host environments (*21*).

The three main posttranslational modifications of proteins that involve carbohydrates are *N*- and *O*-glycosylation and attachment of glycosyl phosphatidyl inositol (GPI) anchors. Recently, a new type of protein glycosylation, namely *C*-mannosylation, has been identified in human RNase 2 (*22*). In this case, the anomeric carbon atom of an α-D-mannopyranose residue is directly linked to the C-2 atom of tryptophan 7 of that polypeptide chain. This diversity of glycosidic linkages in different glycoproteins is particularly challenging from the standpoint of chemical synthesis. Recently, Taylor (*23*) has summarized the present knowledge on chemical syntheses of both naturally occurring and artifical glycopeptides ("glycopeptidomimetics") and glycoproteins.

Examples for the first three types of glycosylation have been found not only in eukaryotic glycoproteins but also in prokaryotic glycopro-

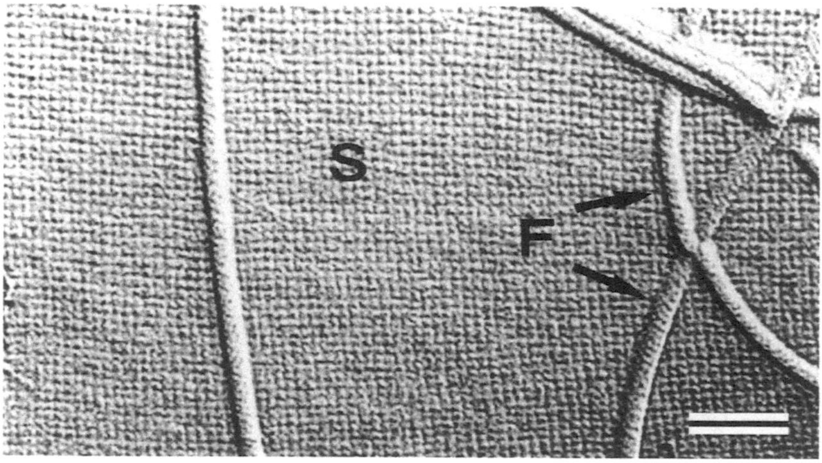

Fig. 1. Electron micrograph of the S-layer glycoprotein lattice from an intact cell of *Thermoanaerobacterium thermosaccharolyticum* D120-70 after freeze-etching and metal-shadowing. Only the protein portion of the S-layer glycoprotein forming the square S-layer lattice can be seen in this preparation. S, S-layer lattice; F, flagellae; Bar = 100 nm [modified from Sleytr & Thorne (*26*)].

teins (Table 1 and Fig. 1). For the latter, Sandercock *et al.* (*13*) have introduced the distinction between surface layer (S-layer) glycoproteins (for review see *11*, *14–16*, *19*, *20*, *24*) and non-S-layer glycoproteins (for review see *12*, *17*, *18*), because S-layer glycoproteins were the first prokaryotic glycoproteins described in detail (*25–27*). Throughout the present review we will use, along with this scheme, the proposed model for presentation of prokaryotic glycoproteins according to their localization in the bacterial cell (for details see *16*). Within the different groups there is frequently no direct connection between the described glycoprotein species; therefore, short introductory statements for each identified glycoconjugate have been included to help present these diverse data in a clear and concise way.

2. S-Layer Glycoproteins

2.1. Uniqueness of S-Layer Glycoproteins

Surface layer (S-layer) glycoproteins are among the best-studied examples of glycosylated prokaryotic proteins. They are widely distri-

Table 1. Prokaryotic Glycoproteins*

Organism	Linkage Region	Detection[a]	References
1. S-Layer Glycoproteins			
Bacteria			
Aneurinibacillus (formerly *Bacillus*) *thermoaerophilus* L420-91	–	BCF	(*74*)
***Aneurinibacillus* (formerly *Bacillus*) *thermoaerophilus* DSM 10155**	**βGalNAc-Thr/Ser**	**AB**	(*76*, ***86***, ***117a***)
***Aneurinibacillus thermoaerophilus* GS4-97**	**βGalNAc-Thr**	**AB**	(***75***, ***112***)
Bacillus smithii	–	F	(*128*)
Clostridium difficile	**–**	**F**	(***93***, ***94***)
Clostridium sp.	–	F	(*132*)
Corynebacterium glutamicum	–	G	(*133*)
Deinococcus radiodurans	–	CF	(*134*)
Desulfotomaculum nigrificans	–	CF	(*131*)
***Geobacillus* (formerly *Bacillus*) *stearothermophilus* NRS 2004/3a**	**βGal-Ser/Thr**	ABE	(*44*, *96*, ***124***, *129–131*)
***Geobacillus stearothermophilus* ATCC 12980, mutant**	**–**	**CE**	(***123a***)
Lactobacillus buchneri	αGlc-Ser	AB	(*97*)
Paenibacillus (formerly *Bacillus*) *alvei*	βGal-Tyr	AB	(*80*, *106*)
Sulfobacillus thermosulfidooxidans	–	CF	(*135*)
Thermoanaerobacter (formerly *Acetogenium*) *kivui*	-Tyr	CE	(*39*, *136*)
Thermoanaerobacter (formerly *Clostridium*) *thermohydrosulfuricus* DSM 568	βGal-Tyr	AB	(*78*, *137*)
Thermoanaerobacter (formerly *Clostridium*) *thermohydrosulfuricus* L77-66	-Tyr?	BCEF	(*137*, *138*)
Thermoanaerobacter (formerly *Clostridium*) *thermohydrosulfuricus* L111-69	βGal-Tyr	AB	(*78*, *139*)
Thermoanaerobacter (formerly *Clostridium*) *thermohydrosulfuricus* S102-70	βGlc-Tyr	AB	(*81*, *98*)
Thermoanaerobacterium (formerly *Clostridium*) *thermosaccharolyticum* D120-70	**βGlc-Tyr**	AB	(***82***, *90*)
Thermoanaerobacterium (formerly *Clostridium*) *thermosaccharolyticum* E207-71	-Tyr	BCEF	(*48*, *73*)

Table 1 (*continued*)

Organism	Linkage Region	Detection[a]	References
***Thermoanaerobacterium thermosulfurigenes* EM1**	**–**	**CE**	(***37***)
Archaea			
Archaeoglobus fulgidus	–	F	(*140*)
Haloarcula japonica	–	F	(*141, 142*)
Halobacterium halobium (formerly *H. salinarium*)	βGlc-Asn, GalNAc-Asn, Gal-Thr	AB	(*10, 54, 143, 144*)
Haloferax volcanii	βGlc-Asn, Gal-Thr	AB**D**	(*14, 56, 65,* ***65a****, 121, 145*)
Hyphomonas jannaschiana	–	F	(*146*)
Methanocorpusculum spp.	–	F	(*147*)
Methanoculleus (formerly *Methanogenium*) *liminatans*	–	F	(*68*)
Methanoculleus (formerly *Methanogenium*) *marisnigri*	–	F	(*148*)
Methanoculleus* (formerly *Methanogenium*) *palmoleii	**–**	**F**	(***66***)
Methanoculleus* (formerly *Methanogenium*) *olentangyi	**–**	**F**	(***68***)
Methanoculleus* (formerly *Methanogenium*) *bourgensis	**–**	**F**	(***66***)
Methanoculleus* (formerly *Methanogenium*) *thermophilus	**–**	**F**	(***67***)
Methanolacinia paynteri	–	F	(*149*)
Methanoplanus limicola	–	EF	(*150*)
Methanosaeta (formerly *Methanothrix*) *soehngenii*	Rha-Asn	ACEF	(*55*)
Methanothermus fervidus	GalNAc-Asn	AB	(*53*)
***Methylobacter alcaliphilus* 20Z**		**CF**	(***70***)
***Methylobacter modestohalophilus* 10S**		**CF**	(***70***)
Methanothermus sociabilis	–	CF	(*122*)
Pyrodictium abyssi	–	F	(*151*)
Staphylothermus marinus	–	CF	(*152*)
Sulfolobus acidocaldarius	–	F	(*153*)
Sulfurococcus mirabilis	–	F	(*154*)
Thermococcus stetteri	–	F	(*155*)
2. Non-S-Layer Glycoproteins			
a) Intracellular Glycoproteins			
Bacteria			
Cellulomonas fimi, β-mannosidase (Man2A)	**–**	**G**	(***158***)
Streptomyces lividans, β-glucosidase	–	F	(*157*)
b) Membrane-associated Glycoproteins			

Table 1 (*continued*)

Organism	Linkage Region	Detection[a]	References
Bacteria			
***Bacillus thuringiensis* subsp. *kurstaki*, crystal toxin**	***N*-glycan**	**DF**	(***165***)
Bacteroides cellulosolvens, cellulosome complex	αGal-Thr/Ser	AB	(*253, 254*)
Bacteroides nodosus, outer membrane protein	–	F	(*255*)
Borrelia burgdorferi, outer membrane protein	GlcNAc-Asn	ADEFG	(*256*)
***Cellulomonas fimi*, mannanase (Man26A)**	**–**	**G**	(***158***)
Chlamydia trachomatis, outer membrane protein	GlcNAc-Asn	AB	(*159-161*)
Chloroflexus aurantiacus, blue copper protein	–	F	(*257*)
***Ehrlichia canis*, outer membrane protein**	**Glc-Ser?**	**CD**	(***173***)
***Ehrlichia chaffeenis*, outer membrane protein**	**Glc-Ser?**	**CD**	(***173***)
***Escherichia coli*, outer membrane protein**	**–**	**F**	(***168***)
Fibrobacter succinogenes, cellulose-binding protein	–	G	(*258*)
Micrococcus luteus (*M. lysodeikticus*), membrane glycoprotein	–	DFG	(*159*)
Myxococcus xanthus, peripheral membrane glycoprotein	–	CF	(*260*)
Pseudomonas syringae, ice nucleation lipoglycoprotein	*O*- and *N*-glycans	ACEFG	(*261, 262*)
***Staphylococcus aureus*, cell wall glycoprotein**	**–**	**G**	(***163***)
Archaea			
Thermoplasma acidophilus, plasma membrane glycoprotein	GlcNAc-Asn	AB	(*263*)
c) *Surface-associated Glycoproteins*			
Bacteria			
Azospirillum brasilense, flagellin	–	EFG	(*17, 264*)
***Burkholderia pseudomallei*, cell surface phosphatase**	**–**	**F**	(***194***)
Campylobacter coli, flagellin	–	DF	(*265, 266*)
Campylobacter fetus, flagellin	–	DF	(*266*)
Campylobacter jejuni, flagellin	**Pse5Ac7Ac-Ser/Thr**	**ABDF**	(***185b***, *266*)
Clostridium thermocellum, cellulosome complex	Gal-Thr	AB	(*79*, ***196***, ***197***, *254, 267*)

Table 1 (*continued*)

Organism	Linkage Region	Detection[a]	References
Clostridium tyrobutyricum, flagellin	–	F	(*175–177*, ***178***)
Escherichia coli, pilin	–	F	(*268*)
***Escherichia coli*, autotransporter adhesin**	**6-deoxy-Tal-Thr**	**C**	(***191c***)
***Mycobacterium smegmatis*, glycophospholipid**	–	**AF**	(***193a***)
Myxococcus xanthus, fimbriae	–	F	(*269*, *270*)
Neisseria gonorrhoeae, pilin	GlcNAc-Ser	AB	(*186*, *187*, ***189***)
Neisseria meningitidis, pilin	2,4-diacetamido-2,4,6-trideoxyhexose-Ser	AB	(*187*, *188*, ***189***, *190*, ***191***, ***191a***, *271*)
Phormidium uncinatum, oscillin	–	CEFG	(*272*)
Pseudomonas aeruginosa, flagellin	–	EF	(*273*)
Pseudomonas aeruginosa 1244, pilin	**βFucNAc-Ser**	**ABC**F	(***185a***, 274)
***Rhodococcus erythropolis*, cell surface glycoprotein**	**–**	**F**	(***192***)
Serpulina hyodysenteriae, flagellin	–	EF	(*275*)
Spirochaeta aurantia, flagellar filament glycoprotein	–	D	(*276*)
Streptococcus salivarius, cell surface fibrils	–	F	(*277*)
***Streptococcus salivarius* mutant A37, fimbriae**	–	**DF**	(***183a***)
Streptococcus sanguis, cell surface fibrils	–	F	(*278*)
***Streptomyces coelicolor*, cell envelope glycoprotein**	**–**	**F**	(***191b***)
***Treponema pallidum*, flagellin**	**–**	**FG**	(***182***)
Archaea			
Halobacterium halobium, flagellin	Glc-Asn	AB	(*279*)
Methanococcus deltae, flagellin	–	FG	(*280*, *281*)
***Methanococcus voltae*, flagellin**	–	**G**	(***185***)
Methanospirillum hungatei, flagellin	–	EF	(*282*, *283*)
Methanothermus fervidus, flagellin	–	F	(*280*)
Natronobacterium magadii, flagellin	–	F	(*284*)
Sulfolobus acidocaldarius, flagellin	–	F	(*285*)
Sulfolobus shibatae, flagellin	–	F	(*286*)
Thermoplasma volcanium, flagellin	–	F	(*286*)

Table 1 (*continued*)

Organism	Linkage Region	Detection[a]	References
d) Extracellular Glycoproteins			
Bacteria			
***Actinobacillus actinomycetemcomitans*, 37-kDa antigen**	**–**	**CF**	(***215***)
Bacillus circulans, β-mannanase	–	C	(*287*)
Bacillus sp., cellulase and xylanase component	–	DF	(*288*)
Cellulomonas fimi, exoglucanase (Cex)	Man/Gal-Thr	ACDEF	(*289*)
Cellulomonas fimi, exoglucanase (CenA)	–	DF	(*290, 291*)
Cellulomonas sp., cellulase	–	DF	(*292*)
Corynebacterium sepedonicum, phytotoxin	Man-Thr	AC	(*293*)
***Cytophaga sp.*, adhesion inhibitor**	**–**	**CF**	(***220***)
Erysipelothrix rhusiopathiae, antigen	–	F	(*294*)
Fibrobacter succinogenes, cellobiosidase	–	F	(*295*)
Flavobacterium meningosepticum, endoglycosidase	Man-Ser/Thr	AB	(*296, 297*)
***Mycobacterium bovis*, several antigens**	**GlcNAc-Asn**	**ACDF**	(*200*, ***203***, *298*)
***Mycobacterium tuberculosis*, 55-kDa antigen**	**–**	**DF**	(*199*, ***202***)
Mycobacterium tuberculosis, 45-kDa antigen	αMan-Thr	AB	(*201, 299*)
***Mycobacterium tuberculosis*, 32 kDa antigen**	**–**	**CDF**	(***202***)
Mycobacterium tuberculosis, 19 kDa antigen	–	DF	(*300*)
***Mycoplasma hyopneumoniae*, adhesin-like glycoprotein**	**–**	**DE**	(***198***)
***Pseudoalteromonas antarctica* NF_3, exopolymer**	**–**	**F**	(***218a, b***)
***Pseudomonas putida*, antifreeze protein**	**–**	**F**	(***217***)
Streptomyces lividans, cellulase	Man/Gal-Thr	ACDEF	(*157, 289*)
Streptomyces lividans, xylanase	–	F	(*301*)
***Thermoanaerobacter thermohydrosulfuricus*, α-glucosidase**	**–**	**F**	(***214***)

Table 1 (continued)

Organism	Linkage Region	Detection[a]	References
Thermoanaerobacterium (formerly *Clostridium*) *thermosaccharolyticum*, extracellular protein complex	–	CF	(*302*)
Thermomonospora fusca, cellulase	–	F	(*303*)
Vibrio sp., β-mannanase	**–**	**F**	(***212***)
Archaea			
Methanobacterium bryantii, copper response protein	–	DE	(*304*)
e) Cellular Glycoproteins			
Bacteria			
Actinomyces naeslundii, aggregation factor	–	**CF**	(***234***)
Bacillus megaterium, protein factor PG-1	–	F	(*305*)
Bacillus thuringiensis, sporangia glycoproteins	*O*-glycans-Ser	ACEFG	(*306*)
Bacillus thuringiensis subsp. *israelensis*, toxin	GlcNAc-Asn	ACDFG	(*307–309*)
***Campylobacter jejuni*, different glycoproteins**	**–**	**EF**	(***227***)
Clostridium acetobutylicum, autolysin	–	F	(*310*)
Enterococcus hirae (formerly *Streptococcus faecium*), N-acetylmuramoylhydrolase	*O*-glycan	CDF	(*311*)
***Flexibacter columnaris*, 60-kDa antigen**	**–**	**F**	(***241***)
***Klebsiella pneumoniae*, glycoprotein fraction P1**	**–**	**F**	(***229***)
***Lactobacillus reuteri*, invertase**	**–**	**F**	(***243***)
***Mycobacterium vaccae*, boiling-water extract**	**–**	**BC**	(***239, 240***)
***Prevotella intermedia*, glycoprotein PGP**	**–**	**CF**	(***233***)
***Streptococcus pyogenes* strain Manfredo, streptococcal acid glycoprotein SAGP**	–	**CF**	(***224, 225***)
Streptococcus pyogenes strain Su, streptococcal acid glycoprotein SAGP	–	CF	(*221-223*, ***224***, ***225***, *226*)
Streptococcus sanguis, platelet aggregation-associated protein	GlcNAc-Asn	ACEFG	(*12*)

Table 1 (*continued*)

Organism	Linkage Region	Detection[a]	References
Archaea			
Methanosarcina mazei, whole cell extract	–	D	(*312*)
f) Synthetic Glycoproteins and Glycopeptides			
High-mannose-type glycopeptide analog	**Glc-Asn**	**ABF**	**(*248*)**
***Bacillus lentus*, modified subtilisin**	**–**	**F**	**(*249*)**
***Geobacillus* (formerly *Bacillus) stearothermophilus* NRS 2004/3a, secondary cell wall polymer**	**βGlcNAc-PP-MurNAc**	**ABC**	**(*36, 42, 252*)**
***Paenibacillus alvei* CCM 2051, secondary cell wall polymer**	**βGlcNAc-PP-MurNAc**	**ABC**	**(*45*)**

* Organisms printed in boldface represent new entries to the list, which have not been considered in previous reviews.
[a] Abbreviations for detection methods: A = linkage region; B = glycan structure; C = monosaccharide composition; D = lectin binding; E = deglycosylation; F = detection of glycoses; G = other methods.

buted in the major lineages of *Archaea*, as well as among *Bacteria*; in the latter lineage, they have been demonstrated exclusively within the low G+C Gram-positive phylum (*e.g.*, *Aneurinibacillus*, *Geobacillus* (formerly *Bacillus*), *Clostridium*, *Desulfotomaculum*, *Paenibacillus*, *Thermoanaerobacter*, *Lactobacillus*) (*15*, *19*).

In general, S-layers are two-dimensional crystalline arrays composed of identical protein or glycoprotein subunits (for an overview see *28*, *29*). They form an outermost envelope layer, 5 to 20 nm thick, which completely covers the prokaryotic cell during all stages of cell growth and division. On growing cells, S-layers represent "dynamic closed surface crystals". The constituting subunits, with apparent molecular masses ranging between 40 and 200 kDa, are aligned in lattices with oblique, square or hexagonal symmetries, exhibiting center-to-center spacings of 2.5 to 35 nm between the morphological units (Fig. 1). S-Layers are planar assemblies of weakly acidic proteins or glycoproteins with isoelectric points, in general, in the range of 4–6 and are typically composed of 40 to 60% hydrophobic amino acids and a rather low, if any, portion of sulfur-containing amino acids. Hydrophobic and hydrophilic amino acids do not form extended clusters; approx. 20% of them are organized as α-helices, approx. 40% occur as β-sheets; aperiodic foldings and the content of β-turns vary between 5 and 45%. High-resolution electron microscopy, scanning force microscopy and three-dimensional image reconstruction studies have revealed that pores of identical size and morphology occupy up to 70% of the S-layer surface. The inherent repetitive physicochemical and structural properties down to the sub-nanometer scale, in combination with the self-assembly characteristics, make S-layers unique organic natural structures.

S-Layers represent the most simple self-assembling membranes that have developed during the evolution of life. They have evolved as a consequence of the direct interaction of prokaryotic cells with different habitats and changing ecological conditions, presumably providing them with a selection advantage. It is recognized that S-layers can act i) as framework to determine cell shape and to aid in the cell division process in some archaea; ii) as structures involved in cell adhesion and surface recognition; iii) as protective coats, molecular sieves, molecule and ion traps or adhesion zones for exoenzymes (*30*, *31*).

Distinct surface properties, such as charge distribution and hydrophobicity, as well as specific interactions with components of the underlying cell envelope layer, are essential for the proper orientation of S-layer (glyco)protein subunits and their ability to assemble into extended crystalline arrays (*32*). The subunits themselves are held together by weak, non-covalent forces. In some species, *e.g.*, in the S-layer proteins SbsA (*33*), SbsB (*34*) and SbsC (*35*) from *Geobacillus (formerly Bacillus) stearothermophilus* (*35a*) wild-type strains, a conserved structural N-terminal element was found to be responsible for anchoring the S-layer subunits via a linker molecule to the underlying peptidoglycan layer (*29*, *36*, *36a*). Recently, an N-terminal so-called S-layer homology (SLH) domain was identified as attachment site for the S-layer glycoprotein of *Thermoanaerobacterium thermosulfurigenes* EM1 to the peptidoglycan layer (*37*). It is proposed that SLH domains mediate attachment via secondary cell wall polymers (SCWPs), covalently bound to peptidoglycan (*38*, *39*). First indications for this mechanism of recognition and association were derived from affinity studies performed with peptidoglycan preparations and the S-layer proteins from different *Geobacillus stearothermophilus* strains (*40*, *41*). Only those proteolytic cleavage fragments of the S-layer proteins that possessed the complete N-terminal region were able to bind and recrystallize into the appropriate lattice type on native peptidoglycan-containing sacculi, provided the secondary cell wall polymer had not been removed. Interestingly, identical SCWPs (*42*) in the peptidoglycans of all *Geobacillus stearothermophilus* wild-type strains investigated so far, seem to be responsible for binding of the S-layer proteins. Initial studies on the structure-function relationships of the S-layer protein from *Geobacillus stearothermophilus* PV72/p2 revealed the coexistence of two binding domains on its N-terminal part, one for peptidoglycan and another for the SCWP. The SCWP-binding domain is located between amino acids 240 and 331 and possesses high serine and glycine contents. The peptidoglycan-binding domain is located between amino acids 1 and 138 of the mature S-layer protein, comprising a typical SLH domain (*36*).

SLH domains have not only been found in S-layer proteins, but also in polysaccharidases and other exocellular proteins from a large variety of bacteria (reviewed in *30*). They are regarded as one of the modes bacteria have evolved to display polypeptides on their cell surface, involving a non-covalent interaction between these domains and various cell wall components. In this context, peptidoglycan (*38*, *39*, *43*) or proteins possessing own SLH domains have also been discussed as interacting with SLH domains through protein-peptidoglycan or protein-protein interactions, respectively. However, the details underlying this cell surface display mechanism have not been fully elucidated.

Besides S-layer glycoproteins and the peptidoglycan, secondary cell wall polymers are one of the major glycosylated cell wall components of Gram-positive bacteria. Thus, they are not only of functional relevance in terms of mediating proper, non-covalent attachment, but also as carbohydrate structures *per se*, contributing to the (physico)chemical properties of the cell wall. The distinct mode of association between SCWP and S-layer (glyco)proteins is currently being investigated. In the course of the characterization of S-layer glycoprotein glycans of various Gram-positive strains, residual amounts of SCWP were frequently co-isolated. As they were readily removed by gel-permeation chromatography using 10 mM sodium chloride as eluent, one can conclude that the binding affinity between the SCWP and the S-layer glycoprotein is of high specificity, but of rather low strength. To unambigously discriminate between S-layer-derived and peptidoglycan-associated polysaccharides, the type of linkage of either glycoconjugate molecule was determined. The first SCWP-peptidoglycan complex whose structure was completely elucidated is the polymer of *Geobacillus stearothermophilus* NRS 2004/3a (*42*, *44*). This anionic polymer, comprising in average five tetrasaccharide repeating units with the structure structure → 4)-β-D-ManA2,3(NAc)$_2$-(1→6)-α-D-Glc-(1→4)-β-D-ManA2,3(NAc)$_2$-(1→3)-α-D-GlcNAc-(1→, represents the SCWP of *G. stearothermophilus* wild-type strains. For *Paenibacillus alvei* CCM 2051, the structure of the complex is proposed to be [(Pyr4,6)-β-D-Man*p*NAc-(1→4)-β-D-Glc*p*NAc-(1→3)]$_{n\sim 11}$-(Pyr4,6)-β-D-Man*p*NAc-(1→4)-α-D-Glc*p*NAc-(1→O)-PO$_2$-O-PO$_2$-(O→6)-MurNAc-. Each disaccharide unit is substituted by 4,6-linked pyruvic acid residues (*45*). The anionic glycan chains of both organisms constituting the SCWP are randomly linked via pyrophosphate groups to C-6 of muramic acid residues of the peptidoglycan layer, with about 20–25% of the muramic acid residues substituted with these polysaccharide chains. In the SCWP of *Bacillus sphaericus* CCM 2177, identical disaccharide motifs were found (*46*). The relevance of anionic polymers for the bacterial cell is frequently discussed in context with their contribution to the polyelectrolyte gel structure of the cell wall and maintenance of the peptidoglycan network in an expanded state, due to charge repulsion (*47*). Besides the mentioned charged polymers, there is strong evidence for the presence of neutral SCWP in various S-layer glycoprotein-covered Bacillaceae, *e.g.*, *Thermoanaerobacterium thermosaccharolyticum* strains E207-71 (*48*) and D120-70 (Schäffer, Kählig, Messner, unpublished observations) and *Aneurinibacillus thermoaerophilus* DSM 10155 (*48a*). From compositional and structural data we conclude that the pyrophosphate linkage is the preferred type of covalent linkage between the SCWP and the peptidoglycan in S-layer-covered, Gram-positive organisms (*42*, *45*). Interestingly, from studies with *Bacillus anthracis* it was only recently proposed that pyruvylated monosaccharides of SCWP constitute the ligand of SLH domains (*49*). Considering that the polymer structures from S-layer carrying organisms elucidated so far include anionic and neutral ones, the modification of such polymers by pyruvylation cannot be regarded as a general requirement for the interaction between SLH domains of several bacterial proteins and SCWP. This assumption is further substantiated by an observation made with *Lactobacillus buchneri* whose S-layer protein is bound to the peptidoglycan sacculus via a neutral polymer (*50*).

Post-translational modifications known to occur in S-layer proteins include cleavage of N- and C-terminal fragments, phosphorylation, and glycosylation, the latter being the main focus of this chapter. Glycosylation is the structurally most diverse modification the newly synthesized protein may undergo (*51*). This diversity is exquisitely exemplified by S-layer glycoprotein glycans, being observed with respect to i) monosaccharide constituents; ii) glycan chain length; iii) glycosidic linkage between carbohydrate and protein portions; iv) interlinking core or pseudo-core region and v) acceptor sequence on the S-layer polypeptide. Most importantly, fundamental differences in structure and biosynthesis have been observed in comparison to the well-investigated eukaryotic glycoproteins.

The detailed analysis of chemical compositions and structures of S-layer glycoproteins, as well as the elucidation of the corresponding pathways constitute the foundations for the construction of tailor-made S-layer (*neo*)glycoproteins, either by organic-chemical synthesis reactions or by genetic engineering techniques. From this knowledge base, new applications based on these glycoproteins will emerge in the future.

In the present chapter, each aspect of prokaryotic protein glycosylation will be addressed in the specific context of S-layer glycoproteins.

2.2. From Chemical Composition to Structural Concept

In Gram-positive bacteria, the (glyco)proteins immobilized on the cell surface are generally of particular interest. As outermost components of the cell envelope, they are involved in crucial physiological processes, and, in the case of pathogenic bacteria, participate in various steps of the infection process, such as adhesion and/or invasion of host cells, binding to host molecules, and protection against phagocytosis (reviewed in *52*).

S-Layer glycoproteins constitute the outermost surface envelope of bacteria. Thus, they obviously play a pivotal role in diversifying the surface properties of the respective organisms for their survival in a competitive habitat. This clearly implies physiological relevance of S-layer glycoprotein glycans. It might be expected that the S-layer glycans, protruding from the cell surface up to 30–40 nm, create a hydrophilic polysaccharide coat similar to that of lipopolysaccharide (LPS) O-antigens of Gram-negative bacteria.

Presently (December 2001), approximately 35 different S-layer glycoprotein glycans from the domains of *Archaea* and *Bacteria* have been characterized (Table 1). For a complete and comprehensive description of these structures the reader is referred to recently published reviews (*19*, *19a*). The objective of this present chapter is to summarize the information on S-layer glycoproteins published since (relevant organisms are written in boldface in Table 1), and to focus on the deduction of some general compositional, structural and eventually biosynthetic concepts of S-layer glycoproteins.

In the case of *Archaea*, most S-layers are glycosylated. Archaeal S-layer glycoprotein lattices have predominantly hexagonal symmetry and are often the only wall component outside the cytoplasmic membrane. The structural data from archaeal investigations have been reviewed previously (*14*). Briefly, archaeal S-layer glycans consist almost exclusively of short heteropolysaccharides with up to ten sugar residues and no repeats (*14*, *15*, *53*). They are bound directly, without an interlinking saccharide core, to the polypeptide portion, predominantly by *N*-glycosidic linkages, such as Glc→Asn (*10*, *54*), GalNAc→Asn

References, pp. 105–124

(*14*, *53*), and Rha→Asn (*55*). On the other hand, short *O*-glycans, linked to threonine, have also been found in clusters close to the membrane-spanning domain in halobacterial S-layer glycoproteins (*10*, *25*, *56*).

The most detailed information is available from the S-layer glycoprotein of *Halobacterium halobium*. This organism is covered with a hexagonally arranged S-layer glycoprotein lattice. A glycosaminoglycan-like heteropolysaccharide from *H. halobium* built of up to 15 repeats is the only long-chain archaeal S-layer glycoprotein glycan known so far (for review see *14*, *27*, *57*). Besides the mentioned S-layer *N*-glycan, another type of sulfated, glucuronic acid-containing, *N*-glycosidically bound saccharide with low molecular mass is present on the same S-layer polypeptide. Thus, the glycoprotein of *H. halobium* presents the interesting situation of being equipped with two different types of *N*-glycosidic linkages, situated within the typical *N*-glycosylation acceptor sequence Asn-X-Ser/Thr. A total of 12 potential *N*-glycosylation sites are found throughout the polypeptide chain. The third S-layer glycan of this organism is a neutral disaccharide with the structure α-D-Glc-(1→2)-Gal-(1→, present in about 20 copies. It occurs in highly clustered arrangements, *O*-glycosidically linked via threonine (*58*).

Identical *O*-glycosidically linked neutral disaccharide chains represent the glycan portion of the hexagonal S-layer lattice of the moderate halophile *Haloferax volcanii*. As is the case with the extreme halophile *H. halobium*, S-layer *O*- and *N*-glycans are simultaneously expressed on the bacterial cell surface (*56*). Concerning archaeal *N*-glycans, it was demonstrated that the glycosaminoglycan chains are necessary for maintaining the structure of the S-layer glycoprotein and, hence, the rod-like shape of either organism. Differences in the net charge of the *N*-saccharides reflect the adaptation to the respective moderately and extremely halophilic environments (for review see *14*). Highly negatively charged loops are required for the stabilization of the S-layer protein in high salt concentrations (*59*). Recently, the S-layer glycoprotein of *H. halobium* and its relation to *Hf. volcanii* was studied in detail by electron microscopy (*60*). As the envelopes of both microorganisms maintain the integrity of the S-layer to only low resolution or disintegrate completely at the salt concentrations suitable for either negative-staining or cryo-electron microscopy, special techniques have been developed. In order to utilize the structural symmetry of the S-layer, isolated cell envelopes were studied as flattened, double-layered two-dimensional arrays (*61*) and cryo-electron microscopy was performed using freeze-substituted cells. Two-dimensional projection maps of both organisms showed striking similarity in array symmetry and subunit organization, at least to a resolution of 2 nm. The morphological units constituting the S-layer glycoprotein lattice have hexagonal arrangement with a center-to-center spacing of 15 nm. This result is not surprising as both organisms require similarly extreme osmotic conditions. In fact, the organization of the S-layer glycoproteins in a number of organisms from extreme environments, *e.g.*, *Sulfolobus shibatae*, *Thermoproteus tenax*, *Staphylothermus marinus*, *Archaeoglobus fulgidus*, have a similar architectural theme (for review see *62*). This arrangement usually consists of a dome-like structure separated from the cell membrane by spacer molecules, resulting in the creation of a quasi-periplasmic space (*63*, *64*). Interestingly, a recent study revealed the existence of several glycosylated, outwardly-orientated *Hf. volcanii* plasma membrane proteins besides the well-investigated 190 kDa S-layer glycoprotein (*65*, *65a*). They possibly exist in the quasi-periplasmic space created by the glycan moiety of the *Hf. volcanii* S-layer glycoprotein with their glycan structures participating in the architecture of this space. Unlike with the S-layer glycoprotein, the membrane association of the 150, 98, 58 and 54 kDa glycoprotein species is Mg^{2+}-independent and very intimate, as suggested by their non-extractability in urea. It was clearly demonstrated by immuno-blotting that the novel glycoproteins are neither precursors nor breakdown products of the 190 kDa S-layer glycoprotein (*65*, *65a*).

Glycosylated S-layer proteins have also been reported for different methanogen species (see Table 1). Only recently, a novel S-layer glycoprotein-covered species was added to the genus *Methanoculleus*. The S-layer of the coccoid methanogen *Methanoculleus palmolei* is composed of hexagonally arranged glycoprotein subunits with a molecular mass of 120 kDa (*66*) and possesses a lattice constant of 15.2 nm. Interestingly, all *Methanoculleus* species studied so far have hexagonal S-layer architecture and glycoprotein subunits (*67*, *68*). However, the glycoprotein nature of the S-layers was only inferred from a positive PAS-staining reaction in SDS-PAGE, and no structural information is available on these S-layer glycan chains.

The pseudomurein sacculus of the hyperthermophilic archaeon *Methanothermus fervidus* was reported to be covered by an S-layer glycoprotein lattice. The heterosaccharide is linked via *N*-acetylgalactosamine to asparagine residues within 20 potential *N*-glycosylation sites of the peptide portion and its primary structure was elucidated several years ago (*53*). A promising first step toward the description of the crystal and molecular structure of this S-layer glycoprotein was taken by Evrard *et al.* (*69*). Until then, structural information from crystalline two-dimensional S-layers was obtained only from electron microscopic and computer image enhancement techniques. In fact, the S-layer glycoprotein of *M. fervidus* was the first S-layer protein to be successfully crystallized. Three-dimensional crystals were formed under microgravity conditions (*69*). The crystallization of the S-layer subunits is a prerequisite for detailed X-ray analysis. The complete elucidation of the three-dimensional structure of this molecule should provide new insights into the molecular principles stabilizing archaeal S-layer glycoproteins in general, and allow the deduction of survival strategies of these organisms which live under extreme environmental conditions such as high temperature, low or high pH value, and high salt concentration.

Concerning the functionality of archaeal S-layer glycoproteins, observations were made with the halotolerant methanotrophs *Methylobacter alcaliphilus* 20Z and *Methylobacter modestohalophilus* 10S that the osmotic pressure of the surrounding medium is balanced by structure/function alterations of the cell envelope and changes of the membrane composition. Regularly arranged S-layer glycoproteins of oblique or hexagonal symmetry, composed of neutral glycoses such as glucose, galactose, mannose, and ribose, were observed on the outer cell walls of *Mb. modestohalophilus* 10S and *Mb. alcaliphilus* 20Z, respectively (*70*). Osmoadaptation at pH 7.2 in the absence of NaCl of the latter coincides with the loss of the tightly packed, cup-shaped S-layer subunits. Thus, the variation of the S-layer might be seen as a mechanism that has evolved in halotolerant bacteria to cope with osmotic stress.

Several differences have been observed between *Archaea* and *Bacteria* regarding general features of S-layer protein glycosylation, indicating once more the wide variety of prokaryotic glycosylation in comparison to eukaryotic glycosylation (the overall occurrence of archaeal and bacterial S-layer glycoproteins is summarized in Table 1). In general, bacterial glycosylation is a rapidly growing field of interest, a substantial body of data having accumulated during the past years. Complete structural analyses of S-layer glycoprotein glycans from Gram-positive bacilli, with the main focus on different mesophilic and thermophilic *Geobacillus* (formerly *Bacillus*) and *Thermoanaerobacter* (formerly *Clostridium*) strains, have been described only by our group, so far. For this present review, certain organisms with glycosylation characteristics fully analyzed only recently have been chosen for the detailed explanation of particular and typical features of S-layer glycoprotein glycans. The methods applied are mentioned briefly in each context.

An important aspect of S-layer glycoprotein glycan analysis concerns the nature of the glycosidic linkage. Identificaton of the glycosylated linkage amino acid constitutes the only unequivocal demonstration of the glycoprotein nature of a given molecule. For glycopeptide purification and analysis, some general methods have been adapted to S-layer glycoprotein research and new ones have been developed. For a survey of methods the paper by Schäffer *et al.* (*20*) is recommended. Briefly, S-layer glycoproteins are obtained from continuously fermented bacteria after breaking the cells by sonication, extraction of the cytoplasmic membrane with Triton X-100, and treatment of the resulting cell walls with the chaotropic agent guanidinium hydrochloride, resulting in disintegration of the S-layer into constituting glycoprotein subunits. Upon dialysis so-called "S-layer self assembly products" are formed, which are the starting material for the production of S-layer glycopeptides by controlled proteolytic degradation. All characterizations of glycan chains, including structure elucidation and identification of linkage regions and acceptor sequences on the S-layer polypeptide are performed on glycopeptide species which have been purified to homogeneity by several chromatographic and/or electrophoretic steps. Ideally, such glycopeptides possess a rather short peptide portion. Finally, structural data are obtained by a combination of nuclear magnetic resonance spectroscopy (NMR) techniques and selective chemical degradation experiments, the latter affecting either the peptide or the carbohydrate portion.

The monosaccharide constituents of most bacterial S-layer glycan chains include a wide range of neutral hexoses, rarely pentoses, 6-deoxy hexoses, and amino sugars. This spectrum is further extended by unusual glycoses, some of which typically occur in lipopolysaccharide (LPS) O-antigens of Gram-negative bacteria (for review see *71*, *72*). Examples for unusual monosaccharides in bacterial S-layer glycans are 3-acetamido-3,6-dideoxy-D-glucose (quinovosamine), observed in *Thermoanaerobacterium thermosaccharolyticum* E207-71 (*73*), D-rhamnose and 3-acetamido-3,6-dideoxy-D-galactose (*N*-acetyl-D-fucosamine), both occurring in *Aneurinibacillus thermoaerophilus* L420-91 (*74*) and GS4-97 (*75*) as well as D-*glycero*-D-*manno*-heptose in the S-layer glycan of *Aneurinibacillus thermoaerophilus* DSM 10155 (*76*). Typically, linear or branched S-layer glycan repeating units are composed of two to six monosaccharide residues with an overall degree of polymerization of up to 50 units in the mature S-layer glycoprotein. This corresponds to a length of approximately 30–40 nm of the outwardly oriented glycan chain. In some cases, additional non-carbohydrate substituents such as phosphate or glycerol are present in the S-layer glycan repeats (*16*).

The square S-layer lattice of *A. thermoaerophilus* GS4-97, exhibiting a center-to-center spacing between the morphological units of approx. 10 nm, consists of identical glycoprotein monomers with a molecular mass of 109 kDa (*75*). After thorough proteolytic degradation of the purified S-layer glycoprotein with pronase E and purification of the reaction mixture by gel permeation, cation-exchange chromatography, chromatofocusing, and reversed-phase (C18) HPLC, a homogeneous glycopeptide fraction was obtained. Monosaccharide analysis by HPAEC-PED revealed rhamnose and *N*-acetylfucosamine in a molar ratio of approx. 2:1 as main constituents of the S-layer glycan, together with a small amount of *N*-acetylgalactosamine. On the basis of integration of NMR signals corresponding to the constituting monosaccharides, an average of 15 hexasaccharide repeating units was estimated for the S-layer glycan chain. Additional signals of rhamnose residues were assigned to the core linkage region between the repeats and the polypeptide-associated β-D-GalNAc residue. The composition of the core varies between one and three α-D-Rha residues, exhibiting the sequence-[α-D-Rha]$_{n=0-2}$-(1→3)-α-D-Rha-(1→3)-β-D-GalNAc. Generally, due to their small content, the elucidation of core structures is more complicated than that of the repeats. Besides these signals, minor signals corresponding to the non-reducing terminal sugar moiety and to the amino acids present in the sample were resolved. By N-terminal sequencing, the polypeptide portion was determined to be Ser-**Thr**-Gly-Asp-Thr, with threonine (in boldface) serving as glycosylation site. With the *A. thermoaerophilus* GS4-97 S-layer glycan, for the first time in glycoproteins, a β-D-GalNAc-(1→*O*)-Thr linkage region was identified. The β-configuration was determined by means of one-bond carbon-proton coupling constants. This finding is in contrast to the common α-D-GalNAc-(1→*O*)-Thr/Ser connections found in eukaryotic glycoproteins (for review see *77*). Another remarkable feature of the *A. thermoaerophilus* GS4-97 S-layer glycan is the occurrence of an *O*-methyl group on the non-reducing terminal rhamnose. Interestingly, the same type of capping, a terminal 3-*O*-methylrhamnose, was observed in the S-layer glycans of *A. thermoaerophilus* L420-91 (*74*) and *Thermoanaerobacter thermohydrosulfuricus* strains L111-69 and L110-69 (*78*). Very recent investigations of the S-layer glycoprotein of *Geobacillus stearothermophilus* NRS 2004/3a revealed that the terminal unit is modified by a 2-*O*-methyl group (*124*). Thus, one might speculate that the presence of *O*-methylated glycose residues plays a role as termination signal for chain elongation during S-layer glycan biosynthesis, as was also discussed with other carbohydrate chains (*79*).

The glycan chains of *Thermoanaerobacter thermohydrosulfuricus* strains L111-69 and L110-69 have identical constituents, and the number of disaccharide repeats of the structure →3)-α-L-Rha-(1→4)-α-D-Man-(1→ varies between 23 and 33, with a maximum at 27 repeats (*78*). In *Tb. thermohydrosulfuricus* L111-69 as well as in *Paenibacillus alvei*

CCM 2051, the S-layer glycan chain is bound via almost identical →3)-α-L-Rha-(1→3)-α-L-Rha-(1→3)-α-L-Rha-(1→3)- core oligosaccharides to the β-D-GalNAc(1→*O*)-Tyr glycosidic linkage unit (*80*). This finding indicates that cores of S-layer glycoprotein glycans are frequently composed of rhamnose residues. Another remarkable feature particular to these glycans is the presence of an *O*-glycosidic linkage via tyrosine. This linkage type was a novelty when first described with strain *Tb. thermohydrosulfuricus* S102-70 (*81*), but in the meantime several other species from the genera *Thermoanaerobacter* and *Thermoanaerobacterium* (*73*, *82*) have been shown to possess this type of linkage (see Table 1).

Another member of the newly described species *Aneurinibacillus* (*83*), *Aneurinibacillus thermoaerophilus* DSM 10155, possesses the only heptose-containing S-layer glycan known so far, giving rise to a total carbohydrate content of the S-layer glycoprotein of 15% (*76*). There are speculations that heptose has been introduced into this organism by lateral gene transfer (for review see *84*, *85*). The glycosylated protomers constituting the square S-layer lattice have a molecular mass of 153 kDa and the disaccharide repeating unit structure of the S-layer glycan was elucidated as →4)-α-L-Rha-(1→3)-β-D-*glycero*-D-*manno*-Hep-(1→(*76*). With this organism, the structural heterogeneity of the linkage region occurring in S-layer glycoprotein glycans is very well exemplified (*86*). This includes both the structure of the core oligosaccharides and their carbohydrate-protein linkage regions. Using modified Edman-degradation protocols (*87*), alkali-labile *O*-glycosidic linkages of GalNAc to serine and threonine residues have been identified. Again, as in *A. thermoaerophilus* GS4-97, *N*-acetylgalactosamine appeared to be linked to the 75 kDa S-layer polypeptide backbone through a β-*O*-linkage. Most likely, both types of glycosidic linkages, β-D-GalNAc(1→*O*)-Thr and β-D-GalNAc(1→*O*)-Ser, occur only once on the S-layer polypeptide. The corresponding peptide sequences derived from MALDI-TOF-MS are Thr-**Thr**-Asn-Xaa-Lys and **Ser**-Gly-Glu-Thr-Xaa-Xaa, respectively (glycosylated amino acids are shown in boldface; Xaa represents unidentified amino acids). Combined evidence from Edman-degradation on Smith-degraded, fluorescently labelled glycopeptides, in combination with liquid chromatography/mass spectrometry analyses of the cleaved glycosylated amino acids and NMR revealed that both glycopeptide species contain equal amounts of the complete core structure α-L-Rha-(1→3)-α-L-Rha-(1→3)-β-D-GalNAc(1→*O*)-Thr/Ser and the truncated forms α-L-Rha-(1→3)-β-D-GalNAc(1→*O*)-Thr/Ser and D-GalNAc(1→*O*)-Thr/Ser. In the S-layer glycoprotein glycan, each core structure is modified by a varying number of repeating unit disaccharides. The variation of S-layer glycan chain length resulting from different degrees of polymerization of identical repeating units can be regarded a common feature of S-layer glycoprotein glycans and is referred to as "glycan chain microheterogeneity". Concerning functional aspects of the S-layer glycoprotein lattice, it was demonstrated with *A. thermoaerophilus* DSM 10155, among other S-layer-carrying *A. thermoaerophilus* and *Geobacillus stearothermophilus* strains, that S-layer(glyco)proteins on Gram-positive bacteria do not form a defensive barrier against the lytic action of membrane vesicles from *Pseudomonas aeruginosa*. During the attack, a peptidoglycan hydrolase is liberated from the vesicles which digests the underlying peptidoglycan-containing cell wall, but the S-layer itself is not affected and remains intact (*88*).

In the course of the elucidation of the complete S-layer glycan structure of *Thermoanaerobacterium thermosaccharolyticum* D120-70, the phenomenon of glycan chain microheterogeneity was, for the first time, fully evaluated (*82*). Until then, the length of S-layer glycans was calculated as an average value obtained from NMR integrals or MALDI analyses of intact (*89*) or degraded S-layer glycoproteins (*78*). The S-layer protein of *T. thermosaccharolyticum* D120-70 bears glycosidically bound glycan chains with a previously established branched hexasaccharide repeating unit structure formed from neutral glycoses (*90*). Examination of the three glycopeptide fractions obtained after pronase E treatment of the glycoprotein indicated the presence of the glycosylated amino acid tyrosine at the N-terminus of each peptide. From sequencing data it can be concluded that there are at least three different glycosylation sites present on the intact S-layer polypeptide of that organism. The corresponding sequences are **Tyr**-Thr, **Tyr**-Ser-Pro-Ala, and **Tyr**-Asp. Native polysaccharide PAGE was used to identify individual glycoforms, *i.e.*, identical peptide portions exhibiting differences in the degree of polymerization of hexasaccharide repeats. The electrophoretic pattern reflected glycoforms that differed in chain length by single hexasaccharide units, supporting the proposed biosynthetic concept of a stepwise assembly of S-layer glycoprotein glycans. The overall degree of polymerization varied between three and ten, but differed within the three glycopeptide fractions. These data were confirmed by MALDI-TOF MS of the respective, gel-eluted and purified glycopeptide fractions. For examination of the carbohydrate-protein linkage region, the S-layer glycan chain was partially hydrolyzed with trifluoroacetic acid. With the aid of 1D and 2D NMR spectroscopy, including a novel diffusion-edited experiment, in combination with the MALDI-TOF evidence, the *O*-glycosidic linkage region was shown to be β-D-Glc-

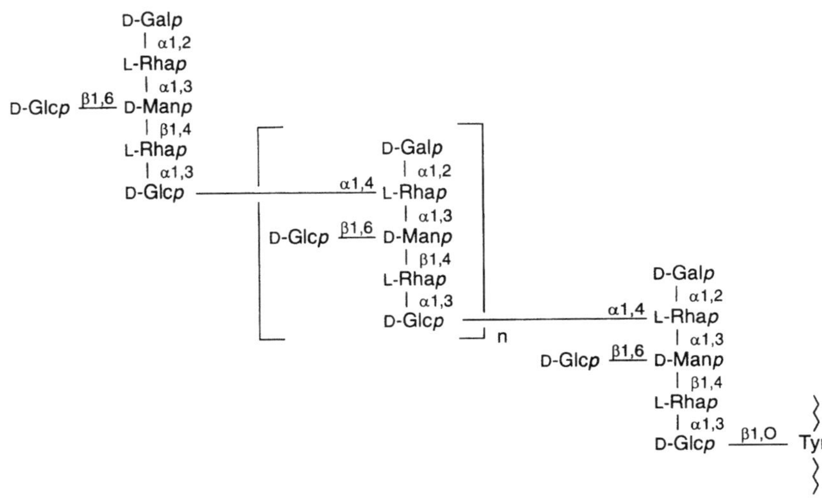

Fig. 2. Schematic representation of the complete S-layer glycoprotein glycan structure from *Thermoanaerobacterium thermosaccharolyticum* D120-70 [modified from Schäffer *et al.* (*82*)].

(1→*O*)-L-Tyr. The β-D-glucose residue is a constituent of the first repeating unit of the S-layer glycan chain, but exhibits a different anomeric configuration (β-) from all other glucose (α-) residues of the repeating unit backbone structure (Fig. 2). This is the first observation of a bacterial S-layer glycan composed of repeating units without a core region connecting the carbohydrate moiety with the polypeptide portion. The change in the anomeric configuration of the linkage glycose obviously implies the action of glycosyl transferases with different specificities. Similar observations have recently been made for the reducing end sugars of the LPS O-antigens of *Pseudomonas aeruginosa* (*91*) and *Salmonella enterica* (*92*). Thus, we suggest to regard this phenomenon as that of a "pseudo-core-region" and not as the total absence of a core-region.

Recently, the glycoprotein nature of the S-layer of different *Clostridium difficile* isolates (*93*, *93a*, *93b*) was demonstrated. *C. difficile* is the most common cause of antibiotic-associated diarrhea and pseudomembranous colitis in humans. The S-layer glycoprotein covers the entire cell surface and is the immunodominant antigen. Therefore, it may contribute to the virulence and represent an important mechanism to escape the host defense system. Two superimposed, structurally different lattices were detected on the cell surface of six *C. difficile* strains by using freeze-etching and negative-staining techniques. The inner lattice exhibits square and the outer lattice hexagonal symmetry. There seem to be two distinct families of S-layer glycoproteins present, possessing conserved N-terminal structural motifs. Compositional data indicate the presence of hexoses and *N*-acetylneuraminic residues in the P36 and P47 kDa glycoproteins. In other strains, however, only one of the two S-layers is glycosylated (*93b*). For an exact investigation of molecular masses, liquid chromatography coupled to electrospray ionization mass spectrometry was applied (*94*).

From the examples listed, wide compositional differences among S-layer glycoprotein glycans are evident, even among phylogenetically closely related species. Despite the reflected complexity, three principal building plans are found in S-layer glycoproteins. A representation of those is given in Fig. 3. The most common scheme is the tripartite architecture (Fig. 3a), comprising a variable carbohydrate chain composed of identical repeating units, which is linked via a core oligosaccharide to the S-layer polypeptide backbone. This structure compares to the assembly principle of LPS O-antigens; however, instead of the lipid A anchor, the glycosidic linkage is present in S-layer glycoproteins (*95*). Organisms for which this principle has been observed include *A. thermoaerophilus* GS4-97 (*75*), *G. stearothermophilus* NRS 2004/3a (*96*, *124*), *P. alvei* CCM 2051 (*80*) and *Tb. thermohydrosulfuricus* L111-69 (*78*). In some cases, capping of the terminal glycose residue with an *O*-methyl group was observed. If the oligosaccharide linker is missing in the S-layer glycan (Fig. 3b), the variable carbohydrate chain is directly bound to the glycosylated amino acid via the first repeating unit. In this scheme, the anomeric configuration of the linkage sugar might remain unchanged (*L. buchneri*; *97*) or be inverted and thus represent a "pseudo-core" (*T. thermosaccharolyticum* D120-70; *82*). Besides the structures comprising extended polysaccharide chains, a truncated form was also observed in one case (Fig. 3c) (*Tb. thermohydrosulfuricus* S102-70; *81*, *98*). This structure comprises only one hexasaccharide as glycan portion, possibly representing a sole core oligosaccharide, which originates either from a mutation or an enzyme defect.

In Gram-positive bacilli, only one distinct scheme of glycoprotein architecture has been found in the bacterial S-layer glycoproteins of each organism. This also includes the occurrence of only one linkage type. Unlike in halobacteria, the simultaneous occurrence of *O*- and *N*-glycans has not been reported for S-layer polypeptides. On the other hand, different acceptor sequences for S-layer glycans may be present on one S-layer polypeptide. Two to six glycosylation sites have been determined per S-layer protein monomer, giving rise to an overall carbohydrate content of S-layer glycoproteins between two and 15%. This is also in

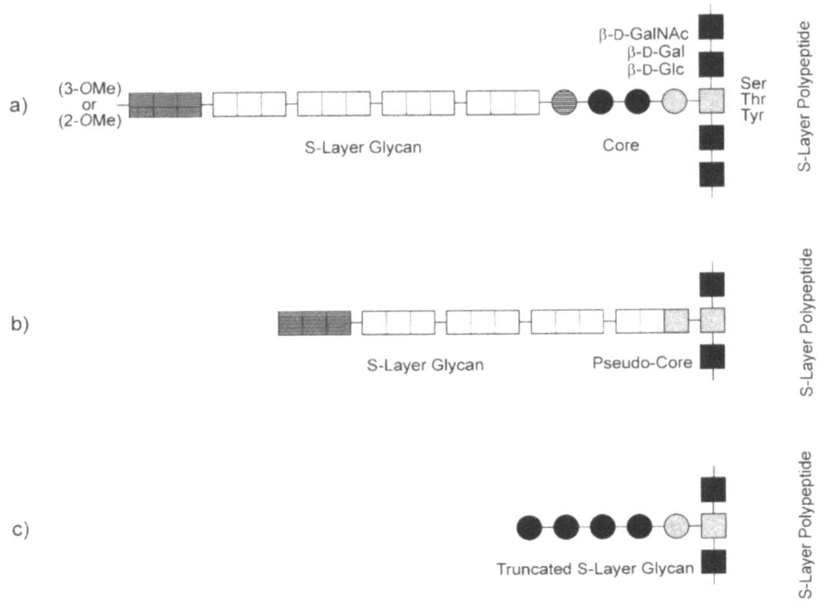

Fig. 3. Schematic representation of the three architectural building plans of S-layer glycoprotein glycans. a) Extended structure, comprising a core-region and the S-layer glycan chain; b) Simplified extended structure, comprising a pseudo-core and the S-layer glycan chain; c) Truncated structure, representing a core region only. (Repeating units composed of individual monosaccharides are represented by open squares; full circles indicate residues of the core-oligosaccharides; possible variations in length, either in the core or the glycan chain, are indicated by striped symbols; the S-layer polypeptide is built up by individual amino acids, symbolized by full squares; grey shaded symbols indicate the sugar and amino acid residues participating in the glycosidic linkage).

contrast to archaea, because in halobacteria, up to 25 glycosylation sites per S-layer monomer may be present (*14*).

An outstanding feature of bacterial S-layer glycoproteins is the occurrence of novel and, in comparison to eukaryotic glycoproteins, unusual glycosidic linkage units. In bacterial S-layer glycans these include, so far, only *O*-glycosidic bonds via serine (β-D-Gal→Ser; β-D-GalNAc →Ser; α-D-Glc→Ser), threonine (β-D-Gal→Thr; β-D-GalNAc→Thr) and tyrosine (β-D-Gal→Tyr; β-D-Glc→Tyr). In archaea, by contrast, the *N*-glycosidic linkage type preponderates.

These principal architectural schemes together with the phenomenon of microheterogeneity in glycan chain length and core composition produce the considerable heterogeneity inherent to S-layer glycoprotein glycans (for review see *19*).

References, pp. 105–124

2.3. Biosynthesis

The diversity in glycan structures and compositions raises interesting questions regarding the biosynthesis of prokaryotic glycoproteins. The first studies of S-layer glycoprotein biosynthesis were performed on archaea. In general, archaea possess many eukaryotic-like properties, including the ability to glycosylate proteins (*14*). Although current understanding of archaeal protein glycosylation is limited, parallels to the eukaryotic protein glycosylation system are apparent. In both eukaryotes and archaea, *N*-glycosylation occurs at asparagine residues of the highly conserved Asn-X-Ser/Thr consensus motif and dolichyl-linked oligosaccharides, similar to those involved in eukaryotic protein glycosylation, have been detected in archaea (*99–101*). So far, only one exception has been observed; the S-layer glycoprotein of *H. halobium* becomes *N*-glycosylated, even if the serine residue in the specific consensus motif site is replaced by valine, leucine, or asparagine, suggesting the presence of different *N*-glycosyltransferases in this organism (*102*). For the biosynthesis of the two different *N*-glycans of the *H. halobium* S-layer, the involvement of different saccharide precursors has been described (*14*). These are C_{55}- and C_{60}-phosphodolichol, respectively. The proposed mechanism includes the presence of a non-glycosylated protein core which is translocated through the cell membrane, and the corresponding glycans traverse the membrane in the lipid-linked form. Similar to the model for the biosynthesis of LPS O-antigens, the constituting repeating unit saccharide chain is assembled at the lipid-linked stage in the cytosol prior to its transfer to the nascent polypeptide portion. Finally, S-layer glycoprotein biosynthesis is completed by the generation of the *N*-glycosyl linkage at the cell surface (for review see *14*). In contrast to the well-investigated *N*-glycan biosynthesis in halobacteria, the mechanism of the assembly of the *O*-glycosyl units is still unknown.

Similar lipid-activated oligosaccharides with C_{55} dolichol species were also confirmed in the S-layer glycoprotein biosynthesis of *Haloferax volcanii* (*65a*, *101*) and *Methanothermus fervidus* (*103*). Interestingly, during the biosynthetic pathway of the S-layer glycan of *M. fervidus*, methylation and transient glycosylation with six or eight glucose residues takes place at the UDP-linked and lipid-linked levels, respectively.

Taken together, the use of dolichol as saccharide carrier seems to be a common feature of archaea and eukaryotes. There are differences, however. In archaea, oligosaccharides are frequently linked via a monophosphate bridge, whereas in eukaryotes they are exclusively linked via pyrophosphate. Archaeal cells also possess a limited spectrum of dolichol species, containing only 11 or 12 isoprene units, again in contrast to the wide spectrum of long-chain dolichols (18–22 isoprene units) present in eukaryotes. On the other hand, many of the enzymes involved in the eukaryotic protein glycosylation machinery have counterparts in archaea (*104*, *105*). The transient occurrence of glycosyl units during glycoprotein processing is another feature common to archaea and eukaryotes.

Recently, studies on the biosynthesis of S-layer glycans of bacteria have been initiated with *Paenibacillus alvei* CCM 2051. Characterization of the isolated intermediates revealed that, in addition to nucleotide-bound monosaccharides, nucleotide-activated oligosaccharides play a role during the assembly of the S-layer glycan chain. In the activated precursors additional sugar constituents, such as GlcNAc residues have been found, which are not present in the mature S-layer glycan. This might indicate the occurrence of trimming reactions in bacteria (*106*, *107*), comparable to those well known in eukaryotic systems (*99*).

A similar route of assembly of the S-layer glycan was suggested for *Thermoanaerobacterium thermosaccharolyticum* E207-71 (*95*). In the first dedicated step of this pathway, sugar-1-phosphates are converted into the corresponding nucleoside diphosphate sugars. So far, only the nucleotide-bound hexasaccharide repeating unit of that organism has been isolated, which is linked to GDP via a reducing mannose residue of the S-layer glycoprotein glycan backbone (*95*). Concerning the lipid intermediate occurring in the course of bacterial S-layer glycan biosynthesis of *T. thermosaccharolyticum* E207-71, preliminary data indicate the presence of a short chain dolichol rather than of undecaprenol (Schäffer, Messner, unpublished observation). Detailed investigations of lipid-linked precursors in S-layer glycoprotein glycan biosynthesis are in progress.

From the data available, we propose the following biosynthesis concept for S-layer glycoproteins. The S-layer glycan chain is assembled in the cytoplasm by the stepwise addition of single repeating units. These, in turn, are formed from the nucleoside diphosphate-activated derivatives of the constituting monosaccharides and assembled on a nucleotide acceptor producing activated oligosaccharide repeats. The occurrence of nucleotide-activated oligosaccharides is in contrast to the well-investigated routes of eukaryotic glycoprotein biosynthesis (*99*). Although not yet demonstrated experimentally, the separate, independent biosynthesis of a (pseudo)core-region is assumed. The activated biosynthetic intermediate would subsequently be transferred to a lipid carrier and the glycolipid intermediate transported across the cytoplasmic membrane. On the outer face of the membrane, the transfer of the

S-layer glycan chains to specific amino acid residues on the S-layer polypeptide would occur to complete the S-layer glycoprotein. It is not known whether the assembly of the core region takes place on the S-layer glycan chain or, independently on the S-layer polypeptide, in a co- or post-translational reaction.

A new approach for elucidating the biosynthesis of S-layer glycoprotein glycans of Gram-positive bacteria benefits from the compositional similarity between S-layer glycans and LPS O-antigens (*95*) and from the progress made in the molecular characterization of the LPS biosynthesis of Gram-negative organisms over recent years (for review see *108–111*).

Initial genetic studies have been performed on the S-layer glycoprotein of *Anenrinibacillus thermoaerophilus* L420-91 to elucidate the biosynthesis of the constituent monosaccharide residues. The constituents of the repeating unit hexasaccharide of this organism are the rare glycoses D-rhamnose and *N*-acetyl-D-fucosamine (*74*), both of which are also found in LPS. GDP-D-rhamnose is the activated form required for the incorporation of D-rhamnose into the S-layer glycan chain. The two genes *gmd* and *rmd* required for the biosynthesis of this nucleotide-activated precursor from GDP-D-mannose have recently been cloned, sequenced and overexpressed in *E. coli* (GenBank accession number AF317224) by Kneidinger *et al.* (*112*). Furthermore, the corresponding enzymes were purified to homogeneity and functional studies were performed. With *A. thermoaerophilus* L420-91, for the first time, the two-step reaction sequence for the conversion of GDP-D-mannose to GDP-D-rhamnose was fully characterized. The genes encoding enzymes for the biosynthesis of a nucleotide sugar precursor are usually clustered together within the corresponding gene cluster for the bacterial polysaccharide (*108*). In fact, the D-rhamnose operon was found to consist only of the two open reading frames for Gmd and Rmd. Both enzymes were classified as novel members of the short chain dehydrogenase/reductase protein family. The first step in the precursor biosynthesis, the dehydration of GDP-D-mannose, is catalyzed by Gmd. Subsequently, the intermediate is reduced in a cofactor-dependent way by Rmd, yielding the final product GDP-D-rhamnose, characterized by chemical and NMR-analysis. It was observed that Gmd catalyzes both the dehydration and, to a lesser extent than Rmd, the reduction step. Gmd from *A. thermoaerophilus* L420-91 is the first bifunctional enzyme reported to display these activities. Possibly, with Gmd alone, insufficient amounts of GDP-D-rhamnose are produced for complete S-layer protein glycosylation. Interestingly, Gmd is highly conserved throughout the domains *Archaea*, *Bacteria* and higher eukaryotic organisms, which might reflect the importance of this enzyme during evolution. The knowledge of the reaction sequence of GDP-D-rhamnose biosynthesis is not only important in the context of S-layer glycan biosynthesis. By the use of enzyme inhibitors, targeted against the first reaction step, it should be possible to prevent the biosynthesis of D-rhamnose-containing LPS structures in pathogenic bacteria. Studies in that direction might provide a promising therapeutic approach.

The data from GDP-D-rhamnose biosynthesis indicated reaction sequence analogies in glycan biosynthesis between Gram-positive and Gram-negative bacteria. Based on these data, biosynthesis studies on the →4)-α-L-Rha-(1→3)-β-D-*glycero*-D-*manno*-Hep-(1→ disaccharide repeat unit from the S-layer glycan of *A. thermoaerophilus* DSM 10155 are currently under way (*76*, *86*). Again, both monosaccharide residues are typical LPS constituents. The nucleotide-activated form of L-rhamnose is dTDP-L-rhamnose, which was found to be formed from dTDP-D-glucose by the sequential action of the RmlABCD enzymes in enterobacteria (*113*, *114*). Detailed reaction mechanisms, however, have been proposed only for RmlA (*115*) and RmlB (*116*). Therefore, for comparative reasons, the two remaining enzymes, RmlC and RmlD, of the *rml* operon were first characterized in *Salmonella enterica* (*117*). Recently, the biosynthetic pathways leading to formation of the nucleotide-activated forms of L-rhamnose and D-*glycero*-D-*manno*-heptose have been elucidated in *A. thermoaerophilus* DSM 10155 (GenBank accession number AF324836; *117a*).

2.4. Molecular Biological Insights

An important line of development is directed towards the genetic manipulation of S-layer (glyco)proteins. This includes both the molecular characterization of the enzymes of the S-layer glycan biosynthesis and the knowledge of the nucleotide sequence encoding the S-layer polypeptide to which the glycan is covalently attached. The molecular understanding of all of these proteins will create the possibility of changing the natural properties of S-layer glycoproteins and open up new opportunities for modulating their structural and functional features by genetic engineering techniques.

Although common structural principles exist in S-layer proteins, sequencing of S-layer genes of organisms from all phylogenetic branches has led to the conclusion that sequence identities are rare (*118*, *119*). The list of S-layer proteins with known amino acid sequences is growing steadily (for a recent survey the reader is referred to the review by Sára and Sleytr; *29*). By contrast, knowledge of sequences encoding S-layer glycoproteins is still rather limited (*119a*).

Specific examples include the genes for the S-layer glycoproteins of *H. halobium* (*120*, *121*), *Hf. volcanii* (*120*, *121*), *M. fervidus* (*122*) and *Haloarcula japonica* (*123*).

The genes for the S-layer glycoproteins of *H. halobium* (GenBank accession number J02767) and *Hf. volcanii* (M62816) have been cloned and sequenced by Lechner and Sumper (*120*, *121*). The S-layer genes of *H. halobium* (*csg*) and *Hf. volcanii* possess open reading frames of 852 and 828 amino acids, respectively. These two proteins show distinct regions of homology, particularly a 21-amino acid hydrophobic domain at the C-terminus, which is a presumed membrane anchor. This domain is adjacent to a region of 11 threonine residues, all of which are *O*-glycosylated. The amino acid sequence Asn-X-Ser/Thr constitutes the *N*-glycosylation sites. The homology between the two proteins is less pronounced toward their N-termini, and there are marked differences in the pattern of glycosylation between the two proteins. The overall homology is estimated to be 40%; possibly, the highly conserved sequences code for sites of essential protein-protein interactions (*14*).

The *slgA* gene of the S-layer glycoprotein of *M. fervidus* (X58297) encodes a precursor of the mature S-layer protein containing 593 amino acids and exhibits 20 consensus motifs for *N*-glycosylation (*122*). The S-layer glycoprotein of *Ha. japonica* (D87290) is composed of 828 amino acids, and the mature polypeptide is preceded by a signal sequence of 34 amino acids (*123*).

Concerning sequencing of S-layer glycoproteins from bacteria, the sequence of a temperature-derived, glycosylated mutant of *Geobacillus stearothermophilus* ATCC 12980 (*123a*) has been deposited at GenBank (AF228338). The first complete data on the nucleotide sequence of a structural S-layer gene (*sgsE*; AF328862) from a glycosylated wildtype-strain was obtained from *G. stearothermophilus* NRS 2004/3a by our group (*124*). In addition, the glycosylation sites have been identified on the S-layer polypeptide of that organism (*124*), and the molecular biological characterization of the biosynthesis of its polyrhamnan S-layer glycan is being performed (Novotny, Schäffer, Messner, unpublished observations). From preliminary data on *A. thermoaerophilus* strains DSM 10154 (Pföstl, Schäffer, Messner, unpublished observation) and DSM 10155 (*117a*), there are first indications of a clustered occurrence of the S-layer gene and of the genes encoding the corresponding S-layer glycosylation machinery on the bacterial genome.

2.5. Perspectives

The mechanisms and underlying rules for the creation and assembly of supramolecular units into functional structures may be learned from nature and is expected to lead to new and fascinating S-layer (glyco)protein-based technologies. The detailed knowledge of S-layer protein glycosylation on a molecular level, in combination with the description of the S-layer structural genes and the localization of the glycosylation sites, will allow the generation of tailor-made S-layer *neo*glycoconjugates of medical or agricultural importance.

The principal techniques to accomplish this aim are schematically depicted in Figure 4. For carbohydrate engineering, detailed information concerning the S-layer glycan biosynthesis cluster is required. This includes characterization of biosynthesis enzymes, transporters and ligases, on the one hand, and the S-layer glycoprotein structural gene encoding the protein with its defined glycosylation sites, on the other hand. By application of the methods of genetic engineering such as mutations, knock-outs or replacements, useful S-layer *neo*glycoproteins will eventually be obtained. These novel molecules may contain modifications of the natural glycan structures or completely new glycan structures, *e.g*, capsular polysaccharides of pathogenic organisms.

The recently performed "shuttling" of genes encoding polysaccharides between different Gram-positive organisms, as well as between Gram-negatives and Gram-positives, clearly demonstrates the potential, but also points to some host-inherent limitations of carbohydrate engineering. The 14.5 kb gene cluster responsible for the exopolysaccharide (EPS) biosynthesis in *Streptococcus thermophilus* Sfi6 (*125*) has been transferred into the non-EPS-producing heterologous host, *Lactococcus lactis* MG1363. Interestingly, analysis of the recombinant EPS showed that it had a similar molecular mass but a different structure from the EPS in the native host. The structure found was →3)-β-D-Glc-(1→3)-α-D-Gal-(1→3)-β-D-Gal-(1→ as opposed to the native structure →3)[α-D-Gal-(1→6)]-β-D-Glc-(1→3)-α-D-GalNAc-(1→3)-β-D-Gal-(1→. In the modified EPS, *N*-acetylgalactosamine was substituted by galactose, most probably, because the *L. lactis* strain lacks UDP-*N*-acetylglucosamine C4-epimerase activity, which would provide UDP-GalNAc for GalNAc incorporation into the EPS. The finding of a modified glycan implies that i) bacterial glycosyltransferases may have multiple specificities for glycosyl donors and acceptors and that ii) the repeating unit polymerase can recognize and polymerize a repeating unit that differs from the native substrate both in the backbone and by the absence of a branching Gal unit in the side chain (*126*).

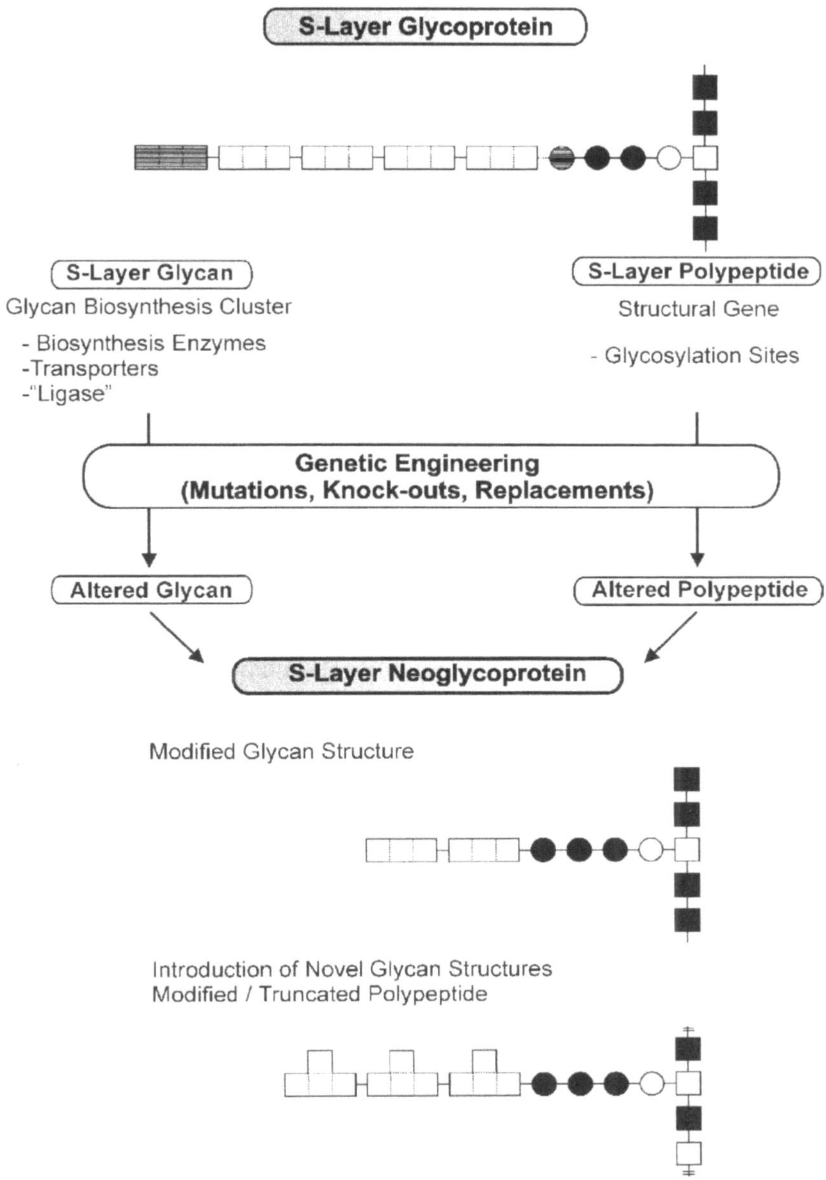

Fig. 4. Principal strategies for the design of S-layer *neo*glycoproteins by "carbohydrate engineering".

In contrast, the type 3 capsular polysaccharide (CPS) from *Streptococcus pneumoniae* was successfully expressed in *L. lactis*, with no structural alterations observed (*127*). Only three of the four CPS biosynthesis genes were found to be necessary for the formation of type 3 CPS, implying a role of the type 3-specific synthase in the extracellular transport of the CPS or the existence of an alternative export system in *L. lactis*. Thus, *L. lactis* has potential as a host for the safe production of capsular antigens of pathogenic organisms.

Both examples listed above demonstrate that the transfer is feasible of entire gene clusters, encoding biosynthesis, polymerization and transport of specific glycans. However, such transfection will not necessarily result in the production of identical polysaccharide structures in different hosts, because differences in host metabolism may affect gene expression and glycan biosynthesis. Therefore, identification of suitable hosts and overcoming of metabolic obstacles will be challenges in the design and construction of S-layer *neo*glycoproteins.

3. Non-S-Layer Glycoproteins

Over the past five years, three reviews have been published in this field (*16–18*). Therefore, in the present contribution, the focus is mainly on those subjects which have not been reviewed thus far. In her recent report, Moens (*18*) comprehensively dealt with flagellar glycoproteins, glycoproteins of pili, fimbriae and cell surface fibrils, glycosylated enzyme complexes, and a few "miscellaneous" glycoproteins of archaeal and bacterial origin.

Even today, in many cases, structural information regarding the sugar composition and the linkage unit is scarce among non-S-layer glycoproteins (for comparison see Table 1). There is, however, an increasing number of reports where medical aspects of these glycoconjugates are discussed. Interestingly, many of the described organisms with glycosylated proteins belong to the Gram-negative bacteria. In general, it is believed that *Escherichia coli* is unable to glycosylate its own cellular proteins. However, the recent sequencing of the entire *E. coli* K-12 genome has led to the identification of glycosyltransferases that may be involved in protein glycosylation (*156*). Nonetheless, the inability of *E. coli* to glycosylate eukaryotic recombinant proteins is explained by differences in glycan biosynthesis existing between eukaryotes and prokaryotes.

With the enormous progress of genetic techniques, an increasing number of sequence data are now available from genes which code for proteins involved in the biosynthesis of glycoprotein glycans. As far as data base entries are known, the available accession numbers are listed in the text. In addition, a novel section (Section 3.6) is included in this review, dealing with enzymatic and chemical synthesis of glycopeptides and glycoproteins and their potential as inhibitors or for binding studies.

3.1. Intracellular Glycoproteins

While most of the prokaryotic glycoproteins described in this review are either membrane- or surface-associated or shed into the surrounding medium of a bacterial cell, only little is known about intracellular glycoproteins. One of them is a glycosylated intracellular β-1,4-glucosidase of *Streptomyces lividans* (*157*) which has been discussed in the recent review by Moens (*18*).

Another example was found in a *Cellulomonas fimi* strain (*158*). Members of the genus *Cellulomonas* produce a diverse set of enzymes which enable them to degrade plant material such as cellulose, xylans, glycans, mannans, chitin and starch (for review see *17*, *18*). Usually, purification of these enzymes from polysaccharide-degrading microorganisms is difficult because of the production of a complex set of enzymes with similar activities. Only recently, two novel mannan-degrading enzymes of *C. fimi* have been isolated (*158*). One of them is a surface-associated mannanase which will be described in Section 3.3. The other enzyme, Man2A, is an intracellular β-mannosidase (β-1,4-mannoside mannohydrolase, EC 3.2.1.25) which catalyzes the removal of β-D-mannose residues from the nonreducing end of oligosaccharides (*158*). Upon rescreening of a λ-ZapII library from *C. fimi* in its excised form using 4-methylumbelliferyl-β-D-mannoside, two positive clones have been obtained. One of them carried plasmid pCManI which contained an open reading frame (ORF) of 2,562 bases (GenBank accession number AF126472) encoding a protein of 842 amino acids with a calculated molecular mass of 94.96 kDa. By size-exclusion chromatography of Man2A, a molecular mass of 103 kDa was determined, indicating that the native enzyme is a monomeric glycoprotein. Man2A comprises a catalytic module in a subfamily of family 2 of the glycosyl hydrolases that, at present, contains only β-mannosidases (*158*). A detailed biochemical characterization of this enzyme including its concerted action with the mannanase Man26A (see Section 3.3) is in progress.

3.2. Membrane-Associated Glycoproteins

Many of the membrane-associated prokaryotic glycoproteins described here are implicated in disease and pathogenicity. In many cases, the glycan structures have not been elucidated and the molecular details of their participation are poorly understood or even unknown.

Chlamydia trachomatis is one of the most common causes of blindness and sexually transmitted diseases in humans. In this context, the role of the major outer membrane protein (MOMP) of *C. trachomatis* serovar L_2/434/Bu has been extensively studied in the past (*159–161*). MOMP and two other chlamydial proteins have been identified as glycoproteins by their binding to plant lectins (*159*). After isolation of MOMP from whole-organism lysates, the glycan portion of the 40-kDa glycoprotein was released from the protein by treatment with *N*-glycosidase F. The major fractions consisted of high-mannose-type oligosaccharides containing eight to nine mannose units. Bi- and triantennary complex-type oligosaccharides with terminal galactose have been found as trace components. These glycans were *N*-linked and contained no sialic acid units (*161*). Removal of the glycan chains abolished the ability of MOMP to bind to HeLa cells. On the other hand, [^{3}H]-labeled, isolated glycans are readily attached to HeLa cells, indicating that the carbohydrate is involved in the attachment of chlamydia to these cells (*159*, *160*). In a competition assay, the involvement of specific receptor sites on HeLa cells has been investigated. Among the high-mannose-type oligosaccharides, the strongest inhibition was obtained with those oligosaccharides containing eight mannose units (Fig. 5). These observations indicate that a specific *N*-glycan fraction plays an essential role in the attachment of MOMP to HeLa cells and, possibly, in the infectivity of *C. trachomatis* in humans. The molecule that recognizes chlamydial MOMP oligosaccharides present on host cells, *i.e.*, HeLa cells, has not been identified at that time (*161*). Possibly it is

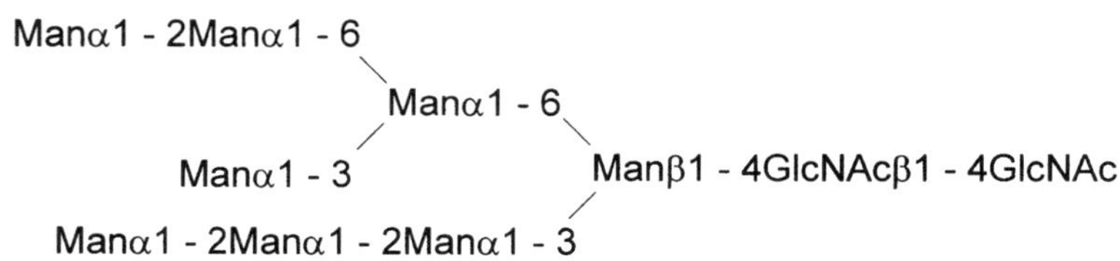

Fig. 5. Triantennary structure of the biologically most active *N*-linked oligosaccharide fraction from the 40 kDa glycoprotein of *Chlamydia trachomatis* [modified from Kuo *et al.* (*161*)].

one of the mannose-binding proteins which are widely distributed on mammalian cells (*162*).

Bacterial cell envelopes contain a number of components with diverse biological properties. Among them is Protein A (PA). In *Staphylococcus aureus* Cowan I, this cell wall glycoprotein is covalently linked to the peptidoglycan layer. It possesses a number of immunomodulatory and antitumor properties (*163*). For the investigation of the primary antibody response in tumor-bearing mice, commercially available preparations of PA (Amersham Biosciences, Uppsala, Sweden) have been used. The mode, dose and time point of administration of PA were found to be critical. PA produced maximum suppression of plaque-forming cell response in normal mice as well as in tumor-bearing mice when administered prior to immunization with sheep red blood cells. In addition, reduced splenic localization of ^{51}Cr-labeled sheep red blood cells was observed after treatment with PA. This effect, caused by the antigen in spleen, may be responsible for reduced response of plaque-forming cells in tumor-bearing mice. In summary, Zaidi *et al.* (*163*) have proposed that suppression of the primary antibody by Protein A helps in tumor regression by reducing the amounts of circulating, soluble immunosuppressive immune complexes.

To study the biochemical nature of the anti-lepidopteran toxin from *Bacillus thuringiensis* subsp. *kurstaki* cells, the toxin was purified after proteolysis of the protoxin by the action of endogenous protease(s) (*164*). This insecticidal crystal toxin is a glycoprotein containing high-mannose-type and hybrid-type oligosaccharide chains (*165*). Lectin binding assays with the purified 66 kDa toxin have shown strong positive reactions to *Galanthus nivalis* agglutinin, indicative of the presence of terminal mannose, α-(1→3)-, α-(1→6)-, or α-(1→2)-linked to mannose. Much weaker reactions were found with peanut agglutinin [β-Gal-(1→3)-GalNAc] and *Sambucus nigra* agglutinin [α-NeuAc-(2→6)-Gal] (*165*). Isolation and purification of integral and peripheral membrane proteins is often facilitated by Concanavalin A (ConA)-Sepharose affinity chromatography (for review see *166, 167*). This glycoprotein has a high binding affinity for Con A and thus could not be eluted with concentrations of methyl α-D-mannopyranoside as high as 0.5 M. Non-specific elution with 0.03% SDS eluted the matrix Con A together with bound glycoprotein. The experimental results indicate that Con A leaching was mainly because of inclusion of detergents in the buffer system and may not be directly related to the nature of the glycoprotein sample (*165*).

In 1996, diarrheal disease was responsible for the death of 2.5 million people worldwide. Most cases of diarrhea occur in developing countries, where especially young children under five years of age die as a result of infection with enterotoxigenic *Escherichia coli*. Strain H10407 was shown to be capable of invading *in vitro* epithelial cell lines derived from the human ileocecum and colon (*168*). From this strain, two separate chromosomally encoded invasion loci (*tia* and *tib*) have been cloned which direct nonadherent and noninvasive laboratory strains to adhere to and invade cultured human intestinal epithelial cells. The *tib* locus encodes the synthesis of TibA, a 104-kDa outer membrane protein that is synthesized as a 100-kDa precursor (preTibA) and must be modified for biological activity. Analysis of the *tib* locus DNA sequence revealed an open reading frame with similarity to RfaQ, a glycosyltransferase. TibA is the only glycoprotein produced by strain H10407. Its glycoprotein nature has been demonstrated on intact cells of strain H10407 by oxidation of the glycans with periodate and labeling with hydrazide-conjugated digoxigenin (DIG). To show that the glycan chains on TibA are surface-exposed and involved in binding to epithelial cells, the DIG-labeled bacteria were then treated with different concentrations of trypsin or proteinase K. Though trypsin treatment did not entirely remove carbohydrates, as little as 10 μg per ml of proteinase K completely eliminated carbohydrate-staining glycopeptides. Using proper controls, the authors have demonstrated that the proteases did not penetrate and destroy the outer membrane, causing contamination with lipopolysaccharides. These results suggest that the TibA glycoprotein was surface exposed and acted as an adhesin that may participate in the disease process (*168*).

Ehrlichia chaffeensis and *Ehrlichia canis* are obligate intracellular bacteria and show tropism for macrophages and monocytes. They are responsible for the diseases human monocytotropic ehrlichiosis and canine ehrlichiosis (*169*). Recently, the 120-kDa protein (P120) of *E. chaffeensis* and the 140-kDa protein (P140) of *E. canis* have been cloned, expressed, and characterized (*170*, *171*). The recombinant *E. chaffeensis* P120 and *E. canis* P140 are useful as serodiagnostic antigens (*171*, *172*). The genes of *E. chaffeensis* (1,644 bp) and *E. canis* (2,064 bp) encode proteins of 548 and 688 amino acids, respectively, with predicted molecular masses of only 61 and 73 kDa, but with electrophoretic mobilities corresponding to 120 kDa (P120) and 140 kDa (P140), respectively (*173*). Similar discrepancies between the molecular masses and the observed electrophoretic mobilities are also known from archaeal and bacterial S-layer glycoproteins (for review see *14*, *89*). Large portions of the P120 and P140 genes have been expressed in *Escherichia coli* and the recombinant proteins exhibited molecular masses ranging from 1.6 to 2 times larger than those predicted by the

References, pp. 105–124

amino acid sequences. Native proteins with molecular masses of 120 and 140 kDa from *E. chaffeensis* and *E. canis*, respectively, reacted with antisera against the corresponding homologous recombinant glycoproteins. When analysed for potential glycosylation sites, the *E. chaffeensis* P120 recombinant protein revealed two *N*-linkage sites (Asn-Xaa-Ser/Thr) and 91 *O*-linkage sites (Ser or Thr). The *E. canis* P140 recombinant protein also comprises two possible *N*-linked glycosylation sites (amino acids 7 and 32) and 147 possible *O*-linkage sites. Tandem repeat units in both genes have been considered as possible glycosylation sites, particularly for *O*-glycosylation (*173*). Incubation of the *E. chaffeensis* and *E. canis* recombinant proteins with *N*-glycosidase F (*N*-linked glycans), *O*-glycosidase [β-Gal-(1$\rightarrow$3)-GalNAc], neuraminidase (sialic acid residues), β-galactosidase (β-linked 1,4-terminal galactose), or glucosaminidase (β-linked *N*-acetylglucosamine) did not result in deglycosylation of the recombinant proteins. Moreover, these proteins did not react with lectins specific for a number of conserved glycan core structures (*173*). Monosaccharide analysis of the glycans on the recombinant glycoproteins of either strain have shown glucose: galactose: xylose in molar ratios of approximately 44 : 22 : 1 (P120, *E. chaffeensis*) and 30 : 1 : 1 (P140, *E. canis*), respectively. The molar ratio of carbohydrate:protein was 18 : 1 (*E. chaffeensis*) and 17 : 1 (*E. canis*). The sugar compositions of the recombinant glycoproteins suggested the presence of unusual glycan modifications (xylose-glucose *O*-linkages to serine residues) which have been described in different eukaryotic glycoproteins and appeared to be similar to the composition and linkages observed with the recombinant glycoproteins of *Ehrlichia* strains (*174*).

In Section 3.1, one of the mannan-degrading enzymes from *Cellulomonas fimi*, the β-mannosidase Man2A, has been described (*158*). Another enzyme from *C. fimi* is the β-mannanase Man26A (mannan endo-1,4-β-mannosidase; EC 3.2.1.78) which catalyzes the random hydrolysis of β-1,4 mannosidic linkages within the backbones of mannans and related polysaccharides. A library of *C. fimi* genomic DNA was screened for mannanase-positive clones; two clones, CMan2 and CMan4, were obtained. The sequence of the putative mannanase gene is deposited at GenBank (accession number AF126471) (*158*). Man26A, like all the extracellular glycosyl hydrolases produced by *C. fimi*, is a modular protein, comprising a catalytic module and at least one other module. Man26A is the only one with an SLH (surface layer homology) module (*38*, *39*). A significant fraction of the enzyme activity appeared to be cell-associated, possibly through binding to the cell surface by the SLH domain. SLH domains are present on different polysaccharidases where they may be involved in attaching the enzymes to the cell surface (for review see *175* and Section 2.1). Man26A produced by *C. fimi* appears to be transiently cell-associated but does not bind to peptidoglycan. In addition, it appears to contain two additional modules of unknown function (*158*). The β-mannanase produced by *E. coli* has a molecular mass of ~100 kDa and the enzyme produced in *C. fimi* has a slightly lower molecular mass. This result was taken to indicate that either *E. coli* or *C. fimi* utilize different leader peptide processing sites in proMan26A, or that they use the same site and *C. fimi* removes additional amino acids from the N-terminus. Obviously, this enzyme is the only mannanase produced by *C. fimi*, with multiple bands on SDS-PAGE arising by proteolysis of the 100-kDa protein in the culture (*158*).

3.3. Surface-Associated Glycoproteins

Besides S-layers (see Section 2), cell surface appendages such as flagella and pili currently represent the best-investigated systems of prokaryotic glycoproteins. Glycans covalently bound to flagellins and pilins have been found in both prokaryotic lineages, the *Archaea* and the *Bacteria* (for detailed review see *18*, *176*). Since the glycosylated flagellin of *Clostridium tyrobutyricum* (for review see *177–180*) has been recently reviewed (*18*), it is not discussed in detail here.

For a better understanding of the processes triggered by the microflora in the human oral cavity, the distinct populations of colonizing prokaryotic organisms need to be characterized. In patients with periodontal diseases, the microflora of subgingival plaque is often dominated by highly heterogeneous populations of spirochetes. Other organisms frequently found in the oral cavity are described in Sections 3.4 and 3.5. With the specific monoclonal antibody H9-2 against *Treponema pallidum*, the type species of the genus *Treponema*, cross-reactive populations of treponemas have been detected in plaque samples from sites of periodontitis and gingivitis (*181*). Using this antibody and a specific staining reaction for the detection of glycoproteins, Wyss (*182*) has demonstrated that the wild-type flagellar core proteins (FlaBs) of cultured oral treponemes are glycosylated. Following periodate oxidation, these proteins were stained by the digoxigenin (DIG)-hydrazide labeling procedure (*183*). Depending on the treponemal species tested, between two and five bands in the molecular range of flagellar core proteins (28 to 35 kDa) were labeled in all strains. However, no glycan staining reaction has been observed for the FlaA-proteins, indicating that the endoflagellar sheath proteins are not glycosylated (*182*).

A very recent paper on a hyperfimbriated mutant of *Streptococcus salivarius* ATCC 29975 demonstrated the glycoprotein nature of the filamentous structure by lectin binding studies (*183a*).

Recent biochemical and genetic analyses have indicated that flagella and pili of a number of prokaryotic organisms, particularly those of pathogens, are posttranslationally modified by glycosylation. This unusual covalent modification may affect interactions between the pathogen and the host. Flagellation in archaea has been extensively studied in the mesophilic methanogen *Methanococcus voltae* (for review see *176*). *M. voltae* has been shown to possess four highly similar flagellin genes (*flaA*, *flaB1*, *flaB2*, and *flaB3*) with *flaB2* as one of the major flagellins (*184*). These flagellins migrate on SDS-PAGE as proteins of approximately 31 and 33 kDa. This molecular mass is approximately 10 kDa larger than the predicted molecular mass of 22.8 kDa (*184*). So far, posttranslational modifications such as glycosylation have not been described for the flagellins of *M. voltae* (*176*). However, the deduced flagellin amino acid sequences comprise a number of carbohydrate attachment consensus sequences. These findings suggested that the flagellins may be glycosylated by glycans that resist conventional detection methods. Recent attempts to express the *M. voltae flaB2* flagellin gene in *E. coli* and *Pseudomonas aeruginosa* resulted in an overexpressed flagellin which migrated at 26.5 kDa (*185*). This molecular mass is 5 to 7 kDa smaller than that of the flagellins isolated from purified *M. voltae* flagella. Although the discrepancy in migration behaviour could be partly explained, the nature of the putative modification of the *M. voltae* flagellin is not fully understood (*185*).

Very recent NMR studies revealed the pilin amino glycan of *Pseudomonas aeruginosa* 1244 to be composed of the trisaccharide α-5-*N*-β-hydroxybutyryl-7-*N*-formyl-pseudaminic acid(α-5NβOH$C_4$7-NfmPse)-(2$\rightarrow$4)-β-Xyl-(1$\rightarrow$3)-β-FucNAc *O*-linked to serine (*185a*). A different type of substitution of pseudaminic acid has been observed on the glycosylated flagellin of *Campylobacter jejuni* (GenBank accession numbers AF345999 and AY034084; *185b*).

Pilus-mediated adhesion is essential in the pathogenesis of *Neisseria meningitidis* and *N. gonorrhoeae*. Pili are assembled from glycoprotein subunits termed pilins. Pilins are the key virulence factors and the major adhesins of these capsulate organisms and contribute to their specificity for a human host. However, the detailed mechanism of glycosylation of pilins has not been elucidated. Since the glycosylated region of pilin is surface-exposed and has a relatively conserved amino acid sequence, it was suggested that glycosylation may provide masking against the host immune response (*186*). The X-ray crystal structure analysis of pilin from *N. gonorrhoeae* with a resolution of 2.6 Å revealed that galactose-α-1,3-*N*-acetylglucosamine is *O*-linked to Ser63 (*186*). Further, Stimson *et al.* (*187*) have demonstrated that an *O*-linked trisaccharide is present between amino acid residues 45 and 73 of *N. meningitidis* pilin. This structure contains nonreducing, terminal 1,4-linked digalactose covalently linked to a 2,4-diacetamido-2,4,6-trideoxyhexose. Nassif and coworkers have investigated the role of glycosylation at Ser63 in the production of soluble pilin in pathogenic *Neisseria* (for review see *188*). Pilin variants of either *N. gonorrhoeae* and *N. meningitidis*, responsible for high and low adhesiveness, have been created either by introducing transcriptional fusions or by site-directed mutagenesis. Production of truncated soluble pilin (S-pilin) was more abundant in *N. gonorrhoeae* rather than in *N. meningitidis* (*189*). These experiments indicate that for *N. gonorrhoeae*, but not for *N. meningitidis*, glycosylation at Ser63 is not required for S-pilin production, suggesting that the mechanisms of production of this pilin are different in these organisms (*189*). In a recent study in *N. meningitidis* a gene, *pglA* (GenBank accession number U73942), involved in pilin glycosylation, has been identified. It codes for a galactosyltransferase and shows a high degree of homology to a group of glycosyl transferase genes involved in the biosynthesis of either lipopolysaccharide or capsular polysaccharide (*190*). In continuation of these efforts, a random plasmid library of *N. meningitidis* strain MC58 has been screened against combined data bases to identify genes involved in polysaccharide biosynthesis (*191*). From a positive clone (GenBank accession number AF014804), a 6.5 kb fragment has been sequenced, showing four open reading frames (ORFs), which were designated *pglB*, *pglC*, *pglD*, and *avtA*. The putative PglB protein consists of 414 amino acids and the calculated molecular mass is 44.5 kDa. It shows closest similarity with a family of presumed glycosyltransferases. The second ORF codes for PglC, a putative protein of 392 amino acids and a molecular mass of 43.3 kDa. The putative PglC protein has high similarity with a number of pyridoxal-binding dehydrases and aminotransferases. Translation of the third ORF yielded a peptide of 637 amino acids with a calculated molecular mass of 71 kDa. The putative PglD protein was highly similar to a family of proteins proposed to be dehydratases involved in LPS biosynthesis. AvtA is an alanine-to valine aminotransferase and has not been implicated in polysaccharide biosynthesis so far (*191*). Insertional mutations in a strain with well-defined LPS and pilin-linked glycan structures have been shown by Western analysis to result in increased migration of the pilin-subunit molecules of *pglB*, *pglC*, and *pglD*. These data indicate that the gel migration differences are due to the alteration or absence of the pilin-linked trisaccharide structure in these mutants. Sequence homology comparisons suggest specific roles for *pglB*, *pglC*, and *pglD* in the biosynthesis of the 2,4-diacetamido-2,4,6-trideoxyhexose. The predicted functions of these genes were based on amino acid sequence homologies with genes in databases that are involved in the formation of similar chemical structures. To verify the proposed gene functions, biochemical and structural analyses are currently under way (*191*). Very recent investigations revealed polymorphism in the pilin glycosylation locus *plg* of *N. meningitidis* strains expressing class II pili (*191a*).

Another very recent genetic analysis is dealing with effects of glycosylation of a *Streptomyces coelicolor* A3(2) cell envelope protein that is required for infection by bacteriophage ϕC31 (*191b*).

Very recently, it could be shown that glycosylation with heptose residues, mediated by the *aah* gene product, is essential for the diffuse adherence of *Escherichia coli* strain 2787 (O126:H27) by the autotransporter adhesin AIDA-I (GenBank accession number AJ304444; *191c*).

Microbial production of emulsifiers commonly accompanies biodegradation of hydrophobic compounds and may be necessary for interaction of the cell with the substrate. *Rhodococcus erythropolis* strain ST-2 emulsifies and grows on crude oil as sole carbon and energy source. The emulsifying agent is produced constitutively in the absence of oil and is associated with the cell surface. Chemical analysis of the polymer showed a composition of about 73% protein and 27% carbohydrate (*192*). When the extracted polymer was treated with lysozyme, a 30-kDa protein was precipitated. Inactivation of the emulsifier by lysozyme indicated the presence of peptidoglycan-like structures in the carbohydrate component (*193*). The high-molecular-weight glycoprotein polymer of *R. erythropolis* ST-2 appears to be noncovalently bound to the cell surface (*192*).

Analysis of *Mycobacterium smegmatis* mc^2155 indicated that changes in glycopeptidolipid acetylation affect sliding motility and biofilm formation (GenBank accession number AF192151; *193a*).

The acid phosphatase complex of *Burkholderia pseudomallei* is a glycoprotein (*194*). In order to elucidate the complex nature of enzymatic activity of acid phosphatase on *B. pseudomallei* cells, the results from several lines of investigation have been compared. Active enzyme fractions were obtained from cell-free extracts, representing the content of the cytoplasm, the outer membrane fraction, and the culture filtrate after column chromatography on Sepharose 6B and DEAE cellulose. Protein analysis (Lowry method) and sugar determination (phenol-H_2SO_4-method) of the column effluents indicated that the highest enzyme activity was present in the outer membrane fractions. To verify the glycoprotein nature of the acid phosphatase from *B. pseudomallei*, the effect of tunicamycin as a glycosylation inhibitor was investigated on growing cells (*195*). In the presence of tunicamycin, the levels of sugar and the acid phosphatase activity were significantly reduced. The surface location of the glycoprotein acid phosphatase was confirmed by immunofluorescence staining with antisera against the phosphatase-active fractions. From these results the conclusion was drawn that the acid phosphatase of *B. pseudomallei* is a glycoprotein tyrosine phosphatase located at the cell surface and that the complex nature of enzymatic activity may reflect the evolutionary stage of the enzyme (*194*).

One of the best examples for surface-associated prokaryotic glycoproteins is the cellulosome complex of *Clostridium thermocellum* which contains, in addition to the major polypeptide scaffoldin, a number of glycosylated and nonglycosylated enzymes (for review see *196*, *197*). Since this multimodular enzyme complex has been described in a number of recent reviews, it is not discussed here (for review see *16–18*, *196*, *197*).

3.4. Extracellular Glycoproteins

A number of organisms have been identified which contain glycosylated proteins not only bound to a membrane or another cell envelope component, but shed considerable amounts of these components into the surrounding medium (for example see *158*, *194*, *198*). The reasons for shedding can be manifold and are not always fully understood. In a number of cases, ageing of the organism and degradation of linkage units is the cause. Other organisms, however, shed exoenzymes into their surrounding environment to degrade nutrients into smaller pieces, suitable for proper processing by the "intracellular" catabolic machinery.

Identification and characterization of mycobacterial antigens are critical for studies of diagnosis, vaccination and pathogenesis of mycobacterial diseases. In culture filtrates of *Mycobacterium tuberculosis*, Concanavalin A (ConA)-binding glycoproteins with apparent molecular masses of 55, 50, and 38 kDa have been described (*199*). By immunoblot with monoclonal antibodies TB71 and TB72, a carbohydrate-dependent and carbohydrate-independent epitope has been identified on the 38 kDa glycoprotein. In a ConA binding assay, the relationship between the *M. tuberculosis* 50- to 55-kDa glycoprotein and the *Mycobacterium bovis* BCG 45- to 47-kDa antigen has been established (*200*). Deglycosylation with jack bean α-D-mannosidase produced a reduction of the molecular masses of each species by 1.5 to 2 kDa which indicated the presence of α-linked mannosyl residues. This observation suggested the presence of short α-linked mannose chains on these mycobacterial antigens. Dobos *et al.* (*201*) have determined the full extent of glycosylation on the *M. tuberculosis* 45-kDa glycoprotein antigen. The sites of glycosylation have been identified in proline-rich domains near the N- and the C-terminus of the 45-kDa protein. The threonine residues at positions 10 and 18 are substituted with α-D-Man*p*-(1→2)-α-D-Man*p*, Thr27 is substituted with a single α-D-Man*p*, and Thr277 is substituted with either α-D-Man*p*, α-D-Man*p*-(1→2)-α-D-Man*p*, or α-D-Man*p*-(1→2)-α-D-Man*p*-(1→2)-α-D-Man*p*. A different sugar composition is obviously present in the 32-kDa antigen from *M. tuberculosis* strain H37Rv, which has been purified by affinity chromatography on immobilized *Lens culinaris* lectin, and electroelution. In this antigen, the sugar composition (molar ratio) of the glycan chains is galactose: xylose: mannose: *N*-acetylglucosamine 5 : 2 : 3 : 1 (*202*). In addition, Montaño and coworkers have identified glycoconjugates from *Mycobacterium bovis* culture filtrates by biotin-hydrazide labeling (*203*). Purification of mycobacterial antigens has been traditionally performed by ConA chromatography (*204*), but the specificity of the lectin and the fact that lectin-binding does not necessarily indicate the absence of a particular carbohydrate residue limits the usefulness of that approach. After periodate oxidation of the sugar moieties, the aldehydes generated were reacted with biotin-hydrazide. To determine if biotin-labeled mycobacterial glycoproteins possessed *N*-linked saccharides, *N*-glycosidase F treatment was performed. This enzyme acts specifically on the inner *N*-acetylglucosamine of biantennary oligosaccharides (*205*) and has no effect on *O*-glycosidically linked glycoconjugates (*203*). Purification of sufficient amounts of these mycobacterial antigens was performed for the purpose of elucidating defense mechanisms against mycobacteria, and to identify those antigenic determinants that are able to confer protection in context with more effective vaccines (for review see *206*, *207*). In a study on biotin-hydrazide-labeled *M. bovis* antigens, 11 carbohydrate-containing protein species with molecular masses between 15 and 62 kDa have been identified (*203*). The 52-kDa antigen was found to represent approximately 26% of the total glycoconjugate content, whereas the 32-kDa glycoprotein corresponded to 1% of the total content. The latter antigen has the same molecular mass as that of *M. tuberculosis* which has been isolated by *Lens culinaris* lectin affinity chromatography (*202*). In the 52-kDa antigen, the biotin label disappeared completely after *N*-glycosidase F treatment, indicating the presence of *N*-linked biantennary oligosaccharides. In contrast, persistence of the biotin label after endo-F treatment would suggest that the glycans of these glycoproteins are either *O*-linked or attached in a different manner (*203*).

Mycoplasma hyopneumoniae is the causative agent of mycoplasmal pneumonia in swine. Severe economic losses in the hog industry can be attributed to infection of animals by this organism. Therefore, studies of the pathogenesis of the infection are expected to contribute toward the development of measures preventing the outbreak of mycoplasmal pneumonia. Adherence of *M. hyopneumoniae* at the ciliated epithelial cells of porcine trachaeas, bronchi, and bronchioles is a prerequisite for colonization and infection by this pathogen (*208*). In this disease, the functions of mucociliary clearance are damaged. Therefore, investigation of the receptor-ligand interaction was considered of crucial importance. Several adhesins have been defined which resemble the epitopes of P1; however, the P97 adhesin of *M. hyopneumoniae* has a different structure (*209*). In a recent study, mycoplasmal proteins separated by SDS-PAGE and transferred to polyvinylidene difluoride (PVDF) membranes, were used to interact with biotinylated epithelial cell surface proteins to detect adhesin candidates. Three predominant protein subunits, P80, P54 and P28, have been purified by Mab affinity chromatography, of which the first two, P80 and P54 are glycoproteins. One P54 and two P28 subunits are able to form a 110 kDa glycoprotein complex (*198*). The glycoprotein nature of these antigens has been identified by deglycosylation of the respective antigens. Following pretreatment and denaturation of the antigenic proteins with 1% SDS, glycans have been removed by treatment with *N*-glycosidase F at 37°C for 6 and 16 h, respectively (*198*). The quantities of glycoproteins expressed remained unchanged, even after 100 serial passages of *M. hyopneumoniae* in the course of the experiments. However, as described in a recent report, *M. hyopneumoniae* had lost its pathogenicity after 39 broth passages (*210*). Therefore, whether the functional groups of amino acids or the glycosylation sites had changed during the previous repeated serial passages remains undecided.

Polysaccharide-degrading enzymes have been isolated from a great variety of lower eukaryotic and prokaryotic microorganisms. Examples from *Cellulomonas fimi* are described in Sections 3.1 and 3.3. Mannans have also been found frequently in plants and marine algae. Recently, the mannanase-producing bacterium *Vibrio* sp. strain MA-138 was isolated from a marine environment (*211*). The organism secretes β-mannanases into the growth medium in the presence of an inducer molecule such as β-mannan (*212*). In Section 3.3, another β-mannanase-secreting organism, *Cellulomonas fimi*, is described whose mannanase seems, however, to be membrane-attached (*158*). The *Vibrio* β-mannanase was purified to homogeneity by chromatographic procedures and native PAGE. By SDS-PAGE, an apparent molecular mass of 49 kDa was determined, and periodic acid-Schiff staining gave a positive colour reaction, indicating the presence of a glycoprotein (*212*). Comparison of the sequences of 23 N-terminal amino acids between the β-mannanases of *Vibrio* sp. strain MA-138 and the terrestric strain *Streptomyces lividans* 66 indicated 60% homology between these enzymes (*213*).

Thermophilic anaerobic bacteria belonging to the clostridia have attracted considerable interest due to their ability to convert a wide range of carbohydrate substrates into technically important products. Recently, a novel type of thermostable α-D-glucosidase from *Thermoanaerobacter thermohydrosulfuricus* DSM 567 exhibiting maltodextrinohydrolase activity has been reported (*214*). The enzyme is apparently a glycoprotein with a molecular mass of 160 kDa and released maltodextrins from starch in addition to glucose. Identification of the single protein band detected in purified preparations as the α-glucosidase was confirmed by activity staining. The α-glucosidase gave a positive staining reaction with a glycoprotein detection kit (*214*).

Similar to other colonizers of the human oral cavity (compare with *Treponema pallidum*, Section 3.3, *Porphyromonas gingivalis* and *Prevotella intermedia*, Section 3.5) *Actinobacillus actinomycetemcomitans* is a Gram-negative, capnophilic coccobacillus which has frequently been isolated from patients with periodontitis. The extracellular antigens of *A. actinomycetemcomitans* Y4 (serotype b) include a 37-kDa glycoprotein which is a major target for IgGs from patients suffering from severe alveolar bone loss (*215*). This glycoprotein forms immune complexes with monoclonal antibodies against a rhamnose-fucose polysaccharide. The 37-kDa antigen has been collected from *A. actinomycetemcomitans* culture supernatants. Purification included chromatofocussing and gel filtration on Sephadex 200. The sample was composed of 72.3% (w/w) protein, 13.3% hexoses, and 2.8% hexosamine. LPS contaminations were ruled out by quantification of the heptose content and analysis of the endotoxin content of the antigen, both of which were negligible. From the reaction with mAb S5, a monoclonal antibody against rhamnose-fucose polysaccharide (*216*), the authors concluded that the glycoprotein contains rhamnose and fucose. The 37-kDa glycoprotein is a powerful inducer of inflammatory cytokines, inducing TNF-α, IL-1β, and IL-6 in mouse macrophage cultures (*215*). Interestingly, the IL-6 inducing activity of the 37-kDa antigen exceeds that of LPS. IL-6 was also produced from macrophages which were stimulated with a heat-treated sample. Therefore, the level of IL-6 production stimulated by the heat-treated sample is likely to be due to hexose- and hexosamine-containing glycans in the sample (*215*).

To cope with hostile environmental conditions such as low temperatures prokaryotic organisms have adopted a variety of survival strategies. The plant-growth-promoting rhizobacterium *Pseudomonas putida* strain GR12-2 secretes an antifreeze glycoprotein into the growth medium at 5°C growth temperature (*217*). That material was concentrated from the culture supernatant by $(NH_4)_2SO_4$ precipitation and purified to apparent homogeneity. SDS-PAGE indicated a molecular mass of 164 kDa. This band gave a bright staining reaction with the periodate-Schiff reagent. In addition, it gave a positive reaction with a lipid staining reagent. To elucidate whether the carbohydrate moiety in the 164-kDa antifreeze glycoprotein plays a role in its antifreeze activity, boric acid and periodic acid treatments were performed. The 164-kDa band disappeared on periodic acid oxidation, while a new band appeared at 72 kDa. This band essentially retained the antifreeze activity of the native glycoprotein. In summary, these results indicated that the carbohydrate component of the 164-kDa polypeptide was less important than the protein moiety in its modification properties of ice crystal growth (*217*). Up to date, carbohydrate in such large amount has only been found associated with antifreeze glycoproteins where it contributes up to 60% of the molecular mass (*218*).

Recently, a heterogeneous glycoprotein fraction was found to be excreted by *Pseudoalteromonas antarctica* NF_3 whose composition was dependent on the substrate composition (*218a*). This biopolymer showed interesting application potentials such as the protection of liposomes against the action of surfactants (*218b*).

There is an increasing interest in the structure and biology of biofilms. Due to the specific growth conditions, fierce competition for space and nutrients appears to take place among biofilm bacteria (*219*). In aquatic habitats, a sizeable portion of the diverse microbial flora is found in biofilms. Two species of gliding bacteria of the genus *Cytophaga* have been isolated from a marine biofilm. One of them produced an extracellular inhibitor which was characterized as a glycoprotein with an apparent molecular mass of 60 kDa (*220*). Carbohydrate analysis was performed on FPLC-purified material after hydrolysis with HCl at 100°C. In the inhibitor the sugars mannose, glucose, glucosamine, galactose, and galactosamine have been detected in a molar ratio of 1.0 : 2.7 : 2.0 : 1.7 : 1.0 (*220*). If this glycoprotein indeed functioned as an inhibitor of adhesion of an organism and its mobility in biofilms, the high-molecular mass would prove advantageous over a low-molecular-mass secondary metabolite. In a biofilm, the larger molecule would be kept in proximity of the producing organism because of relatively slow diffusion through the biofilm's channels, in comparison to smaller molecules (*219*).

3.5. "Cellular" Glycoproteins

Many glycoproteins have been isolated from intact or disrupted cells by various extraction and isolation methods. Subsequently, a topographic localization of their origin within the prokaryotic cell is no longer possible. Therefore, these glycoconjugates are considered as "cellular" glycoproteins, although *in vivo* they would belong to one of the previously mentioned groups (see Sections 3.1 to 3.4).

In our recent review (*16*), we have discussed the isolation of the acidic glycoprotein (SAGP) from *Streptococcus pyogenes* Su strain (*221*). At that time an anti-tumor activity of that substance on various tumors had been proposed and was investigated *in vitro* and *in vivo*, in comparison to the whole cell preparation OK-432 of the same organism, which has been used clinically as a potent antitumor agent (*222*). The growth of sensitive cells such as HeLa cells and sarcoma 180 cells has been proposed to be inhibited by very low doses of SAGP. Recombinant SAGP was obtained after expression in *E. coli* JM103. However, instead of the formation of tetramers, as with the native SAGP, only dimers were formed by the recombinant SAGP (*223*). The influence of glycosylation was not investigated in that study. The results of recent studies on SAGP, or its assumed homolog in *S. pyogenes* strain Manfredo (group A *Streptococcus*) are, however, at variance with the previous assumptions on the mode of action of SAGP (*224*). These authors purified a protein from the group A strain Manfredo cell extract which inhibited proliferative responses to human peripheral blood mononuclear cells (PBMCs) and exhibited, in addition to an N-terminus identical to that of SAGP, a number of other biochemical and size similarities with the latter molecule. In addition, it possessed arginine deimidase activity (*224*). It is obviously the activity of this enzyme that is responsible for inhibition of the proliferation of PBMCs (*225*). SAGP was originally isolated from *S. pyogenes* Su following the observation that the lyophilized preparation, termed OK-432, had antitumor properties (*222*, *226*). Since OK-432 is a lyophilized preparation of the Su strain of *S. pyogenes*, it would be expected to contain SAGP and inhibit rather than induce proliferative responses of human PBMCs. Therefore it is assumed that the activity of SAGP is destroyed during the preparation of OK-432, perhaps during lyophilisation, and it might be the remaining mixture of bacterial proteins that elicits its antitumor effect (*224*, *225*).

Recently, in *Campylobacter jejuni* 81–176 (serogroups O:23, O:36), a general system of protein glycosylation has been identified by which numerous soluble and membrane proteins including flagellin are glycosylated (*227*). The genetic locus of the genes was termed *pgl*. There was no obvious indication of a role of the *pgl* genes in LPS biosynthesis of *C. jejuni*. Interestingly, *pglA* was present in a large number of immunotypes of *Neisseria meningitidis* and other species of *Neisseria*. The high level of conservation of these genes among different *C. jejuni* strains suggests that they are responsible for the addition of glycosyl residues to flagellin. However, flagellins isolated from *pgl* mutants have shown the same glycoform patterns as wild-type flagellins (*227*). Antigenicity differences observed with O:23 and O:36 antiserum suggest, however, that some variation in the glycan chains may occur with post-translational modifications encoded by the *pgl* genes. The number of proteins that are affected by the *pgl* mutations and the ubiquity of their cellular locations is unusual. Their glycoprotein nature has been demonstrated by deglycosylation with trifluoromethanesulfonic acid (*228*). Most of the affected proteins appeared to be highly immunogenic. With the exception of flagellin of *C. jejuni* (*185b*), they were all present in relatively low concentrations. The fact that almost no antibodies were elicited against these modified proteins suggests that the glycosylation could offer an unusual mechanism of immune evasion through masking of primary amino acid sequences (*227*).

The immune system produces its own regulatory factors (*229*). In this context, immunomodulation refers to the ability to stimulate or suppress immune reactions. An immunomodulatory factor, RU 41740, has been extracted from *Klebsiella pneumonia* strain K2 O1. This material is composed of two glycoconjugates, one of which is a lipopolysaccharide-like molecule (F1) and the other, a non-endotoxin fraction (P1), is possibly a glycoprotein (*230*, *231*). However, no detailed biochemical characterization of the P1 fraction has been reported in this work. In view of the bacterial origin of RU 41740, the possible involvement of LPS-like molecules and of molecules unrelated to LPS has been considered. Both molecules were active on immunocompetent cells, but presumably through different cellular receptors. The data indicate that at least the glycoprotein fraction P1 modulates the immune system in an endotoxin-independent manner. RU 41740 has been found mitogenic for B lymphocytes from LPS nonresponder C3H/HeJ mice. This effect was attributed to the P1 fraction (*229*).

References, pp. 105–124

In the course of the characterization of the human dental plaque a number of organisms residing in this complex system have been investigated (compare with Sections 3.3 and 3.4). *Porphyromonas gingivalis* and *Prevotella intermedia* are dominant bacteria in periodontal pockets and have been implicated as pathogens in periodontal diseases (*232*). Recently, a novel immunobiologically active fraction has been prepared from a phenol-water extract of *P. intermedia* ATCC 25611 by Sephadex G-100 column chromatography (*233*). The fraction designated *Prevotella* glycoprotein (PGP) contains rhamnose, galactose, glucose, small amounts of glucosamine and 11% of polypeptide, suggesting that this fraction is a protein-carbohydrate complex. The material showed mitogenicity against splenocytes of C3H/HeN and C3H/HeJ mice, and is devoid of *Limulus* activity. The mitogenic activity of the PGP fraction is resistant to heating at 100°C for 60 min and to pronase E and proteinase K, but is almost completely abolished by periodate oxidation. This observation indicates the importance of the glycan portion of this fraction (*233*).

Another important system identified in human dental plaque is an aggregation factor of *Actinomyces naeslundii* (AnAF) that mediates coaggregation with *Porphyromonas gingivalis* (*234*). Colonization of pathogenic bacteria in periodontal pockets is generally recognized as an essential first step for the development of periodontal disease (*235*). Within the plaque, streptococci, actinomycetes and Gram-positive bacilli are dominant in deeper layers, whereas Gram-negative bacteria predominate in the surface areas (*236*). *A. naeslundii* is often found in the dental plaque of healthy humans and periodontal patients and is therefore considered an early colonizer. By a poorly understood mechanism, cells of *P. gingivalis* adhere to bacteria that have previously colonized the teeth. Thus, this organism is classified as a secondary colonizer. The aggregation factor AnAF has been isolated from sonicated cell suspensions of *A. naeslundii* (*237*). After concentration in an ultraconcentration cell (with a molecular-mass cut-off of 10,000 Da) and column chromatography over Sephacryl S-400 HR, each fraction was tested for chemical composition and inhibitory activity (*234*). The total carbohydrate content was measured by the phenol-H_2SO_4 method, and sugar analysis was performed after HPLC and post-column fluorescence labeling with 1% arginine in 3% boric acid. In the eluate of the Sephacryl S-400 HR column, four distinct peaks were detected, but only the high-molecular-mass substance was further investigated. This included hydrophobic interaction chromatography on HighTrap Octyl sepharose 4FF and anion exchange chromatography on a HighTrap Q column. The collected material showed a single band by SDS-PAGE with a molecular mass of >200 kDa. In AnAF, a sugar content of 0.7 μg hexose per 100 μg of protein was detected. Serine, threonine and asparagine accounted for 6.8 mol% of the total amino acid residues. Carbohydrate analysis indicated that AnAF contains rhamnose (1.2 mol%), galactose (92.6 mol%), and glucose (6.2 mol%), but no pentose (*234*). These results indicate that the coaggregation factor of *A. naeslundii*, AnAF, is a high-molecular-mass glycoprotein. However, it is not yet clear whether AnAF was derived from the fimbriae-like material or from the cell wall (*234*). When pretreated with sodium metaperiodate, AnAF slightly inhibited the coaggregation. However, heat- or protease-treated AnAF preparations had an inhibiting activity identical to native AnAF. These data indicate that the coaggregative factor of *A. naeslundii* is the sugar component of the glycoprotein and that the aggregation is a very strong receptor-adhesin interaction. Bacteria causative of periodontal disease may universally recognize the active site of AnAF. On the other hand, the coaggregation between *A. naeslundii* and streptococci which are not causative agent of periodontal diseases was independent of AnAF, indicating coaggregation to be mediated by another mechanism (*234*).

To study the interaction between the outermost constituents of the cell envelope and host cells, pathogenic and non-pathogenic mycobacteria have been compared. The outermost cell envelope components consist of polysaccharides and proteins (*238*). A polysaccharide of about 3.6 kDa has been identified as the major carbohydrate portion of PS4A, a protein-polysaccharide complex from *Mycobacterium vaccae* with antineoplastic activity (*239*). PS4A comprises ~5–6% (w/w) of the dried *M. vaccae* biomass, with a protein-to-sugar ratio of 1.5 : 1 (*240*). Although the presence of an antineoplastic glycoconjugate in the cell envelope of *M. vaccae* has been proposed, the covalent linkage of the glycan chains of this complex to the polypeptide has not been demonstrated so far. Combined evidence from chemical degradation experiments and ^{1}H- and ^{13}C NMR spectroscopy indicates that the polymer consists of poly-α-(1→4)-linked 3-*O*Me-D-mannopyranose. The antineoplastic activity was tested *in vivo* with the aid of an S-180 murine sarcoma model, and the immunostimulatory activity was tested *in vitro* using murine peritoneal macrophages isolated from BALB/c mice. Only the high-molecular-mass fractions of PS4A showed antineoplastic activity and in general, these fractions possessed better immunostimulating potential (*239*).

For the formulation of efficient fish vaccines, different preparations of immunodominant whole-cell lysates of *Flexibacter columnaris* have been compared. *F. columnaris* is the causative agent of columnaris disease in more than 36 species of fish. The organism causes tremendous losses in the commercial catfish farming industry in the U.S., but wild fish populations are also affected (*241*). From 16-hour cultures, both whole-cell pressure-killed and formalin-killed lysates have been prepared. On SDS-PAGE, up to 22 separate protein bands were detected. The 60-kDa band was identified on Western blots of either type of lysate as a glycoprotein containing sialic acid residues (*241*, *242*). Formalin treatment apparently caused the sialic acid moiety of the 60-kDa protein to be lost or rendered unrecognizable to the fish immune system. Therefore, these results supported the notion that a formalin-killed *F. columnaris* preparation may not be suitable as a protective vaccine for channel catfish (*241*).

Glycosylated enzymes are not restricted to cellulolytic organisms (compare, for example, with mannan-degrading enzymes of *Cellulomonas fimi*, Sections 3.1 and 3.3), but can also be isolated from other prokaryotic species. Many microbial species used today in the fermented food industry belong to the lactic acid bacteria. In sourdough fermentations, it is important to degrade sucrose which represents 1% of the soluble carbohydrates in wheat flour. To metabolize this sugar, microorganisms should possess invertase activity (β-D-fructofuranoside fructohydrolase). This enzyme has been thoroughly investigated in yeasts, but little is known about prokaryotic invertase. Only recently, an invertase was isolated and characterized from *Lactobacillus reuteri* CRL 1100 (*243*). The enzyme was purified to homogeneity by $(NH_4)_2SO_4$ precipitation followed by several chromatographic purification steps. On SDS-PAGE it migrates as a single band with a molecular mass of 58 kDa. Gel filtration of the native enzyme indicated the same molecular mass. The carbohydrate content determined by the phenol-H_2SO_4 method, of the purified invertase of *L. reuteri* CRL 1100 is 48% (*243*).

3.6. Synthetic Glycopeptides and Glycoproteins

Glycopeptides contain a variable carbohydrate domain and a short peptide domain. The carbohydrate chain can be a monosaccharide or a complex, possibly branched oligosaccharide containing a large number of monosaccharide units (see Section 1, Introduction). Little is known about the formation of the glycosidic linkage between the carbohydrate and the peptide during enzymatic glycopeptide biosynthesis. A key approach for addressing this question will be the study of synthetic compounds in biological systems (*23*). The recognition of glycoconjugates as mediators of important biological processes has stimulated investigations of their therapeutic potential. New approaches for the simplification of glycoconjugate synthesis are overcoming the limitations of existing methods. As their accessibility increases, synthetic *neo*glycoproteins are becoming available as medicinal agents in a range of therapies (*244*, *245*).

Glycoamidases [peptide-N^4-(*N*-acetyl-β-D-glucosaminyl)-asparagine amidase, PNGase] have been found in a number of organisms across all kingdoms of life. The enzyme catalyzes the release of intact sugars from *N*-linked glycopeptides and glycoproteins via hydrolysis of the β-amide of the linking asparagine. To probe the mechanism and substrate requirements for glycoamidases, a number of natural and artificial mono- and disaccharide glycopeptides have been synthesized and tested for substrate specificity. In one of them, the linking sugar GlcNAc was replaced by Glc. The unusual Glc-Asn linkage has been found in nature notably in archaeal S-layer glycoproteins (*54*, *121*, *246*). Additionally, Glc-Asn in α-linkage has been found in a nephritogenic glycopeptide (*247*). The synthesis of this artificial glycopeptide was performed by a combination of organic synthesis and enzymatic methods. Synthesis of the enzymatic acceptor **2** (Fig. 6) started from glucose (**1**). After acetylation of the OH-groups introduction of an azide was performed, yielding

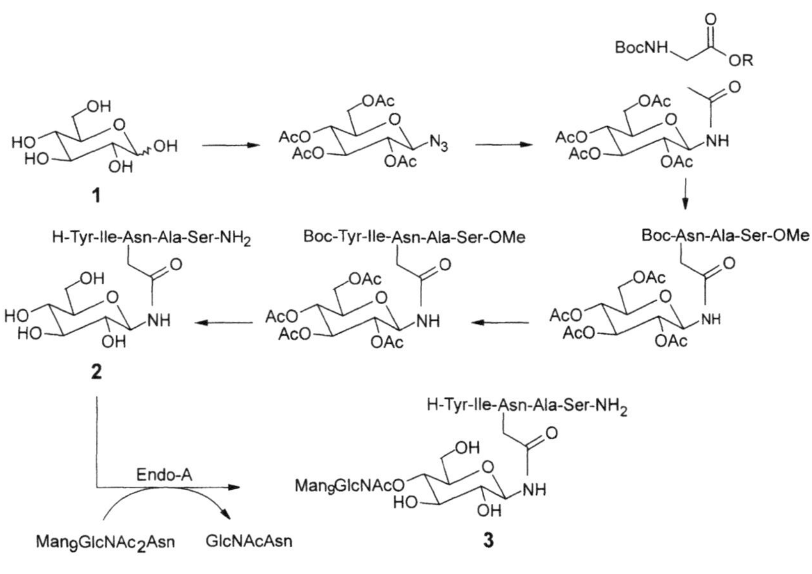

Fig. 6. Chemoenzymatic synthesis of a high-mannose-type glycopeptide analog containing a glucose-asparagine linkage [(*248*), with permission of Elsevier Science Publ., Amsterdam].

glucosyl azide tetra-*O*-acetate. Protected glucosyl asparagine was prepared by different methods, followed by elongation to the glucosyl pentapeptide acceptor molecule (**2**). Enzymatic transfer of the $Man_9GlcNAc$ structure to **2** was performed according to the literature (*248*). After HPLC-purification the desired product was obtained in 39% yield. ^1H NMR, amino acid and monosaccharide analysis confirmed the expected structure of the product **3**. The high-mannose-type glycopeptide analog **3** was tested as an inhibitor for glycoamidases using glycopeptide from a thermolysin digest of bovine asialofetuin as substrate. Glycoamidases from almond (GAA) or from *Flavobacterium meningosepticum* (GAF) and concentrations of 0.5–1000 μM of **3** in the reaction mixture eventually showed K_i values for **3** of 2 μM for GAA and 53 μM for GAF. The lack of turnover and the inhibition results suggested that the acetamido groups of the innermost sugar play an important role in recognition and cleavage of the GlcNAc-Asn bond (*248*).

In nature, glycoproteins are usually present as complex mixtures of glycoforms which are difficult to separate. To explore their individual properties, there is a need for homogeneous sources of carbohydrate-protein conjugates. Recently, a new method for site-selective glycosylation of proteins has been developed (*249*). The potential of the method has been developed by site-selective glycosylation of the serine protease subtilisin of *Bacillus lentus* as a model protein (*250*). A representative library of thio-modified mono- and disaccharide reagents have been synthesized from their parent carbohydrates and were used to modify cysteine mutants of subtilisin from *B. lentus* at specific sites. Site-directed mutagenesis combined with chemical modification reactions permitted preparation of homogeneous *neo*glycoproteins in which both the site of glycosylation and the structure of the introduced glycan have been predetermined. An expansion of this versatile method was possible by the introduction of glycans containing different numbers of acetate groups in combination with carefully controlled pH conditions. These improvements have resulted in a precisely controlled route of preparation methods of *neo*glycoconjugates that is virtually unlimited in the scope of sites and glycans that may be conjugated. It opens up opportunities for the systematic determination of properties of glycosylated proteins (*249*). As part of an ongoing research program to tailor the activity of subtilisin mutants, a library of 48 of their glycosylated forms have been synthesized as single, pure species. In order to use them in peptide synthesis, these enzymes have been screened against a synthetic esterase substrate. Depending on the glycosylation site, the different glycoforms showed remarkable increases in enzyme activity. The tailored specificity and the high activity of the glycoforms can also be reflected by molecular modelling. Therefore, these glycosylated enzymes are ideal candidates, for example, as catalysts in peptide synthesis as they will favour the formation of amide bonds over hydrolysis (*251*).

To investigate the mode of interaction between the S-layer protein of Gram-positive bacteria and the secondary cell wall polymer (SCWP) of the corresponding organism (compare with Section 2.1), chemical synthesis of structurally defined components has been initiated (*252*). Two distinct secondary cell wall polymers are responsible for strain-specific anchoring of the respective S-layer proteins of *Geobacillus stearothermophilus* PV72/p6 (wild type strain) and PV72/p2 (variant strain) to the cell wall of either the wild type or the variant strain (*36*, *40*, *41*). The SCWP of the PV72/p6 variant is a regular polysaccharide containing 2,3-diacetamido-2,3-dideoxy-D-mannuronic acid: *N*-acetylglucosamine : glucose in a ratio of 2 : 1 : 1 (*42*, *44*). In order to derive insights into protein-carbohydrate interactions between the SCWP and the S-layer proteins, it is necessary to synthesize oligosaccharide derivatives corresponding to partial structures of the PV72/p6-related polysaccharide. In a first set of experiments, the target compound 2,3-di-*N*-acetyl-2,3-diacetamido-mannuronic acid β-D-methylglycoside has been synthesized (*252*). This component will be used in syntheses of further oligosaccharide probes as tools in elucidating the nature of the interaction between the S-layer protein and the SCWP in that organism.

4. Conclusions

Innovations in chemical and genetic characterization of prokaryotic glycoproteins have led to a tremendous progress in the general understanding of prokaryotic glycosylation in the past decade.

With S-layer glycoproteins (Section 2), a great diversity of glycan structures has been observed. This points to a general concept of specific adaptation of individual organisms to the environmental conditions in the natural habitats by variation of sugar molecules. In addition to the structural genes of S-layer proteins, genes involved in the biosynthesis of S-layer glycans have been cloned, sequenced and overexpressed. These results will open up the possibility for carbohydrate engineering not only by chemical but also by genetic means in the future. In this manner, the production of tailored glycan and glycoproteins for medical and biotechnological applications will be possible.

With regard to non-S-layer glycoproteins (Section 3), the situation is more diverse. Whereas impressive progress has been made with the characterization of specific prokaryotic glycoproteins such as pilins, flagella, exoenzymes, or shed antigens, the information is still fragmentary with some other glycoproteins. It is interesting to note that most reports from non-S-layer glycoproteins originate from pathogenic microorganisms. So, the awareness of the existence of this type of glycoconjugate among biologists and organic chemists is increasing. In this context it should be noted that specific glycoproteins have been found even in clinical isolates of *Escherichia coli*. For a long time, it had been assumed that enteric bacteria do not synthesize true glycoproteins.

Although carbohydrates of prokaryotic glycoproteins are still poorly understood when compared to other biomolecules, general themes in glycoconjugate function are beginning to emerge. Further progress in prokaryotic glycobiology will be greatly aided by the introduction and application of techniques that include a combination of enzymatic and synthetic methods for production of specific glycan structures for applied aspects.

Acknowledgements

The authors appreciate the critical reading of the manuscript by Prof. Frank M. Unger (Institute of Pharmaceutical Technology and Biopharmacy, University of Vienna).

The work was supported by grants from the Austrian Science Fund (projects P12966-MOB and P14209-MOB to P.M.), the Hochschuljubiläumsstiftung der Stadt Wien (project H096-2000 to C.S.), and the Federal Ministry of Education, Science and Culture.

References

1. Allen HJ, Kisailus EC (eds) (1992) Glycoconjugates: Composition, Structure and Function. Marcel Dekker, New York
2. Varki A (1993) Biological Roles of Oligosaccharides: All of the Theories are Correct. Glycobiology **3**: 97
3. Fukuda M, Kobata A (eds) (1993) Glycobiology. A Practical Approach. IRL Press, Oxford
4. Fukuda M, Hindsgaul O (eds) (1994) Molecular Glycobiology. IRL Press, Oxford
5. Montreuil J, Vliegenthart JFG, Schachter H (eds) (1995) Glycoproteins. Elsevier, Amsterdam
6. Rudd PM, Dwek RA (1997) Glycosylation: Heterogeneity and the 3D Structure of Proteins. Crit Rev Biochem Mol Biol **32**: 1
7. Van den Steen P, Rudd PM, Dwek RA, Opdenakker G (1998) Concepts and Principles of O-Linked Glycosylation. Crit Rev Biochem Mol Biol **33**: 151
8. Apweiler R, Hermjakob H, Sharon N (1999) On the Frequency of Protein Glycosylation, as Deduced from Analysis of the SWISS-PROT Database. Biochim Biophys Acta **1473**: 4
9. Varki A, Cummings R, Esko J, Freeze H, Hart G, Marth J (eds) (1999) Essentials of Glycobiology. Cold Spring Harbor Laboratory Press, Cold Spring Harbor, NY
10. Lechner J, Wieland F (1989) Structure and Biosynthesis of Prokaryotic Glycoproteins. Annu Rev Biochem **58**: 173

11. Messner P, Sleytr UB (1991) Bacterial Surface Layer Glycoproteins. Glycobiology **1**: 545
12. Erickson PR, Herzberg MC (1993) Evidence for the Covalent Linkage of Carbohydrate Polymers to a Glycoprotein from *Streptococcus sanguis*. J Biol Chem **268**: 23780
13. Sandercock LE, MacLeod AM, Ong E, Warren RAJ (1994) Non-S-layer Glycoproteins in Eubacteria. FEMS Microbiol Lett **118**: 1
14. Sumper M, Wieland FT (1995) Bacterial Glycoproteins. In: Montreuil J, Vliegenthart JFG, Schachter H (eds) Glycoproteins. Elsevier, Amsterdam, p 455
15. Messner P (1996) Chemical Composition and Biosynthesis of S-Layers. In: Sleytr UB, Messner P, Pum D, Sára M (eds) Crystalline Bacterial Cell Surface Proteins. RG Landes/Academic Press, Austin, TX, p 35
16. Messner P (1997) Bacterial Glycoproteins. Glycoconjugate J **14**: 3
17. Moens S, Vanderleyden J (1997) Glycoproteins in Prokaryotes. Arch Microbiol **168**: 169
18. Moens S (2000) Non-S-Layer Glycoproteins. A Review. In: Doyle RJ (ed) Glycomicrobiology. Kluwer Academic/Plenum Publishers, New York, p 1
19. Messner P, Schäffer C (2000) Surface Layer Glycoproteins of Bacteria and Archaea. In: Doyle RJ (ed) Glycomicrobiology. Kluwer Academic/Plenum Publishers, New York, p 93
19a. Schäffer C, Messner P (2001) Glycobiology of Surface Layer Proteins. Biochimie **83**: 591
20. Schäffer C, Graninger M, Messner P (2001) Prokaryotic Glycosylation. Proteomics **1**: 248
21. Tuomanen EI (1996) Surprise? Bacteria Glycosylate Proteins Too. J Clin Invest **98**: 2659
22. Vliegenthart JFG, Casset F (1998) Novel Forms of Protein Glycosylation. Curr Opin Struct Biol **8**: 565
23. Taylor CM (1998) Glycopeptides and Glycoproteins: Focus on the Glycosidic Linkage. Tetrahedron **54**: 11317
24. Sleytr UB, Beveridge TJ (1999) Bacterial S-Layers. Trends Microbiol **7**: 253
25. Mescher MF, Strominger JL (1976) Structural (Shape-Maintaining) Role of the Cell Surface Glycoprotein of *Halobacterium salinarium*. Proc Natl Acad Sci USA **73**: 2687
26. Sleytr UB, Thorne KJI (1976) Chemical Characterization of the Regularly Arranged Surface Layers of *Clostridium thermosaccharolyticum* and *Clostridium thermohydrosulfuricum*. J Bacteriol **126**: 377
27. Sumper M (1987) Halobacterial Glycoprotein Biosynthesis. Biochim Biophys Acta **906**: 69
28. Sleytr UB, Messner P, Pum D, Sára M (eds) (1996) Crystalline Bacterial Cell Surface Proteins. RG Landes/Academic Press, Austin TX
29. Sára M, Sleytr UB (2000) S-Layer Proteins. J Bacteriol **182**: 859
30. Beveridge TJ, Pouwels PH, Sára M, Kotiranta A, Lounatmaa K, Kari K, Kerosuo E, Haapasalo M, Egelseer EM, Schocher I, Sleytr UB, Morelli L, Callegari M-L, Nomellini JF, Bingle WH, Smit J, Leibovitz E, Lemaire M, Miras I, Salamitou S, Béguin P, Ohayon H, Gounon P, Matuschek M, Sahm K, Bahl H, Grogono-Thomas R, Dworkin J, Blaser MJ, Woodland RM, Newell DG, Kessel M, Koval SF (1997) Functions of S-Layers. FEMS Microbiol Rev **20**: 99
31. Sleytr UB, Messner P, Pum D, Sára M (1999) Crystalline Bacterial Cell Surface Layers (S-layers): From Supramolecular Cell Structure to Biomimetics and Nanotechnology. Angew Chemie, Int Ed, **38**: 1034

32. Sleytr UB, Sára M, Pum D (2000) Crystalline Bacterial Cell Surface Layers (S-Layers): A Versatile Self-Assembly System. In: Ciferri A (ed) Supramolecular Polymerization. Marcel Dekker, New York, p 177
33. Kuen B, Sleytr UB, Lubitz W (1994) Sequence Analysis of the *sbs*A Gene Encoding the 130-kDa Surface-Layer Protein of *Bacillus stearothermophilus* Strain PV72. Gene **145**: 115
34. Kuen B, Koch A, Asenbauer E, Sára M, Lubitz W (1997) Molecular Characterization of the *Bacillus stearothermophilus* PV72 S-Layer Gene *sbsB* Induced by Oxidative Stress. J Bacteriol **179**: 1664
35. Jarosch M, Egelseer EM, Mattanovich D, Sleytr UB, Sára M (2000) S-Layer Gene *sbs*C of *Bacillus stearothermophilus* ATCC 12980: Molecular Characterization and Heterologous Expression in *Escherichia coli*. Microbiology **146**: 273
35a. Nazina TN, Tourova TP, Poltaraus AB, Novikova EV, Grigoryan AA, Ivanova AE, Lysenko AM, Petrunyaka VV, Osipov GA, Belyaev SS, Ivanov MV (2001) Taxonomic Study of Aerobic Thermophilic Bacilli: Descriptions of *Geobacillus subterraneus* gen. nov., sp. nov. and *Geobacillus uzenensis* sp. nov. from Petrol Reservoirs and Transfer of *Bacillus stearothermophilus*, *Bacillus thermocatenulatus*, *Bacillus thermoleovorans*, *Bacillus kaustophilus*, *Bacillus thermoglucosidasius* and *Bacillus thermodenitrificans* to *Geobacillus* as New Combinations *G. stearothermophilus*, *G. thermocatenulatus*, *G. thermoleovorans*, *G. kaustophilus*, *G. thermoglucosidasius* and *G. thermodenitrificans*. Int J Syst Evol Microbiol **51**: 433
36. Sára M, Egelseer EM, Dekitsch C, Sleytr UB (1998) Identification of Two Binding Domains, One for Peptidoglycan and Another for a Secondary Cell Wall Polymer, on the N-Terminal Part of the S-Layer Protein SbsB from *Bacillus stearothermophilus* PV72/p2. J Bacteriol **180**: 6780
36a. Sára M (2001) Conserved Anchoring Mechanisms Between Crystalline Cell Surface S-Layer Proteins and Secondary Cell Wall Polymers in Gram-Positive Bacteria. Trends Microbiol **9**: 47
37. Brechtel E, Matuschek M, Hellberg A, Egelseer EM, Schmid R, Bahl H (1999) Cell Wall of *Thermoanaerobacterium thermosulfurigenes* EM1: Isolation of Its Components and Attachment of the Xylanase XynA. Arch Microbiol **171**: 159
38. Fujino T, Béguin P, Aubert JP (1993) Organization of a *Clostridium thermocellum* Gene Cluster Encoding the Cellulosomal Scaffolding Protein CipA and a Protein Possibly Involved in Attachment of the Cellulosome to the Cell Surface. J Bacteriol **175**: 1891
39. Lupas A, Engelhardt H, Peters J, Santarius U, Volker S, Baumeister W (1994) Domain Structure of the *Acetogenium kivui* Surface Layer Revealed by Electron Crystallography and Sequence Analysis. J Bacteriol **176**: 1224
40. Ries W, Hotzy C, Schocher I, Sleytr UB, Sára M (1997) Evidence that the N-Terminal Part of the S-Layer Protein from *Bacillus stearothermophilus* PV72/p2 Recognizes a Secondary Cell Wall Polymer. J Bacteriol **179**: 3892
41. Egelseer EM, Leitner K, Jarosch M, Hotzy C, Zayni S, Sleytr UB, Sára M (1998) The S-Layer Proteins of Two *Bacillus stearothermophilus* Wild-Type Strains Are Bound via Their N-Terminal Region to a Secondary Cell Wall Polymer of Identical Chemical Composition. J Bacteriol **180**: 1488
42. Schäffer C, Kählig H, Christian R, Schulz G, Zayni S, Messner P (1999) The Diacetamidodideoxyuronic-Acid-Containing Glycan Chain of *Bacillus stearothermophilus* NRS 2004/3a Represents the Secondary Cell Wall Polymer of Wild-Type *B. stearothermophilus* Strains. Microbiology **145**: 1575

43. Schneewind O, Mihaylova-Petkov D, Model P (1993) Cell Wall Sorting Signals in Surface Proteins of Gram-Positive Bacteria. EMBO J **12**: 4803

44. Messner P, Sleytr UB, Christian R, Schulz G, Unger FM (1987) Isolation and Structure Determination of a Diacetamidodideoxyuronic Acid-Containing Glycan Chain from the S-Layer Glycoprotein of *Bacillus stearothermophilus* NRS 2004/3a. Carbohydr Res **168**: 211

45. Schäffer C, Müller N, Mandal PK, Christian R, Zayni S, Messner P (2000) A Pyrophosphate Bridge Links the Pyruvate-Containing Secondary Cell Wall Polymer of *Paenibacillus alvei* CCM 2051 to Muramic Acid. Glycoconjugate J **17**: 681

46. Ilk N, Kosma P, Puchberger M, Egelseer EM, Mayer H, Sleytr UB, Sára M (1999) Structural and Functional Analyses of the Secondary Cell Wall Polymer of *Bacillus sphaericus* CCM 2177 That Serves as an S-Layer-Specific Anchor. J Bacteriol **181**: 7643

47. Marquis RE (1973) Immersion Refractometry of Isolated Bacterial Cell Walls. J Bacteriol **116**: 1273

48. Altman E, Schäffer C, Brisson J-R, Messner P (1996) Isolation and Characterization of an Amino Sugar-Rich Glycopeptide from the Surface Layer Glycoprotein of *Thermoanaerobacterium thermosaccharolyticum* E207-71. Carbohydr Res **295**: 245

48a. Steindl C, Schäffer C, Wugeditsch T, Graninger M, Müller N, Messner P, The First Biantennary Secondary Cell Wall Polymer of the Domain Bacteria, Isolated from *Aneurinibacillus thermoaerophilus* DSM 10155, and its Influence on the Assembly of the S-Layer Glycoprotein. Biochem J, submitted

49. Mesnage S, Fontaine T, Mignot T, Delepierre M, Mock M, Fouet A (2000) Bacterial SLH Domain Proteins Are Non-Covalently Anchored to the Cell Surface via a Conserved Mechanism Involving Wall Polysaccharide Pyruvylation. EMBO J **19**: 4473

50. Masuda K, Kawata T (1985) Reassembly of a Regularly Arranged Protein in the Cell Wall of *Lactobacillus buchneri* and Its Reattachment to Cell Walls: Chemical Modification Studies. Microbiol Immunol **29**: 927

51. Lis H, Sharon N (1993) Protein Glycosylation. Structural and Functional Aspects. Eur J Biochem **218**: 1

52. Navarre WW, Schneewind O (1999) Surface Proteins of Gram-Positive Bacteria and Mechanisms of Their Targeting to the Cell Wall Envelope. Microbiol Mol Biol Rev **63**: 174

53. Kärcher U, Schröder H, Haslinger E, Allmaier G, Schreiner R, Wieland F, Haselbeck A, König H (1993) Primary Structure of the Heterosaccharide of the Surface Glycoprotein of *Methanothermus fervidus*. J Biol Chem **268**: 26821

54. Wieland F, Heitzer R, Schaefer W (1983) Asparaginyl-Glucose: Novel Type of Carbohydrate Linkage. Proc Natl Acad Sci USA **80**: 5470

55. Pellerin P, Fournet B, Debeire P (1990) Evidence for the Glycoprotein Nature of the Cell Sheath of *Methanosaeta*-Like Cells in the Culture of *Methanothrix soehngenii* Strain FE. Can J Microbiol **36**: 631

56. Mengele R, Sumper M (1992) Drastic Differences in Glycosylation of Related S-Layer Glycoproteins from Moderate and Extreme Halophiles. J Biol Chem **267**: 8182

57. Wieland F, Lechner J, Bernhardt G, Sumper M (1981) Sulphation of a Repetitive Saccharide in Halobacterial Cell Wall Glycoprotein. Occurrence of a Sulphated Lipid-Linked Precursor. FEBS Lett **132**: 319

58. Wieland F, Lechner J, Sumper M (1982) The Cell Wall Glycoprotein of Halobacteria: Structural, Functional and Biosynthetic Aspects. Zbl Bakt Hyg, I Abt Orig C **3**: 178

59. Sumper M (1993) S-Layer Glycoproteins from Moderately and Extremely Halophilic Archaeobacteria. In: Beveridge TJ, Koval SF (eds) Advances in Bacterial Paracrystalline Surface Layers. Plenum, New York, p 109
60. Trachtenberg S, Pinnick B, Kessel M (2000) The Cell Surface Glycoprotein Layer of the Extreme Halophile *Halobacterium salinarium* and Its Relation to *Haloferax volcanii*: Cryo-Electron Tomography of Freeze-Substituted Cells and Projection Studies of Negatively Stained Envelopes. J Struct Biol **130**: 10
61. Kessel M, Wildhaber I, Cohen S, Baumeister W (1988) Three-Dimensional Structure of the Regular Surface Glycoprotein Layer of *Halobacterium volcanii* from the Dead Sea. EMBO J **7**: 1549
62. Baumeister W, Lembcke G (1992) Structural Features of Archaebacterial Cell Envelopes. J Bioenerg Biomembr **24**: 567
63. Messner P, Pum D, Sára M, Stetter KO, Sleytr UB (1986) Ultrastructure of the Cell Envelope of the Archaebacteria *Thermoproteus tenax* and *Thermoproteus neutrophilus*. J Bacteriol **166**: 1046
64. Wildhaber I, Baumeister W (1987) The Cell Envelope of *Thermoproteus tenax*: Three-Dimensional Structure of the Surface Layer and its Role in Shape Maintenance. EMBO J **6**: 1475
65. Eichler J (2000) Novel Glycoproteins of the Halophilic Archaeon *Haloferax volcanii*. Arch Microbiol **173**: 445
65a. Eichler J (2001) Post-Translational Modification of the S-Layer Glycoprotein Occurs Following Translocation Across the Plasma Membrane of the Haloarchaeon *Haloferax volcanii*. Eur J Biochem **268**: 4366
66. Zellner G, Messner P, Winter J, Stackebrandt E (1998) *Methanoculleus palmolei* sp. nov., an Irregularly Coccoid Methanogen from an Anaerobic Digester Treating Wastewater of a Palmoil Plant in North-Sumatra, Indonesia. Int J Syst Bacteriol **48**: 1111
67. Zabel HP, König H, Winter J (1985) Emended Description of *Methanogenium thermophilicum*, Rivard and Smith, and Assignment of New Isolates to This Species. System Appl Microbiol **6**: 72
68. Zellner G, Sleytr UB, Messner P, Kneifel H, Winter J (1990) *Methanogenium liminatans* spec. nov., a New Coccoid, Mesophilic Methanogen Able to Oxidize Secondary Alcohols. Arch Microbiol **153**: 287
69. Evrard C, Declerq JP, Debaerdemaeker T, König H (1999). The First Successful Crystallization of a Prokaryotic Extremely Thermophilic Outer Surface Layer Glycoprotein. Z. Kristallogr **214**: 427
70. Khmelenina VN, Kalyuzhnaya MG, Sakharovsky VG, Suzina NE, Trotsenko YA, Gottschalk G (1999) Osmoadaptation in Halophilic and Alkaliphilic Methanotrophs. Arch Microbiol **172**: 321
71. Knirel YA, Kochetkov NK (1994) The Structure of Lipopolysaccharides of Gram-Negative Bacteria. III. The Structure of O-Antigens. Biochemistry (Moscow) **59**: 1325
72. Rocchetta HL, Burrows LL, Lam JS (1999) Genetics of O-Antigen Biosynthesis in *Pseudomonas aeruginosa*. Microbiol Mol Biol Rev **63**: 523
73. Altman E, Schäffer C, Brisson J-R, Messner P (1995) Characterization of the Glycan Structure of a Major Glycopeptide from the Surface Layer Glycoprotein of *Clostridium thermosaccharolyticum* E207-71. Eur J Biochem **229**: 308
74. Kosma P, Neuninger C, Christian R, Schulz G, Messner P (1995) Glycan Structure of the S-Layer Glycoprotein of *Bacillus* sp. L420-91. Glycoconjugate J **12**: 99

75. Schäffer C, Müller N, Christian R, Graninger M, Wugeditsch T, Scheberl A, Messner P (1999) Complete Glycan Structure of the S-Layer Glycoprotein of *Aneurinibacillus thermoaerophilus* GS4-97. Glycobiology **9**: 407
76. Kosma P, Wugeditsch T, Christian R, Zayni S, Messner P (1995) Glycan Structure of a Heptose-Containing S-Layer Glycoprotein of *Bacillus thermoaerophilus*. Glycobiology **5**: 791; Erratum: Glycobiology (1996) **6**: 5
77. Vliegenthart JFG, Montreuil J (1995) Primary Structure of Glycoprotein Glycans. In: Montreuil J, Vliegenthart JFG, Schachter H (eds) Glycoproteins. Elsevier, Amsterdam, p 13
78. Bock K, Schuster-Kolbe J, Altman E, Allmaier G, Stahl B, Christian R, Sleytr UB, Messner P (1994) Primary Structure of the O-Glycosidically Linked Glycan Chain of the Crystalline Surface Layer Glycoprotein of *Thermoanaerobacter thermohydrosulfuricus* L111-69. Galactosyl Tyrosine as a Novel Linkage Unit. J Biol Chem **269**: 7137
79. Gerwig GJ, Kamerling JP, Vliegenthart JFG, Morag (Morgenstern) E, Lamed R, Bayer EA (1991) Primary Structure of O-Linked Carbohydrate Chains in the Cellulosome of Different *Clostridium thermocellum* Strains. Eur J Biochem **196**: 115
80. Messner P, Christian R, Neuninger C, Schulz G (1995) Similarity of "Core" Structures in Two Different Glycans of Tyrosine-Linked Eubacterial S-Layer Glycoproteins. J Bacteriol **177**: 2188
81. Messner P, Christian R, Kolbe J, Schulz G, Sleytr UB (1992) Analysis of a Novel Linkage Unit of O-Linked Carbohydrates from the Crystalline Surface Layer Glycoprotein of *Clostridium thermohydrosulfuricum* S102-70. J Bacteriol **174**: 2236
82. Schäffer C, Dietrich K, Unger B, Scheberl A, Rainey FA, Kählig H, Messner P (2000) A Novel Type of Carbohydrate-Protein Linkage Region in the Tyrosine-Bound S-Layer Glycan of *Thermoanaerobacterium thermosaccharolyticum* D120-70. Eur J Biochem **267**: 5482
83. Meier-Stauffer K, Busse H-J, Rainey FA, Burghardt J, Scheberl A, Hollaus F, Kuen B, Makristathis A, Sleytr UB, Messner P (1996) Description of *Bacillus thermoaerophilus* sp. nov., to Include Sugar Beet Isolates and *Bacillus brevis* ATCC 12990. Int J Syst Bacteriol **46**: 532
84. Lan R, Reeves PR (1996) Gene Transfer is a Major Factor in Bacterial Evolution. Mol Biol Evol **13**: 47
85. Lan R, Reeves PR (2000) Intraspecies Variation in Bacterial Genomes: The Need for a Species Genome Concept. Trends Microbiol **8**: 396
86. Wugeditsch T, Zachara NE, Puchberger M, Kosma P, Gooley AA, Messner P (1999) Structural Heterogeneity in the Core Oligosaccharide of the S-Layer Glycoprotein from *Aneurinibacillus thermoaerophilus* DSM 10155. Glycobiology **9**: 787
87. Zachara NE, Gooley AA (2000) Identification of Glycosylation Sites in Mucin Peptides. In: Corfield AP (ed) Glycoprotein Methods and Protocols. The Mucins. Methods in Molecular Biology, vol 125. Humana Press, Totowa, NJ, p 121
88. Kadurugamuwa JL, Mayer A, Messner P, Sára M, Sleytr UB, Beveridge TJ (1998) S-Layered *Aneurinibacillus* and *Bacillus* spp. are Susceptible to the Lytic Action of *Pseudomonas aeruginosa* Membrane Vesicles. J Bacteriol **180**: 2306
89. Allmaier G, Schäffer C, Messner P, Rapp U, Mayer-Posner FJ (1995) Accurate Determination of the Molecular Weight of the Major Surface Layer Protein Isolated from *Clostridium thermosaccharolyticum* by Time-of-Flight Mass Spectrometry. J Bacteriol **177**: 1402
90. Altman E, Brisson J-R, Messner P, Sleytr UB (1990) Chemical Characterization of the Regularly Arranged Surface Layer Glycoprotein of *Clostridium thermosaccharolyticum* D120-70. Eur J Biochem **188**: 73

91. Sadovskaya I, Brisson J-R, Thibault P, Richards JC, Lam JS, Altman E (2000) Structural Characterization of the Outer Core and the O-Chain Linkage Region of Lipopolysaccharide from *Pseudomonas aeruginosa* Serotype O5. Eur J Biochem **267**: 1640

92. Olsthoorn MMA, Petersen BO, Duus J, Haverkamp J, Thomas-Oates JE, Bock K, Holst O (2000) The Structure of the Linkage Between the O-Specific Polysaccharide and the Core Region of the Lipopolysaccharide from *Salmonella enterica* Serovar Typhimurium Revisited. Eur J Biochem **267**: 2014

93. Cerquetti M, Molinari A, Sebastianelli A, Diociaiuti M, Petruzelli R, Capo C, Mastrantonio P (2000) Characterization of Surface Layer Proteins from Different *Clostridium difficile* Clinical Isolates. Microb Pathog **28**: 363

93a. Karjalainen T, Waligora-Dupriet AJ, Cerquetti M, Spigaglia P, Maggioni A, Mauri P, Mastrantonio P (2001) Molecular and Genomic Analysis of Genes Encoding Surface-Anchored Proteins from *Clostridium difficile*. Infect Immun **69**: 3442

93b. Calabi E, Ward S, Wren B, Parton T, Panico M, Morris H, Dell A, Dougan G, Fairweather N (2001) Molecular Characterization of the Surface Layer Proteins from *Clostridium difficile*. Mol Microbiol **40**: 1187

94. Mauri PL, Pietta PG, Maggioni A, Cerquetti M, Sebastianelli A, Mastrantonio P (1999) Characterization of Surface Layer Proteins from *Clostridium difficile* by Liquid Chromatography/Electrospray Ionization Mass Spectometry. Rapid Commun Mass Spectrom **13**: 695

95. Schäffer C, Wugeditsch T, Neuninger C, Messner P (1996) Are S-Layer Glycoproteins and Lipopolysaccharides Related? Microb Drug Resist **2**: 17

96. Messner P, Sleytr UB (1988) Asparaginyl-Rhamnose: A Novel Type of Protein-Carbohydrate Linkage in a Eubacterial Surface-Layer Glycoprotein. FEBS Lett **228**: 317

97. Möschl A, Schäffer C, Sleytr UB, Messner P, Christian R, Schulz G (1993) Characterization of the S-Layer Glycoproteins of Two Lactobacilli. In: Beveridge TJ, Koval SF (eds) Advances in Bacterial Paracrystalline Surface Layers. Plenum, New York, p 281

98. Christian R, Schulz G, Schuster-Kolbe J, Allmaier G, Schmid ER, Sleytr UB, Messner P (1993) Complete Structure of the Tyrosine-Linked Saccharide Moiety from the Surface Layer Glycoprotein of *Clostridium thermohydrosulfuricum* S102-70. J Bacteriol **175**: 1250

99. Kornfeld R, Kornfeld S (1985) Assembly of Asparagine-Linked Oligosaccharides. Annu Rev Biochem **54**: 631

100. Lechner J, Wieland F, Sumper M (1985) Biosynthesis of Sulfated Saccharides N-Glycosidically Linked to the Protein via Glucose. Purification and Identification of Sulfated Dolichyl Monophosphoryl Tetrasaccharides from Halobacteria. J Biol Chem **260**: 860

101. Kuntz C, Sonnenbichler J, Sonnenbichler I, Sumper M, Zeitler R (1997) Isolation and Characterization of Dolichol-Linked Oligosaccharides from *Haloferax volcanii*. Glycobiology **7**: 897

102. Zeitler R, Hochmuth E, Deutzmann R, Sumper M (1998) Exchange of Ser-4 for Val, Leu or Asn in the Sequon Asn-Ala-Ser Does Not Prevent N-Glycosylation of the Cell Surface Glycoprotein from *Halobacterium halobium*. Glycobiology **8**: 1157

103. Hartmann E, König H (1989) Uridine and Dolichyl Diphosphate Activated Oligosaccharides are Intermediates in the Biosynthesis of the S-Layer Glycoprotein of *Methanothermus fervidus*. Arch Microbiol **151**: 274

104. Zhu BCR, Laine RA (1996) Dolichyl-Phosphomannose Synthase from the Archae *Thermoplasma acidophilum*. Glycobiology **6**: 811

105. Burda P, Aebi M (1999) The Dolichol Pathway of N-Linked Glycosylation. Biochim Biophys Acta **1426**: 239

106. Altman E, Brisson J-R, Messner P, Sleytr UB (1991) Chemical Characterization of the Regularly Arranged Surface Layer Glycoprotein of *Bacillus alvei* CCM 2051. Biochem Cell Biol **69**: 72

107. Hartmann E, Messner P, Allmeier G, König H (1993) Proposed Pathway for Biosynthesis of the S-Layer Glycoprotein of *Bacillus alvei*. J Bacteriol **175**: 4515

108. Reeves PR (1994) Biosynthesis and Assembly of Lipopolysaccharide. New Comp Biochem **27**: 281

109. Whitfield C (1995) Biosynthesis of Lipopolysaccharide O-Antigens. Trends Microbiol **3**: 178

110. Raetz CRH (1996) Bacterial Lipopolysaccharides: A Remarkable Family of Bioactive Macroamphiphiles. In: Neidhart FC, Curtis III R, Ingraham JL, Lin ECC, Low KB, Magasanik B, Reznikoff WS, Riley M, Schaechter M, Umbarger HE (eds) *Escherichia coli* and *Salmonella enterica*. Cellular and Molecular Microbiology, 2nd ed, vol 1. ASM Press, Washington, DC, p 1035

111. Heinrichs DE, Yethon JA, Whitfield C (1998) Molecular Basis for Structural Diversity in the Core Regions of the Lipopolysaccharides of *Escherichia coli* and *Salmonella enterica*. Mol Microbiol **30**: 221

112. Kneidinger B, Graninger M, Adam G, Puchberger M, Kosma P, Zayni S, Messner P. (2001) Identification of Two GDP-6-Deoxy-D-*Lyxo*-4-Hexulose Reductases Sythesizing GDP-D-Rhamnose in *Aneurinibacillus thermoaerophilus* L420-91^{T}. J Biol Chem **276**: 5577.

113. Kornfeld S, Glaser L (1961) The Enzymatic Synthesis of Thymidine-Linked Sugars. I. Thymidine Diphosphate Glucose. J Biol Chem **236**: 1791

114. Glaser L, Kornfeld S (1961) The Enzymatic Synthesis of Thymidine-Linked Sugars. II. Thymidine Diphosphate L-Rhamnose. J Biol Chem **236**: 1795

115. Lindquist L, Kaiser R, Reeves PR, Lindberg AA (1993) Purification, Characterization and HPLC Assay of *Salmonella* Glucose-1-Phosphate Thymidylyl-Transferase from the Cloned *rfbA* Gene. Eur J Biochem **211**: 763

116. Stein A, Kula M-R, Elling L (1998) Combined Preparative Enzymatic Synthesis of dTDP-6-Deoxy-4-*keto*-D-Glucose from dTDP and Sucrose. Glycoconjugate J **15**: 139

117. Graninger M, Nidetzky B, Heinrichs DE, Whitfield C, Messner P (1999) Characterization of dTDP-4-Dehydrorhamnose 3,5-Epimerase and dTDP-4-Dehydrorhamnose Reductase, Required for dTDP-L-Rhamnose Biosynthesis in *Salmonella enterica* Serovar Typhimurium LT2. J Biol Chem **274**: 25069

117a. Kneidinger B, Graninger M, Puchberger M, Kosma P, Messner P (2001) Biosynthesis of Nucleotide-Activated D-*glycero*-D-*manno*-Heptose. J Biol Chem **276**: 20935

118. Messner P, Sleytr UB (1992) Crystalline Bacterial Cell-Surface Layers. Adv Microb Physiol **33**: 213

119. Kuen B, Lubitz W, Barton GJ (1993) Structural and Functional Analysis of the S-Layer Protein from *Bacillus stearothermophilus*. In: Beveridge TJ, Koval SF (eds) Advances in Bacterial Paracrystalline Surface Layers. Plenum, New York, p 143

119a. Sleytr UB, Sára M, Pum D, Schuster B, Messner P, Schäffer C. Self Assembly Protein Systems: Microbial S-Layers. In: Steinbüchel A (ed) Biopolymers, vol. 3, Polyamides and Complex Proteinaceous Matrices (Part A). Wiley-VCH, Weinheim, in press

120. Lechner J, Sumper M (1987) The Primary Structure of a Procaryotic Glycoprotein. Cloning and Sequencing of the Cell Surface Glycoprotein Gene of Halobacteria. J Biol Chem **262**: 9724
121. Sumper M, Berg E, Mengele R, Strobl I (1990) Primary Structure and Glycosylation of the S-Layer Protein of *Haloferax volcanii*. J Bacteriol **172**: 7111
122. Bröckl G, Behr M, Fabry S, Hensel R, Kaudewitz H, Biendl E, König H (1991) Analysis and Nucleotide Sequence of the Genes Encoding the Surface-Layer Glycoprotein of the Hyperthermophilic Methanogens *Methanothermus fervidus* and *Methanothermus sociabilis*. Eur J Biochem **199**: 147
123. Wakai H, Nakamura S, Kawasaki H, Takada K, Mizutani S, Aono R, Horikoshi K (1997) Cloning and Sequencing of the Gene Encoding the Cell Surface Glycoprotein of *Haloarcula japonica* Strain TR-1. Extremophiles **1**: 29
123a. Egelseer EM, Danhorn T, Pleschberger M, Hotzy C, Sleytr UB, Sára M (2001) Characterization of an S-Layer Glycoprotein Produced in the Course of S-Layer Variation of *Bacillus stearothermophilus* ATCC 12980 and Sequencing and Cloning of the *sbsD* Gene Encoding the Protein Moiety. Arch Microbiol **177**: 70
124. Schäffer C, Wugeditsch T, Kählig H, Scheberl A, Zayni S, Messner P (2002) The Surface Layer (S-Layer) Glycoprotein of *Geobacillus stearothermophilus* NRS 2004/3a. Analysis of Its Glycosylation. J Biol Chem **277**: 6230
125. Stingele F, Neeser J-R, Mollet B (1996) Identification and Characterization of the *eps* (Exopolysaccharide) Gene Cluster from *Streptococcus thermophilus* Sfi6. J Bacteriol **178**: 1680
126. Stingele F, Vincent SJF, Faber EJ, Newell JW, Kamerling JP, Neeser J-R (1999) Introduction of the Exopolysaccharide Gene Cluster from *Streptococcus thermophilus* Sfi6 into *Lactococcus lactis* MG1363: Production and Characterization of an Altered Polysaccharide. Mol Microbiol **32**: 1287
127. Gilbert C, Robinson K, Le Page RWF, Wells JM (2000) Heterologous Expression of an Immunogenic Pneumococcal Type 3 Capsular Polysaccharide in *Lactococcus lactis*. Infect Immun **68**: 3251
128. Messner P, Scheberl A, Schweigkofler W, Hollaus F, Rainey FA, Burghardt J, Prillinger H (1997) Taxonomic Comparison of Different Thermophilic Sugar Beet Isolates with Glycosylated Surface Layer (S-Layer) Proteins and Their Affiliation to *Bacillus smithii*. System Appl Microbiol **20**: 559
129. Küpcü Z, März L, Messner P, Sleytr UB (1984) Evidence for the Glycoprotein Nature of the Crystalline Cell Wall Surface Layer of *Bacillus stearothermophilus* Strain NRS 2004/3a. FEBS Lett **173**: 185
130. Christian R, Schulz G, Unger FM, Messner P, Küpcü Z, Sleytr UB (1986) Structure of a Rhamnan from the Surface-Layer Glycoprotein of *Bacillus stearothermophilus* Strain NRS 2004/3a. Carbohydr Res **150**: 265
131. Sleytr UB, Sára M, Küpcü Z, Messner P (1986) Structural and Chemical Characterization of S-Layers of Selected Strains of *Bacillus stearothermophilus* and *Desulfotomaculum nigrificans*. Arch Microbiol **146**: 19
132. Karnauchow TM, Koval SF, Jarrell KF (1992) Isolation and Characterization of Three Thermophilic Aerobes from a St. Lucia Hot Spring. System Appl Microbiol **15**: 296
133. Peyret JL, Bayan N, Joliff G, Gulik-Krzywicki T, Mathieu L, Shechter E, Leblon G (1993) Characterization of the *cspB* Gene Encoding PS2, an Ordered Surface-Layer Protein in *Corynebacterium glutamicum*. Mol Microbiol **9**: 97
134. Peters J, Peters M, Lottspeich F, Schäfer W, Baumeister W (1987) Nucleotide Sequence Analysis of the Gene Encoding the *Deinococcus radiodurans* Surface

Protein, Derived Amino Acid Sequence, and Complementary Protein Chemical Studies. J Bacteriol **169**: 5216

135. Severina LO, Senyushkin AA, Karavaiko GI (1993) Ultrastructure and Chemical Composition of the S-Layer of *Sulfolobus thermosulfidooxidans*. Dokl Akad Nauk **328**: 633

136. Peters J, Rudolf S, Oschkinat H, Mengele R, Sumper M, Kellermann J, Lottspeich F, Baumeister W (1992) Evidence for Tyrosine-Linked Glycosaminoglycan in a Bacterial Surface Protein. Biol Chem Hoppe-Seyler **373**: 171

137. Messner P, Schuster-Kolbe J, Schäffer C, Sleytr UB, Christian R (1993) Glycoprotein Nature of Select Bacterial S-Layers. In: Beveridge TJ, Koval SF (eds) Advances in Bacterial Paracrystalline Surface Layers. Plenum, New York, p 95

138. Altman E, Brisson J-R, Gagné SM, Kolbe J, Messner P, Sleytr UB (1992) Structure of the Glycan Chain from the Surface Layer Glycoprotein of *Clostridium thermohydrosulfuricum* L77-66. Biochim Biophys Acta **1117**: 71

139. Christian R, Messner P, Weiner C, Sleytr UB, Schulz G (1988) Structure of a Glycan from the Surface-Layer Glycoprotein of *Clostridium thermohydrosulfuricum* L111-69. Carbohydr Res **176**: 160

140. Zellner G, Stackebrandt E, Kneifel H, Messner P, Sleytr UB, Conway de Macario E, Zabel H-P, Stetter KO, Winter J (1989) Isolation and Characterization of a Thermophilic, Sulfate Reducing Archaebacterium, *Archaeoglobus fulgidus* Strain Z. System Appl Microbiol **11**: 151

141. Nishiyama Y, Takashina T, Grant WD, Horikoshi K (1992) Ultrastructure of the Cell Wall of the Triangular Halophilic Archaebacterium *Haloarcula japonica* Strain TR-1. FEMS Microbiol Lett **99**: 43

142. Nakamura S, Mizutani S, Wakai H, Kawasaki H, Aono R, Horikoshi K (1995) Purification and Partial Characterization of the Cell Surface Glycoprotein from Extremely Halophilic Archaeon *Haloarcula japonica* Strain TR-1. Biotechnol Lett **17**: 705

143. Mescher MF, Strominger JL (1976) Purification and Characterization of a Prokaryotic Glycoprotein from the Cell Envelope of *Halobacterium salinarium*. J Biol Chem **251**: 2005

144. Paul G, Lottspeich F, Wieland F (1986) Asparaginyl-N-Acetylgalactosamine. Linkage Unit of Halobacterial Glycosaminoglycan. J Biol Chem **261**: 1020

145. Zhu BCR, Drake RR, Schweingruber H, Laine RA (1995) Inhibition of Glycosylation by Amphomycin and Sugar Nucleotide Analogs PP36 and PP55 Indicates that *Haloferax volcanii* β-Glycosylates Both Glycoproteins and Glycolipids Through Lipid-Linked Intermediates: Evidence for Three Novel Glycoproteins and Novel Sulfated Dihexosyl-Archaeol Glycolipid. Arch Biochem Biophys **319**: 355

146. Shen N, Weiner RM (1998) Isolation and Characterization of S-Layer Proteins from a Vent Prosthecate Bacterium. Microbios **93**: 7

147. Zellner G, Stackebrandt E, Messner P, Tindall BJ, Conway de Macario E, Kneifel H, Sleytr UB, Winter J (1989) *Methanocorpusculaceae* fam. nov., Represented by *Methanocorpusculum parvum*, *Methanocorpusculum sinense* spec. nov. and *Methanocorpusculum bavaricum* spec. nov. Arch Microbiol **151**: 381

148. Bayley DP, Koval SF (1994) Membrane Association and Isolation of the S-Layer Protein of *Methanoculleus marisnigri*. Can J Microbiol **40**: 237

149. Zellner G, Messner P, Kneifel H, Tindall BJ, Winter J, Stackebrandt E (1989) *Methanolacinia* gen. nov., Incorporating *Methanomicrobium paynteri* as *Methanolacinia paynteri* comb. nov. J Gen Appl Microbiol **35**: 185

150. Cheong G-W, Cejka Z, Peters J, Stetter KO, Baumeister W (1991) The Surface Protein Layer of *Methanoplanus limicola*: Three-Dimensional Structure and Chemical Characterization. System Appl Microbiol **14**: 209
151. Pley U, Schipka J, Gambacorta A, Jannasch HW, Fricke H, Rachel R, Stetter KO (1991) *Pyrodictium abyssi* sp. nov. Represents a Novel Heterotrophic Marine Archaeal Hyperthermophile Growing at 110°C. System Appl Microbiol **14**: 245
152. Peters J, Nitsch M, Kühlmorgen B, Golbik R, Lupas A, Kellermann J, Engelhardt H, Pfander J-P, Müller S, Goldie K, Engel A, Stetter K-O, Baumeister W (1995) Tetrabrachion: A Filamentous Archaebacterial Surface Protein Assembly of Unusual Structure and Extreme Stability. J Mol Biol **245**: 385
153. Grogan DW (1996) Organization and Interactions of Cell Envelope Proteins of the Extreme Thermoacidophile *Sulfolobus acidocaldarius*. Can J Microbiol **42**: 1163
154. Bashkatova NA, Severina LO, Golovacheva RS, Mityushina LL (1991) Surface Layers of Extremely Thermoacidophilic Archaebacteria of the Genus *Sulfolobus*. Mikrobiologiya **60**: 90
155. Gongadze GM, Kostyukova AS, Miroshnichenko ML, Bonch-Osmolovskaya EA (1993) Regular Proteinaceous Layers of *Thermococcus stetteri* Cell Envelope. Curr Microbiol **27**: 5
156. Blattner FR, Plunkett III G, Bloch CA, Perna NT, Burland V, Riley M, Collado-Vides J, Glasner JD, Rode CK, Mayhew GF, Gregor J, Davis NW, Kirkpatrick HA, Goeden MA, Rose DJ, Mau B, Shao Y (1997) The Complete Genome Sequence of *Escherichia coli* K-12. Science **277**: 1453
157. Mihoc A, Kluepfel D (1990) Purification and Characterization of a β-Glucosidase from *Streptomyces lividans* 66. Can J Microbiol **36**: 53
158. Stoll D, Stålbrand H, Warren RAJ (1999) Mannan-Degrading Enzymes from *Cellulomonas fimi*. Appl Environ Microbiol **65**: 2598
159. Swanson AF, Kuo C-C (1991) Evidence That the Major Outer Membrane Protein of *Chlamydia trachomatis* is Glycosylated. Infect Immun **59**: 2120
160. Swanson AF, Kuo C-C (1994) Binding of the Glycan of the Major Outer Membrane Protein of *Chlamydia trachomatis* to HeLa Cells. Infect Immun **62**: 24
161. Kuo C-C, Takahashi N, Swanson AF, Ozeki Y, Hakomori S-I (1996) An N-linked High-Mannose Type Oligosaccharide, Expressed at the Major Outer Membrane Protein of *Chlamydia trachomatis*, Mediates Attachment and Infectivity of the Microorganism to HeLa Cells. J Clin Invest **98**: 2813
162. Taylor ME, Conary JT, Lennartz MR, Stahl PD, Drickamer K (1990) Primary Structure of the Mannose Receptor Contains Multiple Motifs Resembling Carbohydrate-Recognition Domains. J Biol Chem **265**: 12156
163. Zaidi SIA, Singh KP, Raisuddin S, Jafri A, Saxena AK, Choudhary S, Ray PK (1995) Modulation of Primary Antibody Response by Protein A in Tumor-Bearing Mice. Immunopharmacol Immunotoxicol **17**: 759
164. Kumar NS, Venkateswerlu G (1998) Analysis of 66 kDa Toxin from *Bacillus thuringiensis* subsp. *kurstaki* Reveals Differential Amino Terminal Processing of Protoxin by Endogenous Protease(s). Biochem Mol Biol Int **45**: 769
165. Kumar NS, Reddy ST, Venkateswerlu G (1998) Effect of Sodium Dodecyl Sulfate on Concanavalin A-Sepharose During Affinity Chromatography of High-Affinity Binding Toxin Protein of *Bacillus thuringensis* subsp. *kurstaki*. Anal Lett **31**: 1677
166. Ogata S, Muramatsu T, Kobata A (1975) Fractionation of Glycopeptides by Affinity Column Chromatography on Concanavalin A-Sepharose. J Biochem **78**: 687

167. Krusius T, Finne J, Rauvala H (1976) The Structural Basis of the Different Affinities of Two Types of Acidic N-Glycosidic Glycopeptides for Concanavalin A-Sepharose. FEBS Lett **71**: 117
168. Lindenthal C, Elsinghorst EA (1999) Identification of a Glycoprotein Produced by Enterotoxigenic *Escherichia coli*. Infect Immun **67**: 4084
169. Rikihisa Y (1991) The Tribe *Ehrlichieae* and Ehrlichial Diseases. Clin Microbiol Rev **4**: 286
170. Yu X-J, Crocquet-Valdes P, Cullman LC, Walker DH (1996) The Recombinant 120-Kilodalton Protein of *Ehrlichia chaffeensis*, a Potential Diagnostic Tool. J Clin Microbiol **34**: 2853
171. Yu X-J, McBride JW, Diaz CM, Walker DH (2000) Molecular Cloning and Characterization of the 120-Kilodalton Protein Gene of *Ehrlichia canis* and Application of the Recombinant 120-Kilodalton Protein for Serodiagnosis of Canine Ehrlichiosis. J Clin Microbiol **38**: 369
172. Yu X-J, Crocquet-Valdes P, Walker DH (1997) Cloning and Sequencing of the Gene for a 120-Kilodalton Immunodominant Protein of *Ehrlichia chaffeensis*. Gene **184**: 149
173. McBride JW, Yu X-J, Walker DH (2000) Glycosylation of Homologous Immunodominant Proteins of *Ehrlichia chaffeensis* and *Ehrlichia canis*. Infect Immun **68**: 13
174. Nishimura H, Kawabata S, Kisiel W, Hase S, Ikenaka T, Takao T, Shimonishi Y, Iwanaga S (1989) Identification of a Disaccharide (Xyl-Glc) and a Trisaccharide (Xyl_2-Glc) O-Glycosidically Linked to a Serine Residue in the First Epidermal Growth Factor-Like Domain of Human Factors VII and IX and Protein Z and Bovine Protein Z. J Biol Chem **264**: 20320
175. Béguin P, Lemaire M (1996) The Cellulosome: An Exocellular, Multiprotein Complex Specialized in Cellulose Degradation. Crit Rev Biochem Mol Biol **31**: 201
176. Jarrell KF, Bayley DP, Kostyukova AS (1996) The Archaeal Flagellum: A Unique Motility Structure. J Bacteriol **178**: 5057
177. Bedouet L, Arnold F, Robreau G, Batina P, Talbot F, Binet A (1998) Evidence for an Heterogeneous Glycosylation of the *Clostridium tyrobutyricum* ATCC 25755 Flagellin. Microbios **94**: 183
178. Bedouet L, Arnold F, Robreau G, Batina P, Talbot F, Malcoste R (1998) Partial Analysis of the Flagellar Antigenic Determinant Recognized by a Monoclonal Antibody to *Clostridium tyrobutyricum*. Microbiol Immunol **42**: 87
179. Arnold F, Bedouet L, Batina P, Robreau G, Talbot F, Lécher P, Malcoste R (1998) Biochemical and Immunological Analyses of the Flagellin of *Clostridium tyrobutyricum* ATCC 25755. Microbiol Immunol **42**: 23
180. Arnold F, Bedouet L, Batina P, Robreau G (1999) Cloning and Sequencing of the Central Region of the Flagellin Gene from the Gram-Positive Bacterium *Clostridium tyrobutyricum* ATCC 25755. Microbiol Immunol **43**: 1
181. Riviere GR, Wagoner MA, Baker-Zander SA, Weisz KS, Adams DF, Simonson L, Lukehart SA (1991) Identification of Spirochetes Related to *Treponema pallidum* in Necrotizing Ulcerative Gingivitis and Chronic Periodontitis. N Engl J Med **325**: 539
182. Wyss C (1998) Flagellins, But Not Endoflagellar Sheath Proteins, of *Treponema pallidum* and of Related Pathogen-Related Oral Spirochetes are Glycosylated. Infect Immun **66**: 5751
183. Haselbeck A, Hösel W (1993) Immunological Detection of Glycoproteins on Blots Based on Labeling with Digoxigenin. In: Hounsell EF (ed) Glycoprotein Analysis in Biomedicine. Humana Press, Totowa, NJ, p 161
183a. Lévesque C, Vadeboncoeur C, Chandad F, Frenette M (2001) *Streptococcus salivarius* Fimbriae are Composed of a Glycoprotein Containing a Repeated Motif Assembled into a Filamenous Nondissociable Structure. J Bacteriol **183**: 2724

184. Jarrell KF, Bayley DP, Florian V, Klein A (1996b) Isolation and Characterization of Insertional Mutations in Flagellin Genes in the Archaeon *Methanococcus voltae*. Mol Microbiol **20**: 657
185. Bayley DP, Jarrell KF (1999) Overexpression of *Methanococcus voltae* Flagellin Subunits in *Escherichia coli* and *Pseudomonas aeruginosa*: A Source of Archaeal Preflagellin. J Bacteriol **181**: 4146
185a. Castric P, Cassels FJ, Carlson RW (2001) Structural Characterization of the *Pseudomonas aeruginosa* 1244 Pilin Glycan. J Biol Chem **276**: 26479; Erratum: J Biol Chem **276**: 36058
185b. Thibault P, Logan SM, Kelly JF, Brisson JR, Ewing CP, Trust TJ, Guerry P (2001) Identification of the Carbohydrate Moieties and Glycosylation Motifs in *Campylobacter jejuni* Flagellin. J Biol Chem **276**: 34862
186. Parge HE, Forest KT, Hickey MJ, Christensen DA, Getzoff ED, Tainer JA (1995) Structure of the Fibre-Forming Protein Pilin at 2.6 Å Resolution. Nature **378**: 32
187. Stimson E, Virji M, Makepeace K, Dell A, Morris HR, Payne G, Saunders JR, Jennings MP, Barker S, Panico M, Blench I, Moxon ER (1995) Meningococcal Pilin: A Glycoprotein Substituted with Digalactosyl 2,4-Diacetamido-2,4,6-Trideoxyhexose. Mol Microbiol **17**: 1201
188. Marceau M, Forest K, Beretti J-L, Tainer J, Nassif X (1998) Consequences of the Loss of O-Linked Glycosylation of Meningococcal Type IV Pilin on Piliation and Pilus-Mediated Adhesion. Mol Microbiol **27**: 705
189. Marceau M, Nassif X (1999) Role of Glycosylation at Ser63 in Production of Soluble Pilin in Pathogenic *Neisseria*. J Bacteriol **181**: 656
190. Jennings MP, Virji M, Evans D, Foster V, Srikhanta YN, Steeghs L, van der Ley P, Moxon ER (1998) Identification of a Novel Gene Involved in Pilin Glycosylation in *Neisseria meningitidis*. Mol Microbiol **29**: 975
191. Power PM, Roddam LF, Dieckelmann M, Srikhanta YN, Tan YC, Berrington AW, Jennings MP (2000) Genetic Characterization of Pilin Glycosylation in *Neisseria meningitidis*. Microbiology **146**: 967
191a. Kahler CM, Martin LE, Tzeng Y-L, Miller YK, Sharkey K, Stephens DS, Davies JK (2001) Polymorphisms in Pilin Glycosylation Locus of *Neisseria meningitidis* Expressing Class II Pili. Infect Immun **69**: 3597
191b. Cowlishaw DA, Smith MCM (2001) Glycosylation of a *Streptomyces coelicolor* A3(2) Cell Envelope Protein is Required for Infection by Bacteriophage ϕC31. Mol Microbiol **41**: 601
191c. Benz I, Schmidt MA (2001) Glycosylation with Heptose Residues Mediated by the *aah* Gene Product is Essential for Adherence of the AIDA-I Adhesin. Mol Microbiol **40**: 1403
192. Bailey SA, Ward B (1997) Emulsification of Crude Oil by *Rhodococcus erythropolis* Strain ST-2 via a Cell-Surface, Lysozyme-Sensitive Glycoprotein. System Appl Microbiol **20**: 545
193. Patel MN, Gopinathan KP (1986) Lysozyme-Sensitive Bioemulsifier for Immiscible Organophosphorus Pesticides. Appl Environ Microbiol **52**: 1224
193a. Recht J, Kolter R (2001) Glycopeptidolipid Acetylation Affects Sliding Motility and Biofilm Formation in *Mycobacterium smegmatis*. J Bacteriol **183**: 5718
194. Kondo E, Kurata T, Naigowit P, Kanai K (1996) Evolution of Cell-Surface Acid Phosphatase of *Burkholderia pseudomallei*. Southeast Asian J Trop Med Public Health **27**: 592
195. Kondo E, Wangroongsaub P, Kanai K (1994) Effects of Tunicamycin on the pH-Activity Pattern of Acid Phosphatase in *Pseudomonas pseudomallei*. Southeast Asian J Trop Med Public Health **25**: 144

196. Shoham Y, Lamed R, Bayer EA (1999) The Cellulosome Concept as an Efficient Microbial Strategy for the Degradation of Insoluble Polysaccharides. Trends Microbiol **7**: 275
197. Bayer EA, Shoham Y, Lamed R (2000) The Cellulosome. An Extracellular Organelle for Degrading Plant Cell Wall Polysaccharides. In: Doyle RJ (ed) Glycomicrobiology. Kluwer Academic/Plenum Publishers, New York, p 387
198. Chen J-R, Lin J-H, Weng C-N, Lai S-S (1998) Identification of a Novel Adhesin-Like Glycoprotein from *Mycoplasma hyopneumoniae*. Vet Microbiol **62**: 97
199. Espitia C, Mancilla R (1989) Identification, Isolation and Partial Characterization of *Mycobacterium tuberculosis* Glycoprotein Antigens. Clin Exp Immunol **77**: 378
200. Espitia C, Espinosa R, Saavedra R, Mancilla R, Romain F, Laqueyrerie A, Moreno C (1995) Antigenic and Structural Similarities Between *Mycobacterium tuberculosis* 50- to 55-Kilodalton and *Mycobacterium bovis* BCG 45- to 47-Kilodalton Antigens. Infect Immun **63**: 580
201. Dobos KM, Khoo K-H, Swiderek K, Brennan PJ, Belisle JT (1996) Definition of the Full Extent of Glycosylation of the 45-Kilodalton Glycoprotein of *Mycobacterium tuberculosis*. J Bacteriol **178**: 2498
202. Montaño LF, Massó F, Páez A, Sandoval S, Vázquez L, Sánchez L, Fournet B, Zenteno E (1994) Isolation of a 32 kDa *Mycobacterium tuberculosis* Protein by Lectin Affinity Chromatography. Comp Biochem Physiol **108B**: 265
203. Massó F, Varela E, Páez A, Zenteno E, Montaño LF (1998) Identification of Major Glycoconjugates from *Mycobacterium bovis* Culture Filtrate by Biotin-Hydrazide Labeling. Glycoconjugate J **15**: 843
204. Debray H, Decout D, Strecker G, Spik G, Montreuil J (1981) Specificity of Twelve Lectins Towards Oligosaccharides and Glycopeptides Related to N-Glycosylproteins. Eur J Biochem **117**: 41
205. Trimble RB, Tarentino AL (1991) Identification of Distinct Endoglycosidase (Endo) Activities in *Flavobacterium meningosepticum*: Endo F1, Endo F2, and Endo F3. Endo F1 and Endo H Hydrolyze Only High Mannose and Hybrid Glycans. J Biol Chem **266**: 1646
206. Bonato VLD, Lima VMF, Tascon RE, Lowrie DB, Silva CL (1998) Identification and Characterization of Protective T Cell in hsp65 DNA-Vaccinated and *Mycobacterium tuberculosis*-Infected Mice. Infect Immun **66**: 169
207. Gonzáles-Amaro R, Portales-Pérez DP, Baranda L, Moncada B, Toro C, López-Briones S, Espitia C, Mancilla R (2000) Co-Stimulatory Signals Increase the Reactivity of $\gamma\delta$ T Cells Towards Mycobacterial Antigens. Clin Exp Immunol **120**: 468
208. Zielinski GC, Ross RF (1993) Adherence of *Mycoplasma hyopneumoniae* to Porcine Ciliated Respiratory Tract Cells. Am J Vet Res **54**: 1262
209. Zhang Q, Young TF, Ross RF (1995) Identification and Characterization of a *Mycoplasma hyopneumoniae* Adhesin. Infect Immun **63**: 1013
210. Chen JW, Zhang L, Song J, Hwang F, Dong Q, Liu J, Qian Y (1992) Comparative Analysis of Glycoprotein and Glycolipid Composition of Virulent and Avirulent Strain Membranes of *Mycoplasma hyopneumoniae*. Curr Microbiol **24**: 189
211. Araki T, Tamaru Y, Morishita T (1992) β-1,4-Mannanases from Marine Bacteria *Vibrio* spp. MA-129 and MA-138. J Gen Appl Microbiol **38**: 343
212. Tamaru Y, Araki T, Amagoi H, Mori H, Morishita T (1995) Purification and Characterization of an Extracellular β-1,4-Mannanase from a Marine Bacterium, *Vibrio* sp. Strain MA-138. Appl Environ Microbiol **61**: 4454

213. Arcand N, Kluepfel D, Paradis FW, Morosoli R, Shareck F (1993) β-Mannanase of *Streptomyces lividans* 66: Cloning and DNA Sequence of the *manA* Gene and Characterization of the Enzyme. Biochem J **290**: 857
214. Wimmer B, Lottspeich F, Ritter J, Bronnenmeier K (1997) A Novel Type of Thermostable α-D-Glucosidase from *Thermoanaerobacter thermohydrosulfuricus* Exhibiting Maltodextrinohydrolase Activity. Biochem J **328**: 581
215. Tani Y, Tani M, Kato I (1997) Extracellular 37-kDa Antigenic Protein from *Actinobacillus actinomycetemcomitans* Induces TNF-α, IL-1β, and IL-6 in Murine Macrophages. J Dent Res **76**: 1538
216. Amano K, Nishihara T, Shibuya N, Noguchi T, Koga T (1989) Immunochemical and Structural Characterization of a Serotype-Specific Polysaccharide Antigen from *Actinobacillus actinomycetemcomitans* Y4 (Serotype b). Infect Immun **57**: 2942
217. Xu H, Griffith M, Patten CL, Glick BR (1998) Isolation and Characterization of an Antifreeze Protein With Ice Nucleation Activity from the Plant Growth Promoting Rhizobacterium *Pseudomonas putida* GR12-2. Can J Microbiol **44**: 64
218. DeVries AL, Komatsu SK, Feeney RE (1970) Chemical and Physical Properties of Freezing Point-Depressing Glycoproteins from Antarctic Fishes. J Biol Chem **245**: 2901
218a. de la Maza A, Parra JL, López O, Congregado F, Bozal N, Guinea J (1997) Assembly Properties of a Glycoprotein Produced by *Pseudoalteromonas antarctica* NF_3. J Colloid Interface Sci **192**: 286
218b. Cócera M, López O, Coderch L, Mercadé ME, Parra JL, de la Maza A, Guinea J (2001) Partitioning of SDS in Liposomes Coated by the Exopolymer Excreted by *Pseudoalteromonas antarctica* NF_3 as a Measure of Vesicle Protection Against this Surfactant. J. Biomat Sci, Polymer Ed **12**: 255
219. Costerton JW, Lewandowski Z, Caldwell DE, Korber DR, Lappin-Scott HM (1995) Microbial Biofilms. Annu Rev Microbiol **49**: 711
220. Burchard RP, Sorongon ML (1998) A Gliding Bacterium Strain Inhibits Adhesion and Motility of Another Gliding Bacterium Strain in a Marine Biofilm. Appl Environ Microbiol **64**: 4079
221. Yoshida J, Yoshimura M, Takamura S, Kobayashi S (1985) Purification and Characterization of an Antitumor Principle from *Streptococcus hemolyticus*, Su Strain. Jpn J Cancer Res **76**: 213
222. Kanaoka M, Fukita Y, Taya K, Kawanaka C, Negoro T, Agui H (1987) Antitumor Activity of Streptococcal Acid Glycoprotein Produced by *Streptococcus pyogenes* Su. Jpn J Cancer Res **78**: 1409
223. Kanaoka M, Negoro T, Kawanaka C, Agui H, Nabeshima S (1991) Streptococcal Antitumor Protein: Expression in *Escherichia coli* Cells and Properties of the Recombinant Protein. Agric Biol Chem **55**: 743
224. Degnan BA, Palmer JM, Robson T, Jones CED, Fischer M, Glanville M, Mellor GD, Diamond AG, Kehoe MA, Goodacre JA (1998) Inhibition of Human Peripheral Blood Mononuclear Cell Proliferation by *Streptococcus pyogenes* Cell Extracts is Associated with Arginine Deiminase Activity. Infect Immun **66**: 3050
225. Degnan BA, Fontaine MC, Doebereiner AH, Lee JJ, Mastroeni P, Dougan G, Goodacre JA, Kehoe MA (2000) Characterization of an Isogenic Mutant of *Streptococcus pyogenes* Manfredo Lacking the Ability to Make Streptococcal Acid Glycoprotein. Infect Immun **68**: 2441
226. Yoshida J, Takamura S, Nishio M (1998) Characterization of a Streptococcal Antitumor Glycoprotein (SAGP). Life Sci **62**: 1043

227. Szymanski CM, Yao R, Ewing CP, Trust TJ, Guerry P (1999) Evidence for a System of General Protein Glycosylation in *Campylobacter jejuni*. Mol Microbiol **32**: 1022
228. Edge ASB, Faltynek CR, Hof L, Reichert Jr LE, Weber P (1981) Deglycosylation of Glycoproteins by Trifluoromethanesulfonic Acid. Anal Biochem **118**: 131
229. Guenounou M (1996) Mechanism of Action of the Immunmodulating Agent RU 41740. Curr Therap Res **57**(Suppl A): 16
230. Kol O, Montreuil J, Fournet B, Zalisz R, Smets P (1987) Purification of the Lipopolysaccharide Fraction from *Klebsiella pneumoniae* O1 K2 by High-Performance Liquid Chromatography. J Chromatogr **396**: 281
231. Kol O, Montreuil J, Fournet B, Zalisz R, Smets P (1989) Separation by High-Performance Liquid Chromatography of Oligosaccharides Obtained After Mild Acid Hydrolysis of *Klebsiella pneumoniae* O1 K2 (NCTC 5055) Lipopolysaccharides. J Chromatogr **474**: 452
232. Holt SC, Bramanti TE (1991) Factors in Virulence Expression and Their Role in Periodontal Disease Pathogenesis. Crit Rev Oral Biol Med **2**: 177
233. Iki K, Kawahara K, Sawamura S, Arakaki R, Sakuta T, Sugiyama A, Tamura H, Sueda T, Hamada S, Takada H (1997) A Novel Component Different from Endotoxin Extracted from *Prevotella intermedia* ATCC 25611 Activates Lymphoid Cells from C3H/HeJ Mice and Gingival Fibroblasts from Humans. Infect Immun **65**: 4531
234. Yamaguchi T, Kasamo K, Chuman M, Machigashira M, Inoue M, Sueda T (1998) Preparation and Characterization of an *Actinomyces naeslundii* Aggregation Factor that Mediates Coaggregation with *Porphyromonas gingivalis*. J Periodont Res **33**: 460
235. Moore LVH, Moore WEC, Cato EP, Smibert RM, Burmeister JA, Best AM, Ranney RR (1987) Bacteriology of Human Gingivitis. J Dent Res **66**: 989
236. Listgarten MA (1976) Structure of the Microbial Flora Associated with Periodontal Health and Disease in Man. J Periodont **47**: 1
237. Kolenbrander PE, Andersen RN (1990) Characterization of *Streptococcus gordonii* (*S. sanguis*) PK488 Adhesin-Mediated Coaggregation with *Actinomyces naeslundii* PK606. Infect Immun **58**: 3064
238. Lemassu A, Ortalo-Magne A, Bardou F, Silve G, Laneelle M-A, Daffe M (1996) Extracellular and Surface-Exposed Polysaccharides of Non-Tuberculous Mycobacteria. Microbiology **142**: 1513
239. Tian X-X, Li A, Zhou W, Farrugia IV, Groves MJ (1999) Isolation and Biological Activities of an Antineoplastic Protein-Polysaccharide Complex (PS4A) Obtained from *Mycobacterium vaccae*. Anticancer Res **19**: 237
240. Tian X-X, Li A, Farrugia IV, Mo X, Crich D, Groves MJ (2000) Isolation and Identification of Poly-α-(1$\rightarrow$4)-Linked 3-*O*-Methyl-D-Mannopyranose from a Hot-Water Extract of *Mycobacterium vaccae*. Carbohydr Res **324**: 38
241. Bader JA, Klesius PH, Vinitnantharat S (1997) Comparison of Whole-Cell Antigens of Pressure- and Formalin-Killed *Flexibacter columnaris* from Channel Catfish (*Ictalurus punctatus*). Am J Vet Res **58**: 985
242. Fujihara MP, Nakatani RE (1971) Antibody Production and Immune Responses of Rainbow Trout and Coho Salmon to *Chondrococcus columnaris*. J Fisher Res Board Can **28**: 1253
243. Cuezzo de Ginés S, Maldonado MC, Font de Valdez G (2000) Purification and Characterization of Invertase from *Lactobacillus reuteri* CRL 1100. Curr Microbiol **40**: 181
244. Koeller KM, Wong C-H (2000) Emerging Themes in Medicinal Glycoscience. Nature Biotechnol **18**: 835

245. Koeller KM, Wong C-H (2000) Complex Carbohydrate Synthesis Tools for Glycobiologists: Enzyme-Based Approach and Programmable One-Pot Strategies. Glycobiology **10**: 1157
246. Schreiner R, Schnabel E, Wieland F (1994) Novel N-Glycosylation in Eukaryotes: Laminin Contains the Linkage Unit β-Glucosylasparagine. J Cell Biol **124**: 1071
247. Shibata S, Takeda T, Natori Y (1988) The Structure of Nephritogenoside. A Nephritogenic Glycopeptide with α-N-Glycosidic Linkage. J Biol Chem **263**: 12483
248. Deras IL, Takegawa K, Kondo A, Kato I, Lee YC (1998) Synthesis of a High-Mannose-Type Glycopeptide Analog Containing a Glucose-Asparagine Linkage. Bioorg Med Chem Lett **8**: 1763
249. Davis BG, Lloyd RC, Jones JB (2000) Controlled Site-Selective Protein Glycosylation for Precise Glycan Structure-Catalytic Activity Relationships. Bioorg Med Chem **8**: 1527
250. Stabile MR, Lai WG, DeSantis G, Gold M, Jones JB (1996) Probing the Specificity of the S_1 Binding Site of M222 Mutants of Subtilisin *B. lentus* with Boronic Acid Inhibitors. Bioorg Med Chem Lett **6**: 2501
251. Lloyd RC, Davis BG, Jones JB (2000) Site-Selective Glycosylation of Subtilisin *Bacillus lentus* Causes Dramatic Increases in Esterase Activity. Bioorg Med Chem **8**: 1537
252. Hedenetz A, Schmid W, Unger FM (2000) A Short Synthesis of 2,3-Diacetamido-2,3-Dideoxy-D-Mannuronic Acid-Derivatives. Proc Jahrestagung Österr Biochem Ges Innsbruck, Abstract PO-57
253. Gerwig GJ, Kamerling JP, Vliegenthart JFG, Morag E, Lamed R, Bayer EA (1992) Novel Oligosaccharide Constituents of the Cellulase Complex of *Bacteroides cellulosolvens*. Eur J Biochem **205**: 799
254. Gerwig GJ, Kamerling JP, Vliegenthart JFG, Morag E, Lamed R, Bayer EA (1993) The Nature of the Carbohydrate-Peptide Linkage Region in Glycoproteins from the Cellulosomes of *Clostridium thermocellum* and *Bacteroides cellulosolvens*. J Biol Chem **268**: 26956
255. Emery, DL, Clark BL, Stewart DJ, O'Donnell IJ, Hewish DR (1984) Analysis of the Outer Membrane Proteins of *Bacteroides nodosus*, the Causal Organism of Ovine Footrot. Vet Microbiol **9**: 155
256. Sambri V, Stefanelli C, Cevenini R (1992) Detection of Glycoproteins in *Borrelia burgdorferi*. Arch Microbiol **157**: 205
257. McManus JD, Brune DC, Han J, Sanders-Loehr J, Meyer TE, Cusanovich MA, Tollin G, Blankenship RE (1992) Isolation, Characterization, and Amino Acid Sequences of Auracyanins, Blue Copper Proteins from the Green Photosynthetic Bacterium *Chloroflexus aurantiacus*. J Biol Chem **267**: 6531
258. Gong J, Egbosimba EE, Forsberg CW (1996) Cellulose-Binding Proteins of *Fibrobacter succinogenes* and the Possible Role of a 180 kDa Cellulose-Binding Glycoprotein in Adhesion to Cellulose. Can J Microbiol **42**: 453
259. Doherty H, Condon C, Owen P (1982) Resolution and In Vitro Glycosylation of Membrane Glycoproteins in *Micrococcus luteus* (*lysodeikticus*) FEMS Microbiol Lett **15**: 331
260. Maeba PY (1986) Isolation of a Surface Glycoprotein from *Myxococcus xanthus*. J Bacteriol **166**: 644
261. Kozloff LM, Turner MA, Arellano F (1991) Formation of Bacterial Membrane Ice-Nucleating Lipoglycoprotein Complexes. J Bacteriol **173**: 6528
262. Turner MA, Arellano F, Kozloff LM (1991) Components of Ice Nucleation Structures of Bacteria. J Bacteriol **173**: 6515

263. Yang LL, Haug A (1979) Purification and Partial Characterization of a Procaryotic Glycoprotein from the Plasma Membrane of *Thermoplasma acidophilum*. Biochim Biophys Acta **556**: 265
264. Moens S, Michiels K, Vanderleyden J (1995) Glycosylation of the Flagellin of the Polar Flagellum of *Azospirillum brasilense*, a Gram-Negative Nitrogen-Fixing Bacterium. Microbiology **141**: 2651
265. Guerry P, Doig P, Alm RA, Burr DH, Kinsella N, Trust TJ (1996) Identification and Characterization of Genes Required for Post-Translational Modification of *Campylobacter coli* VC167 Flagellin. Mol Microbiol **19**: 369
266. Doig P, Kinsella N, Guerry P, Trust TJ (1996) Characterization of a Post-Translational Modification of *Campylobacter* Flagellin: Identification of a Sero-Specific Glycosyl Moiety. Mol Microbiol **19**: 379
267. Gerwig GJ, de Waard P, Kamerling JP, Vliegenthart JFG, Morgenstern E, Lamed R, Bayer EA (1989) Novel O-Linked Carbohydrate Chains in the Cellulase Complex (Cellulosome) of *Clostridium thermocellum*. J Biol Chem **264**: 1027
268. Tomoeda M, Inuzuka M, Date T (1975) Bacterial Sex Pili. Progr Biophys Mol Biol **30**: 23
269. Dobson WJ, McCurdy HD (1979) The Function of Fimbriae in *Myxococcus xanthus*. I. Purification and Properties of *M. xanthus* Fimbriae. Can J Microbiol **25**: 1152
270. Dobson WJ, McCurdy HD, MacRae TH (1979) The Function of Fimbriae in *Myxococcus xanthus*. II. The Role of Fimbriae in Cell-Cell Interactions. Can J Microbiol **25**: 1359
271. Virji M, Saunders JR, Sims G, Makepeace K, Maskell D, Ferguson DJP (1993) Pilus-Faciliated Adherence of *Neisseria meningitidis* to Human Epithelial and Endothelial Cells: Modulation of Adherence Phenotype Occurs Concurrently with Changes in Primary Amino Acid Sequence and the Glycosylation Status of Pilin. Mol Microbiol **10**: 1013
272. Hoiczyk E, Baumeister W (1997) Oscillin, an Extracellular, Ca^{2+}-Binding Glycoprotein Essential for the Gliding Motility of Cyanobacteria. Mol Microbiol **26**: 699
273. Brimer CD, Montie TC (1998) Cloning and Comparison of *fliC* Genes and Identification of Glycosylation in the Flagellin of *Pseudomonas aeruginosa* a-Type Strains. J Bacteriol **180**: 3209
274. Castric P (1995) *pilO*, a Gene Required for Glycosylation of *Pseudomonas aeruginosa* 1244 Pilin. Microbiology **141**: 1247
275. Li Z, Dumas F, Dubreuil D, Jacques M (1993) A Species-Specific Periplasmic Flagellar Protein of *Serpulina* (*Treponema*) *hyodysenteriae*. J Bacteriol **175**: 8000
276. Brahamsha B, Greenberg EP (1989) Cloning and Sequence Analysis of *flaA*, a Gene Encoding a *Spirochaeta aurantia* Flagellar Filament Surface Antigen. J Bacteriol **171**: 1692
277. Weerkamp AH, van der Mei HC, Liem RSB (1984) Adhesive Cell Wall-Associated Glycoprotein of *Streptococcus salivarius* (K^{+}) Is a Cell Surface Fibril. FEMS Microbiol Lett **23**: 163
278. Morris EJ, Ganeshkumar N, Song M, McBride BC (1987) Identification and Preliminary Characterization of a *Streptococcus sanguis* Fibrillar Glycoprotein. J Bacteriol **169**: 164
279. Wieland F, Paul G, Sumper M (1985) Halobacterial Flagellins are Sulfated Glycoproteins. J Biol Chem **260**: 15180
280. Faguy DM, Koval SF, Jarrell KF (1992) Correlation Between Glycosylation of Flagellin Proteins and Sensitivity of Flagellar Filaments to Triton X-100 in Methanogens. FEMS Microbiol Lett **90**: 129

281. Bayley DP, Kalmokoff ML, Jarrell KF (1993) Effect of Bacitracin on Flagellar Assembly and Presumed Glycosylation of the Flagellins of *Methanococcus deltae*. Arch Microbiol **160**: 179
282. Southam G, Kalmokoff ML, Jarrell KF, Koval SF, Beveridge TJ (1990) Isolation, Characterization, and Cellular Insertion of the Flagella from Two Strains of the Archaebacterium *Methanospirillum hungatei*. J Bacteriol **172**: 3221
283. Faguy DM, Koval SF, Jarrell KF (1994) Physical Characterization of the Flagella and Flagellins from *Methanospirillum hungatei*. J Bacteriol **176**: 7491
284. Fedorov OV, Paytibratov MG, Kostyukova AS, Osina NK, Tarasov VY (1994) Protofilament as a Structural Element of Flagella of Haloalkalophilic Archaebacteria. Can J Microbiol **40**: 45
285. Grogan DW (1989) Phenotypic Characterization of the Archaebacterial Genus *Sulfolobus*: Comparison of Five Wild-Type Strains. J Bacteriol **171**: 6710
286. Faguy DM, Bayley DP, Kostyukova AS, Thomas NA, Jarrell KF (1996) Isolation and Characterization of Flagella and Flagellin Proteins from the Thermoacidophilic Archaea *Thermoplasma volcanium* and *Sulfolobus shibatae*. J Bacteriol **178**: 902
287. Yoshida S, Sako Y, Uchida A (1998) Cloning, Sequence Analysis, and Expression in *Escherichia coli* of a Gene Coding for an Enzyme from *Bacillus circulans* K-1 that Degrades Guar Gum. Biosci Biotechnol Biochem **62**: 514
288. Paul J, Varma AK (1992) Glycoprotein Components of Cellulase and Xylanase Enzymes of a *Bacillus* sp. Biotechnol Lett **14**: 207
289. Ong E, Kilburn DG, Miller RC Jr, Warren RAJ (1994) *Streptomyces lividans* Glycosylates the Linker Region of a β-1,4-Glycanase from *Cellulomonas fimi*. J Bacteriol **176**: 999
290. Gilkes NR, Langsford ML, Kilburn DG, Miller Jr RC, Warren RAJ (1984) Mode of Action and Substrate Specificities of Cellulases from Cloned Bacterial Genes. J Biol Chem **259**: 10455
291. Gilkes NR, Warren RAJ, Miller Jr RC, Kilburn DG (1988) Precise Excision of the Cellulose Binding Domains from two *Cellulomonas fimi* Cellulases by a Homologous Protease and the Effect on Catalysis. J Biol Chem **263**: 10401
292. Béguin P, Eisen H (1978) Purification and Partial Characterization of Three Extracellular Cellulases from *Cellulomonas* sp. Eur J Biochem **87**: 525
293. Strobel GA, Talmadge KW, Albersheim P (1972) Observations on the Structure of the Phytotoxic Glycopeptide of *Corynebacterium sepedonicum*. Biochim Biophys Acta **261**: 365
294. Meier B, Brunotte CM, Franz B, Warlich B, Petermann M, Ziesenis A, Schuberth H-J, Habermehl GG, Petzoldt K, Leibold W (1992) Isolation of a High-Molecular Mass Glycoprotein from Culture Supernatant of an Arthritogenic Strain of the Bacteria *Erysipelothrix rhusiopathiae* Reacting With "Inductive" Monoclonal Antibodies Derived From Rats With Erysipelas Polyarthritis. Biol Chem Hoppe-Seyler **373**: 715
295. Huang L, Forsberg CW, Thomas DY (1988) Purification and Characterization of a Chloride-Stimulated Cellobiosidase from *Bacteroides succinogenes* S85. J Bacteriol **170**: 2923
296. Plummer Jr TH, Tarentino AL, Hauer CR (1995) Novel, Specific O-Glycosylation of Secreted *Flavobacterium meningosepticum* Proteins. Asp-Ser* and Asp-Thr*-Thr Consensus Sites. J Biol Chem **270**: 13192
297. Reinhold BB, Hauer CR, Plummer TH, Reinhold VN (1995) Detailed Structural Analysis of a Novel, Specific O-Linked Glycan from the Prokaryote *Flavobacterium meningosepticum*. J Biol Chem **270**: 13197

298. Fifis T, Costopoulos C, Radford AJ, Bacic A, Wood PR (1991) Purification and Characterization of Major Antigens from a *Mycobacterium bovis* Culture Filtrate. Infect Immun **59**: 800
299. Dobos KM, Swiderek K, Khoo K-H, Brennan PJ, Belisle JT (1995) Evidence for Glycosylation Sites on the 45-Kilodalton Glycoprotein of *Mycobacterium tuberculosis*. Infect Immun **63**: 2846
300. Garbe T, Harris D, Vordermeier M, Lathigra R, Ivanyi J, Young D (1993) Expression of the *Mycobacterium tuberculosis* 19-Kilodalton Antigen in *Mycobacterium smegmatis*: Immunological Analysis and Evidence for Glycosylation. Infect Immun **61**: 260
301. Kluepfel D, Vats-Mehta S, Aumont F, Shareck F, Morosoli R (1990) Purification and Characterization of a New Xylanase (Xylanase B) Produced by *Streptomyces lividans* 66. Biochem J **267**: 45
302. Van Rijssel M, Gerwig GJ, Hansen TA (1993) Isolation and Characterization of an Extracellular Glycosylated Protein Complex from *Clostridium thermosaccharolyticum* with Pectin Methylesterase and Polygalacturonate Hydrolase Activity. Appl Environ Microbiol **59**: 828
303. Calza RE, Irwin DC, Wilson DB (1985) Purification and Characterization of Two β-1,4-Endoglucanases from *Thermomonospora fusca*. Biochemistry **24**: 7797
304. Kim B-K, Pihl TD, Reeve JN, Daniels L (1995) Purification of the Copper Response Extracellular Proteins Secreted by the Copper-Resistant Methanogen *Methanobacterium bryantii* BKYH and Cloning, Sequencing, and Transcription of the Genes Encoding These Proteins. J Bacteriol **177**: 7178
305. Taku A, Fan DP (1976) Purification and Properties of a Protein Factor Stimulating Peptidoglycan Synthesis in Toluene- and LiCl-Treated *Bacillus megaterium* Cells. J Biol Chem **251**: 1889
306. García-Patrone M, Tandecraz JS (1995) A Glycoprotein Multimer from *Bacillus thuringensis* Sporangia: Dissociation into Subunits and Sugar Composition. Mol Cell Biochem **145**: 29
307. Bulla Jr LA, Kramer KJ, Davidson LI (1977) Characterization of the Entomocidal Parasporal Crystal of *Bacillus thuringiensis*. J Bacteriol **130**: 375
308. Pfannenstiel MA, Muthukumar G, Couche GA, Nickerson KW (1987) Amino Sugars in the Glycoprotein Toxin from *Bacillus thuringiensis* subsp. *israelensis*. J Bacteriol **169**: 796
309. Muthukumar G, Nickerson KW (1987) The Glycoprotein Toxin of *Bacillus thuringensis* subsp. *israelensis* Indicates a Lectinlike Receptor in the Larval Mosquito Gut. Appl Environ Microbiol **53**: 2650
310. Webster JR, Reid SJ, Jones DT, Woods DR (1981) Purification and Characterization of an Autolysin from *Clostridium acetobutylicum*. Appl Environ Microbiol **41**: 371
311. Kawamura T, Shockman GD (1983) Purification and Some Properties of the Endogenous, Autolytic N-Acetylmuramoylhydrolase of *Streptococcus faecium*, a Bacterial Glycoenzyme. J Biol Chem **258**: 9514
312. Yao R, Macario AJL, Conway de Macario E (1992) Immunochemical Differences Among *Methanosarcina mazei* S-6 Morphologic Forms. J Bacteriol **174**: 4683

Carbazole Alkaloids IV

D. P. Chakraborty[1,*] and Shyamali Roy[2]
[1] Institute of Natural Products, Calcutta, India
[2] Calcutta, India

Contents

* Deceased July 24, 2000

I. Introduction

The European dye stuff industry originally provided the main stimulus for development of the chemistry of carbazoles since the discovery of the first member of the group, carbazole (**1**) from coal tar in 1872 by Graebe and Glazer (*35*). The isolation of carbazole as a degradation product of strychnine (**2**) by Clemo and Perkin (*18*) in 1924 provided further impetus for progress in the chemistry of carbazoles. N-Vinylcarbazole (**3**) was subsequently synthesised (*12*) and polymers based on vinyl carbazoles were developed subsequently (*95*) with the dye stuff industries and polymer industry contributing largely to further development of the chemistry of carbazoles up to 1960. The discovery of the first naturally occurring carbazole, the alkaloid murrayanine (**4**)* from *Murraya koinigii* Spreng; may be considered a further landmark in development of the chemistry and biology of carbazoles; the report on its structure was, however, published only in 1965 (*12*). Carbazole, methylcarbazole and dimethylcarbazole were isolated from soil humus in 1967 by Haworth *et al.* (*14*). More recently there has been added interest in the chemistry of carbazoles due to their use in electrophotography and related areas.

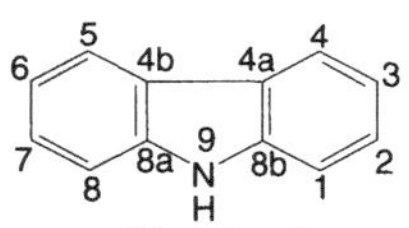

(**1**) carbazole

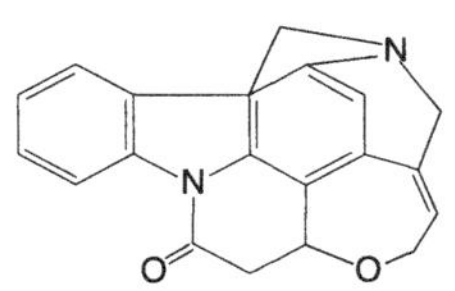

(**2**) strychnine

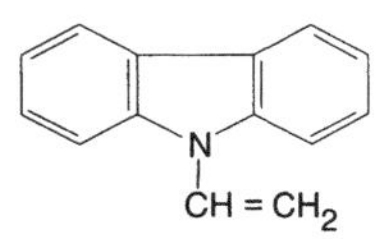

(**3**) N-vinylcarbazole

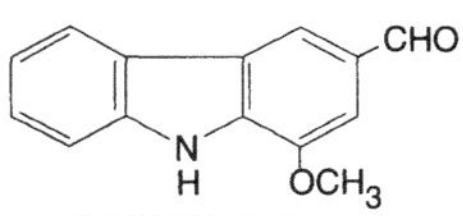

(**4**) murrayanine

* Murrayanine was the first carbazole alkaloid isolated from nature, not girinimbine as mentioned in ref. 9.

The number of carbazole alkaloids reported up to 1991 (*13*) was nearly one hundred if the diazofluorenes of the kinamycin group (**5**, **6**) are included (*86*). These were considered as carbazole alkaloids before 1994 (*34*, *78*).

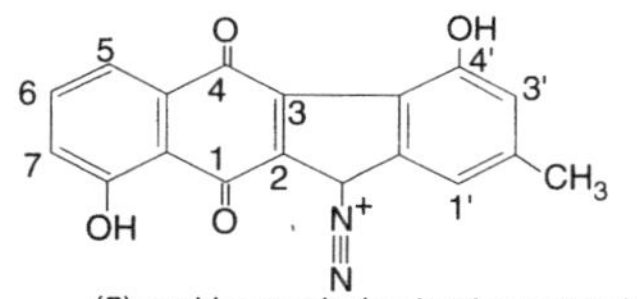

(**5**) prekinamycin (revised structure)

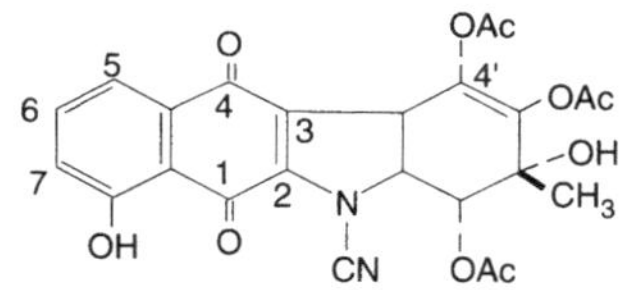

(**6**) kinamycin C

Since then, their number has grown to nearly two hundred. The present review describes work reported after 1991 as well as work not included in the previous reviews of this series (*1*, *10*, *13*).

A. Nomenclature

The nomenclature adopted by Patterson and Capbell (*87*) in the Ring Index has been followed in the present articles. The rings are denoted by A, B and C in the tricyclic system. Abbreviations like C-1 and H-1 have been used to denote carbons and hydrogens at the various positions of the carbazole nucleus (**1**).

B. Occurrence

The genera *Clausena*, *Glycosmis* and *Murraya* belonging to the subtribe Clausanae of the sub-family Aurantoidae of Rutaceae have elaborated the largest number of carbazole alkaloids. The occurrence of carbazole alkaloids in the three genera reflects their chemotaxonomic homogeneity and perhaps provides chemical support for their classification as a separate subtribe. Carbazoles have also been reported from the genera *Micromelum* of Rutaceae and *Ekebergia* of Meliaceae and have recently been isolated from a member of the family Ranunculae (*69*).

Some lower plants of the genera *Nocardiopsis*, *Streptomyces*, *Streptoverticillium* and the alga *Hyella* as well as some mammalian sources like bovine and equine (*72*, *108*) urine have also been found to yield carbazole alkaloids.

The genus *Murraya*, especially the species *Murraya euchrestifolia* Hyata, is the richest source of alkaloids based on C_{13}-, C_{18}-, C_{23}-skeletons and of dimers derived from monomeric units. It is pertinent

Table 1. Biological Sources of Carbazole Alkaloids

A. Higher Plants	1. *Cimicifuga simplex*
	2. *Clausena anisata* (wild) Olive
	3. *C. excavata* Burm f.
	4. *C. harmandiana* Pierre (syn. *C. wampi* Olive)
	5. *C. heptaphylla* Wt. and Arn.
	6. *C. indica* Olive
	7. *C. lansium* (Lour) Skeels
	8. *Ekebergia senegalensis*
	9. *Glycosmis pentaphylla* (Retz.) D.C. (syn. *G. arborea* (Corr.)
	10. *G. mauritiana*
	11. *Micromelum zeylinicum*
	12. *Murraya euchrestifolia* Hyata
	13. *M. gleni*
	14. *M. koenigii* Spreng
	15. *M. paniculata* L. Oack (syn. *M. exotica* Linn
	16. *M. siamensis.*
B. Lower Plants	17. *Actinomadura melliaura*
(Schizomycophyta)	18. *Arcyria denudata*
	19. *Aspergillus flavus*
	20. *A. tubingensis*
	21. *Nocardia aerocoligenes*
	22. *Nocardiopsis dassonville*
	23. *Nocardiopsis* sp.
	24. *Streptomyces chromofuscus* DC 118
	25. *S. exfoliatus* 2419-SVT 2
	26. *Streptoverticillium ehimense*
	27. *S.* sp.
	28. *S. staurosporeus* Anaya.
C. Marine Source	29. *Hyella caespitosa* Bron et Flah
	30. *Tednia ignis.*
D. Mammalian Source	31. *Bovine urine*
	32. *Equine urine*

here to mention that while *Murraya koenigii* Spreng grown in Indian soil has not been reported to yield any dimeric alkaloids, the same species grown in the green house in Japan from seeds collected in Taiwan yielded dimeric carbazoles (*49*). Variation in the alkaloidal contents of a particular species collected from Japan and of the same from Singapore has also been observed by Japanese workers (*47*, *43*).

As mentioned earlier indolocarbazole alkaloids have been reported from lower plants. The biological sources of carbazole alkaloids are listed in Table 1. In addition, abiologic sources such as coal tar, Kuwait petroleum oil, soil humus and river water or mud have been found to yield carbazoles. Carbazoles have also been detected in the polluted atmosphere of industrial areas as well as in cigarette smoke presumably

Table 2. Distribution of Carbazole Alkaloids (for structures consult references (*1, 10 and 13*) in addition to the present review)

Plant Group Type	Name of the alkaloid (The number in parentheses denotes, the source as listed in Table 1)
A. Higher Plants	
i) C_{13}-skeleton	2-Methyl-7-hydroxycarbazole (1), carbazole (9, 3), carbazole-3-carboxylic acid (7), 3-carboxymethylcarbazole (7), carboethoxy-3-methylcarbazole (14), clausenal (5), clausenapins (5), clausenol (2), clausenine (20), clauszoline C, I, J, K, L, M (3), O-demethylmurrayanine (2), euchrestine A, B, C, D, E, (12), eustifoline C (12), 3-formylcarbazole (12), 3-formyl-7-hydroxy-carbazole (12), 3-formyl-6-methoxycarbazole (7), 9-formyl-3-methylcarbazole (14), glycozolidine (9), glycozolidol (9), glycozoline (9), glycozolinine (9), glycozolinon (9), glycozoline, 2-hydroxy-3-methylcarbazole, isomurrayafoline B (12), koeniline (1), lansine (7), 3-carboxymethyl-6-methoxycarbazole, N-methoxy-3-formylcarbazole (12), N-methoxy-3-hydroxymethyl-carbazole (12), 2-methoxy-3-methylcarbazole (14), 7-methoxy-O-methylmukonal (16), 7-methoxymukonal (4), 6-methoxymurrayanine (7), 3-methylcarbazole (5, 6), O-methylmukonal (16), mukoeic acid (14), mukoline (14), mukolidine (14), mukonal (14), mukonidine (14), mukonine (14), murrafoline A, B (12), murrayaline A, B, C, D (12), murrayanine (12, 14), murrayanol (12), murrayastine (12), murrayaquinone A (12)
ii) C_{18}-skeleton	Atanisatin (2), clausenipin (6), murrafoline D, G, H (3), dihydroxygirinimbine (12), ekeberginine (8), eustifoline A, D (12), furostifoline (12), girinimbine (14), glycomaurine (10), heptaphylline (5), heptazolicine (5), heptazolidine (5), heptazoline (5), indizoline (6), isomurrayafoline B (12), koenidine or koenigicine (14), koenigine (14), koenimbine (14), koenine (14), 6-methoxyheptaphylline (6), 7-methoxyheptaphylline (2), 7-methoxymurrayacine (16), mukoenine A, C (14), mukonocine (14), mupamine (2, 9), murrayacine (14), murrafoline B (12), pyrayafoline A, B, C (12), murrayaquinone B (12), pyrayaquinone A, B (12).
iii) C_{23}-skeleton	Bicyclomahanimbicine (14), bicyclomahanimbine (14), clauszoline A, F (3), cyclomahanimbine (14), euchrestine B, C, D, E (12), eustifoline B, C (12), isomahanimbine or mahanimbicine (14), isomahanine (14), isomurrayazoline (14), mahanimbicine (14), mahanimbine (14), *dl*-mahanimbine (14), mahanimbinine (14), mahanimbinol (14), mahanimboline (14), mahanine (14), mukoenine B (12), murrayazolidine (14), murrayazoline (14),

Table 2. (*continued*)

	isomurrayazoline (14), murrayazolinine (14), murrayzolinol (14), murrayaquinone D (12).
iv) Dimeric carbazoles	Bikoeniquinone A (14), bis-7-hydroxygirinimbine A, B (12), bis-2-hydroxy-3-methylcarbazole (14), bis-mahanine (14), bis-murrayazolinol (12), bis-murrayaquinone A (14), chrestifoline A, B, C, D, E, G, H (12), indole dimer (14), murrafoline A, B, C, D, E, F (12), murrastifoline A, B, C, D, E, F (12), oxydimurrafoline (12).
B. Lower Plants	
(i) Marine source (Cyanophyta)	1-Acetylcarbazole (30), 6-chlorohyellazole (29), hyellazole and 6-chlorohyellazole (29), 1-methylcarbazole (30).
(ii) Microbial sources (Schizomicophyta)	Carbazomycine A, B, C, D, E, F, G, H (26), carbazomycinal (26), carazostatin (24), carquinostatin (25), 6-methoxycarbazomycinal (24), tubingensin A (20), antitumor alkaloids AT 2433 A1, A2, B1, B2 (20) arcyriaflavin B, C (18), protein kinase C inhibitor K-252a, 252b, 252c, 252d (23), rebeccamycin (21), staurosporine (28), Tan-999 (22), Tan-1030 A (26), UCN-01 (27), aflavazole (19).
C. Mammalian source	3-Chlorocarbazole (31), carbazole (32).
D. Carbazoles from abiological sources	
i) Coal Tar	Carbazole.
ii) Kuwait petroleum oil	Carbazole, methylcarbazole.
iii) Soil humus	Carbazole, 2-methylcarbazole and dimethylcarbazole
iv) River water and mud	Carbazole.
v) Cigarette smoke	1-, 2-, 3-, 4-methylcarbazoles, 9-alkylcarbazole, 1,9-, 2,9-, 3,7-dimethylcarbazole and several other methyl-, dimethyl- and alkylcarbazoles.

due to combustion of tobacco. The distribution of carbazole alkaloids and carbazoles from biological and abiological sources is listed in Table 2.

II. Methods of Structure Elucidation

A. Physical Methods

Applications of physical methods for structure elucidation and identification of carbazole alkaloids made since 1990 (*13*) are briefly summarised in this section.

1. Ultraviolet Absorption Spectra

Ultraviolet absorption spectra are useful for obtaining information on structures of carbazole alkaloids. UV absorption data of compounds reported after 1990 (*13*) are detailed in Table 3. The "solvent cage effect" due to hydrogen bonding and N-ethylation on the spectral shifts in the electronic UV spectra of carbazoles has been studied by Kadiri *et al.* (*51*).

2. IR Spectra

IR absorption at 1630–1645 cm^{-1}, diagnostic for a formyl group chelated to a hydroxyl group, has been extensively used in structure elucidation while IR bands near 750^{-1} cm have been utilised for determining the substitution pattern in the aromatic rings of the carbazole skeleton.

3. NMR Spectra

a) ^{1}H-NMR Methodology

The Low field H-4 and H-5 signals in the range δ7.17–7.50, appropriately shifted due to adjacent substitution, are utilised for assignment of the protons in ring A or C. Signals of protons at other positions are assigned according to procedures detailed in the previous reviews. The chemical shifts of the olefinic protons at H-3 or H-2 are reported to be helpful for assigning the methyl group of 1:4 carbazolequinones to C-2 or C-3. In the ^{1}H NMR spectrum of 2-methylcarbazolequinone, the signal of the olefinic proton (H-3) appeared at lower field (0.05) than that of H-2 of the corresponding 3-methyl derivatives. Table 4, lists ^{1}H NMR data of some carbazole alkaloids.

b) Application of NOE and INEPT METHODS

The nuclear Overhauser effect has been used to confirm the assignment of the proton signals by irradiating the appropriate proton and observing the enhancement of the signal of the proton under investigation, INEPT, 2DNMR, COSY HETCOSY AND NOESY methods have been utilised to confirm assignment of the signals of the protons in the aromatic rings.

c) ^{13}C NMR Spectra

Salient features for assigning the signals in the ^{13}C NMR spectra have been discussed in a previous review (*13*). The ^{13}C NMR spectra of

Table 3. Ultraviolet Absorption Spectra of Some Carbazole Alkaloids

Compound	λ_{max}
A. 3-Methylcarbazole derivatives	
1. Carbazole-3-carboxylic acid (**248**)	230 (4.18), 237 (4.7), 269 (4.22)
2. 3-Carbomethoxycarbazole (**249**)	241 (4.57), 251 (4.50), 273 (4.57-sh) 287 (4.68-sh.), 320 (3.82), 3.32 (3.73-sh.), 346 (3.30).
3. Carazostatin (**110**)	218 (15300), 235 (141000), 254 (76000) 276 (60000), 303 (83000), 342 (20000)
4. 9-Carboethoxy-3-methyl carbazole (**233**).	229, 250-52, 262, 278 (sh), 287, 304, 315.
5. Clausenal (**20**)	231, (4.45), 257 (4.09), 261 (4.10), 327 (3.44), 354 (3.42).
6. Clausenine (**234**)	226 (4.34), 242 (4.32), 299 (4.00), 304 (3.44), 354 (3.42).
7. Clausenol (**25**)	228, (4.30), 253 (4.36), 303 (4.02), 356 (3.50).
8. Clauszoline C (**250**)	205, 222, 228, 282, 309, 320, 236 (sh).
9. Clauszoline D (**254**)	206, 217, 242, 276, 291, 301 (sh), 356.
10. Clauszoline F (**270**)	206, 216, 242, 279, 291, 300 (sh).
11. Clauszoline I (**252A**)	223 (sh), 233, 247 (sh), 269, 275, 322, 336.
12. Clauszoline J (**252B**)	240, 276, 308, 318.
13. Clauszoline K (**252C**)	226 (sh), 232, 252 (sh), 293, 332.
14. Clauszoline L (**252D**)	218, 238 (sh), 248, 281, 320 (sh).
15. Clauszoline M (**252E**)	217, 240, 275, 290, 356.
16. Euchrestine A (**255**)	213, 236, 265, 295, 313, 318, 329.
17. Euchrestine B (**271**)	212, 238, 264, 310, 318, 331.
18. Euchrestine C (**272**)	212, 237, 266, 306, 320, 329.
19. Euchrestine D (**273**)	212, 238, 265, 312, 321 (sh),
20. Euchrestine E (**274**)	212, 233, 264, 308 (sh), 316.
21. Eustifoline C (**276**)	205, 231, 258, 268, 290, 300, 325, 357.
22. 3-Formyl-7-hydroxycarbazole (**241**)	233, 245 (sh), 273, 288, 326.
23. 3-Formyl-6-methoxycarbazole (**243**)	229 (4.38), 225 (4.16), 281 (4.35), 297 (3.7), 333 (3.80).
24. 9-Formyl-3-methyl-carbazole (**235**)	230 (4.15), 243 (4.08, sh), 255 (3.97, sh), 268 (3.89).
25. Glycomaurol (**259**)	289, (3.79), 298 (3.88), 327 (3.05)
26. 3-Carbomethoxy-6-methoxycarbazole (**251**)	227 (4.39), 241 (sh 4.28), 266 (sh 4.22), 285 (4.50).
27. N-Methoxy-3-hydroxymethylcarbazole (**239**)	237, 263, 283, 287, 293, 334, 350.
28. 7-Methoxy-O-methylmukonal (**245**)	222 (4.27), 238 (4.30), 244 (4.20), 254 (4.30), 302 (5.5), 3.0 (3.80)
29. 7-Methoxymukonal (**244**)	222 (4.27), 238 (4.30), 244 (4.20), 254 (4.33), 302 (4.54), 340 (3.80).
30. 6-Methoxymurrayanine (**246**)	239 (4.35), 251 (4.20), 287 (4.40), 294 (4.36), 335 (3.85), 349 (3.30).

Table 3. (*continued*)

31. Mukoenine A (**264**)	217, 238, 260 (sh), 288, 300, 321 (sh), 333, 356.
32. Mukoenine B (**287**)	222, 238, 284 (sh), 294, 331, 342 (sh), 360 (sh).
33. Murrayaline B (**236**)	223, 259 (sh), 303, 380.
34. Murrayaline C (**247**)	239 (sh), 257 (sh), 287, 330, 357, 364, 368.
35. Murrayaline D (**279**)	243, 255 (sh), 287, 330, 357, 364. 368.
36. Murrayanol (**282**)	208 (4.06), 236 (4.69), 262 (4.88), 296 (4.25) (sh), 316 (4.11 sh), 328 (4.63).
37. Aflavazole (**310**)	219 (15300), 242 (17300), 263 (7500), 297 (7600), 327 (1300), 341 (1600).
B. Furocarbazoles	
38. Eustifoline D (**256**)	206, 204, 250, 268, 298, 340, 354.
39. Furostifoline (**257**)	214, 227, 238, 260, 274, 297, 320, 334.
C. Pyranocarbazoles	
40. Clauszoline A (**269**)	204, 240, 265, 306, 375.
41. Clauszoline B (**253**)	204, 242, 252 (sh), 264 (sh), 291, 304, 376.
42. Clauszoline E	204, 213, 242, 273, 292, 358.
43. Clauszoline G	204, 234, 245 (sh), 278, 300, 361.
44. Clauszoline H (**268**)	230, 258, 288, 310, 323, 348, 363.
45. Eustifoline B (**275**)	220, 230, 268, 304, 324, 362.
46. Glycomaurine (Eustifoline A) (**187**)	215 (4.18), 231 (4.23), 262 (3.90), 270 (3.87), 312 (3.82), 322 (3.85), 378 (3.22).
47. Isomahanine (**277**)	220 (4.51), 238 (4.8-sh), 262 (3.90), 270 (3.87), 288 (3.65-sh), 297 (3.57), 312 (3.85), 322 (3.85), 378 (3.21).
48. 7-Methoxymurrayacine (**261**)	330 (4.53), 335 (3.19).
49. Mukoenine C (**265**)	220, 223, 228 (sh), 278, 288, 325 (sh), 343.
50. Murrayaquinone E (**266**)	232, 262, 306.
51. (+) Murrayazoline	223 (4.39 sh), 246 (4.66), 261 (4.29-sh), 309 (3.1), 341 (3.66).
52. Pyrayafoline B (**176**)	222, 238, 286 (sh), 296 (sh), 329, 353.
53. Pyrayafoline C (**263**)	222, 238, 286 (sh), 296 (sh), 329, 353.
54. Pyrayafoline D (**283**)	222, 238, 268 (sh), 296, 331, 342.
55. Pyrayafoline E (**286**)	233 (sh), 252, 288, 330 (sh), 335.
D. Indolocarbazoles	
56. AT 2433-A1 (**313**)	200 (30781), 235 (3993), 283 (307), 316 (5562), 395 (3865).
57. AT 2433-A2 (**314**)	198 (30478), 234 (38645), 286 (33598), 314 (43890), 394 (3453).
58. AT 2433-B1 (**315**)	202 (28980), 234 (41280), 284 (33617), 400 (4057).
59. AT 2433-B2 (**316**)	201 (30807), 233 (41076), 282 (33390), 315 (46935), 400 (3969).

Table 3. (*continued*)

Compound	λ_{max}
60. Protein kinase C inhibitor K-252a (**317**)	226 (420000), 248 (41000), 264 (sh 44000), 280 (sh 68000), 290 (102000), 320 (sh 19000), 350 (19000), 367 (21000).
61. Protein kinase C inhibitor K-252b (**318**)	232 (29000), 246 (28000), 268 (sh 29000), 280 (sh 42000), 291 (62000), 323 (12000), 337 (16000), 353 (13000).
62. Protein kinase C inhibitor K-252c (**319**)	230 (37000), 238 (30000), 246 (28000 sh), 257 (29000 sh), 287 (86000), 320 (16000), 331 (2000), 341 (1600), 358 (1100).
63. Protein kinase-C (**320**) inhibitor K-252d	223 (40000), 242 (sh 36000), 248 (sh 31000), 260 (sh 34000), 268 (sh 38000), 280 (sh 61000), 290 (95000), 322 (17000), 355 (23000), 347 (16000), 364 (16000).
64. Rebeccamycin (**116**)	238, 256, 293, 314, 382, 390.
65. Staurosporine (**224**)	243, 263 (sh), 292, 322 (sh)
66. Tan-999 (**323**)	245 (31200), 296 (60300), 341 (19500), 362 (16200), 368 (11000).
67. Tan-1030 A (**321**)	233 (29000), 244 (28000), 263 (31000 sh), 275 (4200 sh), 289 (71000), 309 (13400), 333 (17000), 359 (12100), 369 (13400).
E. Biscarbazoles	
68. Bikoeniquinone	227, 240, 257 (sh), 289, 331.
69. Bis-2-hydroxy-3-methylcarbazole (**289**)	214, 237, 269 (sh), 304, 336.
70. Bismahanine (**307**)	223, 245, 297, 331, 341 (sh), 359 (sh).
71. Bismurrayafoline C (**308**)	226, 235, 260 (sh), 285 (sh), 311, 335.
72. Bismurrayafoline D (**310**)	224, 238, 260 (sh), 285 (sh), 310, 334.
73. Bismurrayaquinone A (**290**)	225, 241, 252, 260 (sh), 280 (sh), 291, 330, 333, 392.
74. Chrestifoline A (**291**)	225, 243, 252 (sh), 282, 291, 330, 340, 360.
75. Chrestifoline B (**297**)	226, 241, 252, 260 (sh), 280, 290, 326, 340, 360.
76. Chrestifoline C (**301**)	226, 242, 251, 282, 290, 327, 327, 341, 360.
77. Indole dimer (**288**)	227, 271, 284.
78. Murrafoline G (**298**)	229, 243, 254 (sh), 284, 293, 302, 318, 332, 343 (sh), 348.
79. Chrestifoline D (**292**)	223, 241, 250, 275, 291, 335.
80. Murrafoline H (**302**)	216, 239, 256 (sh), 284, 293, 302, 318, 332, 343 (sh), 348.
81. Murranimbine (**306**)	218, 240, 260, 306, 336.
82. Murrastifoline A (**294**)	227, 243, 285, 291, 334, 347.
83. Murrastifoline B (**295**)	226, 242, 252, 280, 292, 335.
84. Murrastifoline C (**303**)	205, 238, 254, 263, 287, 335, 353.
85. Murrastifoline D (**299**)	205, 238, 254, 263, 287, 335, 353.
86. Murrastifoline E (**300**)	214, 230, 239, 255, 281, 291, 303, 348.
87. Murrastifoline F (**296**)	227, 243, 253 (sh), 282, 292, 332, 346.

References, pp. 224–230

Table 4. ^{1}H NMR Data of Some Carbazole Alkaloids Proton Signals in δ PPM; TYMSδ (O): Coupling Constants Given in Brackets in Some Cases

Alkaloid	NH	H-4	H-5	Other Ar-protons	Protons of substituents on aromatic ring	Other protons
I. Compounds with monocarbazole unit						
1. 1-Acetyl-carbazole					2.8 (3H-OAc g.)	
2. Carbazole-3-carboxylic acid	10.83	9.10 (*brs*)	8.40 (*d*, 7,5))	7.33, 7.30 (4HCom. H-1, H-6, H-8) 8.30 (*dd*, 7.5, 1.85; H-2)	5.83 (*d*, 8; COOH)	
3. 3-Carboxymethyl-carbazole	10.75 (*brs*)	8.82 (*d*, 1.4)	8.23 (*d*, 8.0)	7.57 (*d*, 8.5; H-1) 7.45 (*dd*, 8.0, 6.6; H-7) 8.03 (*dd*, 8.5, 1; H-2), 7.25 (*dd*, 8.0, 6.8; H-6), 7.75 (*d*, 8.0; H-8)	3.91 (*s*, COOMe)	
4. Carazostatin		7.3	7.9 (*dd*, 8.0, 1.5)	7.17 (*dd*, 8.0, 8.0, 1.5 H-6), 7.37 (*dd*, 8.0, 1.5; H-7) 7.42 (*dd*, 8.0, 1.5; H-8)	2.38 (Ar-CMe)	2.87, 1.65, 1.45-1.30 0.91 (*m*, 3H) -heptyl chain with C-Me.
5. Clauszoline C	10.36 (*brs*)	8.4 (*s*)	7.97 (*d*, 7.7)	7.10 (H-1), 7.02 (H-8) 6.99 (*dd*, 7.7, 2)	3.89 (3H-OMe) at (C2), 3.86 (3H; OMe at C-7) 3.82 (3H; COOMe at C-3)	
6. Clauszoline D	10.3 (*brs*)	8.30 (*s*)	7.57 (*d*, 7.7)	7.05 (*t*, 7.7; H-6) 6.88 (*d*, 7.7; H-7)	3.73 (*brd*, 9.5; OH) 2.6 (*dd*, 1.3, 1.8; OH), 2.89 (1.3; H) 4.28 (OH), 3.72	1.33 (3H, *s*, Me) 1.34 (3H, *s*, Me) 8.95 (*br*; 8-OH), 4.28 (OH), 3.72 (1H, *br*; OH)
7. Clauszoline F	8.5 (*brs*)	8.20	7.57 (*d*, 7.7)	7.10 (*t*, 7.7; H-6)	9.9 (1H; CHO at 3) 6.83 (*d*, 7.7; H-7)	1.62 (3H, *s*, Me), 5.35 (H; H-2) 5.07 (*m*; H-6, 3.67 (2H-7.0; H-1) 2.10 (H), 1.92, 1.60, 1.4 (3 Me, *s*), 15.44 (*br*-OH at 8)
8. Clauszoline I	10.7 (*brs*)	8.02 (*d*, 1.1)	8.17 (*d*, 7.7)	7.22 (*t*, 7.77; H-6) 7.3 (*t*, 7.7; H-7) 7.61 (*d*, 7.7; H-8)	3.88 (3H; COOMe)	
9. Clauszoline J	10.26	8.38	7.93 (*d*, 8.0)	7.01 (H-1), 6.76 (H-6) 6.96 (H-8)		3.82 (3H; 7-OCH_2) 3.88 (3H; 2-OCH_3)
10. Clauszoline K	8.29 (*br*)	8.51 (*d*, 1.5)	8.00 (*d*, 8.0)	6.96 (*d*, 2.2; H-8) 6.9 (*dd*, 8; 2.2; H-6) 7.7 (*d*, 8.4); 7.91 (*dd*, 1.5, 8)	10.08 (1H; CHO at 3)	
11. Clauszoline L	8.18	8.70 (1.80)	7.99	7.39 (*d*, 6; H-1), 8.06 (*dd*, 1.8, 6; H-2) 6.19 (*dd*, 8, 2.2; H-6) 6.96 (*d*, 2.2)	3, 97 (3H; $COOCH_3$) 3.91 (3H; OCH_3)	
12. Clauszoline M	10.63 (*br*)	8.0	7.58 (*d*, 7.7)	6.9 (H-1), 7.05 (*t*, 7.7H-6), 6.90 (*d*, 7.7; H-7)	11.4 (OH), 9.99 (CHO), 8.86 (OH)	
13. Euchrestine A	7.72 (*s*)	7.65 (*s*)	7.63 (*d*, 8.4)	6.69 (H-6), 6.83 (*s*; H-1)	2.38 (Ar-CH_3)	1.78, 1.89 (3H, *s*) 3.59, 5.36 (*t*)-DMA chain.
14. Euchrestine B	7.73 (*s*)	7.67 (*s*)	7.70 (*d*, 8.0)	6.80 (*s*; H-1) 6.83 (*d*, 8; H-6)	2.38 (Ar-CH_2)	1.57, 1.62, 1.87 (3Me; *s*) 2.061 (H, *m*), 3.62 (*d*, 6.7) 5.32 (*t*, 6.7; 1H), 5.37 (*t*, 6.7, 1H)-geranyl chain.

Table 4. (continued)

Alkaloid	NH	H-4	H-5	Other Ar-protons	Protons of substituents on aromatic ring	Other protons
15. Euchrestine C	7.73 (*s*)	7.65 (*s*)	7.63 (*d*, 8.4)	6.81 (*d*, 8; H-1) 6.70 (*d*, 8; H-6)	2.38 (Ar-CH_3)	1.58, 1.65, 1.88 (*s*; Me) 2.12 (H, *m*) 3.59 (*d*, 7.12H) 5.37 (*t*, 6.7, 1H), 5.07 (*m*), 3.36 (*t*, 6.71 h)-geranyl chain
16. Euchrestine D	7.73	7.56 (*s*)	7.74 (*dd*, 8, 4.1)	6.82 (*d*, 2; H-8) 6.68 (*dd*, 8; 2.4; H-6)	2.38 (Ar-CH_3)	1.59, 1.66, 1.8 (*s*, 3H each) 2.11, H4 3.59 (*d*, 6.7; 2H), 5.07 (*m*) 3.36 (*t*, 6.7)-geranyl chain
17. Euchrestine E	7.63	7.65	7.66 (*d*, 8.0)	6.81 (H-1) 6.73 (*d*, 8; H-6)	4.90 (OH) 4.50 (-OH) 2.38 (Ar-CH_3)	1.0, 1.56, 1.64, (3H, each) 3.99 (*t*, .4), 5.07 (*m*), 3.10 (*dd*, 9, 16.5) -geranyl chain
18. Euchrestine F	7.81	7.92 (*brs*)		7.29 (*d*, 8.3; H-1) 7.21 (*dd*, 8.3, 1.2; H-2) 6.96 (*d*, 8.6; H-7) 7.15 (*d*, 8.6; H-8)	2.51 (Ar-CH_3)	1.56, 1.63 (3 methyl protons) 2.09 (H, *m*), 4.00 (2H, *d*), 6.90 (*br*, OH), 5.05 (*m*)
19. 3-Formyl-7-hydroxy-carbazole	8.1	8.62	8.15 (*d*, 8.0)	7.98 (*dd*, 0.7, 8.5; H-2)) 7.53 (*d*, 8.5; H-1) 7.48 (*d*, 5; H-8) 7.33 (*d*, 8.0, 5; H-6)	10.10 (Ar-CHO)	
20. 3-Formyl-6-hydroxy-carbazole	10.74 (*br*)	8.66 (*brs*)	7.82 (*d*, 2.0)	7.48 (*d*, 8.8; H-8) 7.10 (*dd*, 8.8, 2; H-7) 7.92 (*dd*, 8.2, 2.0; H-2) 7.58 (*d*, 2.7; H-1)	6.05 (Ar-CHO) 3.90 (Ar-OMe)	
21. Glycomaurrol	7.82	7	7.93 (*brs*)	6.95 (*d*, 8.5; H-1) 7.13 (*d*, 8; H-2) 7.19–7.31 (*m*, H-7, 8)		1.76, 1.95 (3H, *s*, each) 3.99 (*d*, 6.50) 5.0 (*d*, 6.5)
22. 1-Methyl-carbazole					2.56 (Ar CH_3)	
23. O-Methyl-mukonal		8.58	7.12		10.5 (Ar CHO) 4.5 (Ar-OMe)	
24. 3-Carboxy-methyl-6-methoxy-carbazole	10.63 (*brs*)	8.93 (*brs*)	7.70 (*d*, 1.8)	7.2 (*d*, 7.8; H-8) 7.18 (*dd*, 7.8, 1.8; H-7), 8.27 (*dd*, 7.8, 1.8; H-2), 7.48 (*d*, 7.8; H-1)	4.03 (COOMe) 3.96 (*s*, OMe at 6)	
25. 7-Methoxy-heptaphylline						
26. N-Methoxy-3-hydroxymethyl carbazole		8.06 (*brs*)	8.0 (7.88)	7.25 (*t*, 8.8; H-6) 7.50 (H-2)	4.13 (OH), 4.86 (2H-CH_2 OH) 5.15 (*brs*, -OH)	
27. 7-Methoxy-O-methylmukonal			7.97 (*d*, 8.3)	6.83 (*dd*, 8.3, 2.2; H-6)		
28. 7-Methoxy-mukonal					11.56 (OH-chelated to CHO); 9.83 (CHO at 3)	
29. 3-Formyl-carbazole	10.90 (*brs*)	8.67 (*brs*)	8.22 (*d*, 7.6)	7.27 (*dd*, 7.6, 6.8; H-6) 7.6 (*d*, 8.6; H-1), 7.59 (*d*, 8.0; H-8); 7.7 (*dd*, 8.0, 6.8; H-7), 7.96 (*dd*, 8.6, 1.2; H-2)	10.98 (*s*, 3-CHO)	
30. 6-Methoxy-murrayanine (1,6-dimethoxy-carbazole)		8.50 (*brs*)	7.90 (*d*, 1.8)	7.20 (*dd*, 7.8, 1.8; H-7) 7.6 (*d*, 7.9; H-8)	4.0 (*s*-OMe) 4.3 (ArOCH_3) 10.20 (CHO at 3)	

References, pp. 224–230

Table 4. (*continued*)

Alkaloid	NH	H-4	H-5	Other Ar-protons	Protons of substituents on aromatic ring	Other protons
31. Mukoenine A	7.84 (*brs*)	7.68	7.93 (*d*, 7.7)	7.17 (*t*, 7.8; H-6) 7.3 (*t*, 7.7; H-7) 7.38 (*d*, 7.7; H-3)	2.0 (Ar-CH_3)	5.18 (OH) 3.62 (*d*, 7; 2H; 5.37 (1H; *t*, 7.1, 1.91) (3H, prenylchain)
32. Mukoenine B	8.22 (*brs*)	8.07	7.98	7.33 (H-7) 7.4 (*d*, 8; H-8)	9.93 (Ar-CHO)	7.66 (OH); 5.35 (*t*, 6.3) 1.07 (*m*), 3.67 (*d*, 6.8 2H) 2.10 (H, m), 1.91, 1.61, 1.57 (*s*, $3CH_3$) prenyl chain.
33. Murrayaline B	10.79	7.73	8.16 (*d*, 8.7)	7.20 (H-1) 8.2 (OH) 6.89 (*d*, 8.7; H-6)	2.34 (Ar-CH_3) 10.58 (CHO, *m*, 8)	5.0 (*d*, 9, 7; H-2) 6.58 (*d*, 1.7 (*s*, $2CH_3$))
34. Murrayaline C	10.79	8.80	8.10	6.95 (H-1)	11.11 (OH), 9.956 (Ar-CHO) 10.0 (CHO at 8) 4.03 (Ar-OCH_3)	
35. Murrayaline D	8.5	8.08	7.88 (*d*, 8.8)	7.0 (H-1) 6.95 (H-6)	6.17 (OH), 9.98 (Ar-CHO, 3)	1.56, 1.60, 1.91, 2.10 (H, *m*), 3.68 (2 *d*, 7.3, 5.08 (*m*), 5.36 (*t*, 7.3)-geranyl chain
36. Murrayanol B		7.72 (*d*, 8.6)	7.7	6.80 (H-8), 6.83 (H-3)	2.38 (Me at 6) 3.91 (OMe at 7)	1.57, 1.63, 1.88 (3Me) 3.62 (*d*, 7.3, 2H), 5.02–5.12 (1H, s), 2.00 (2H, *m*), 5.33 (1H, *m*) ger. chain
37. Aflavazole (a hexacyclic base)			7.95 (*br*, *d*, 8.0)	7.1 (*dd*, 8, 7.1, H-6) 7.27 (*br*, *dd*, 7.8, 7.1, H-7), 7.12 (*s*, H-1) 7.39 (*br*, *d*, 7.8, H-8)	2.15, 2.75 (*br*, *dd*, 11, 7.1, methylenes) 2.35 (Ar, Me) 1.19 (*d*, 6.8, C-CH_3)	0.6 (C-Me), 1.33 (*s*, C-CH_3), 2.89 (-CH_3)
II. Furo- and pyrano-carbazoles						
38. Clauszoline A	8.31 (*brs*)	7.98 (*s*)	7.44 (*d*, 7.7)	6.90 (*d*, 7.7, H-6)	11.64 (s, OH-2) 9.89 (Ar-CHO)	6.48, 5.63 (*d*, 9.9, 1H each) 1.53 (6H-gem-diMe) 5.3 (*d*, 7.3), 3.15 (2H)
39. Clauszoline B	8.3 (*brs*)	8.07 (*s*)	7.3 (*d*, 7.7)	6.90 (*d*, 7.7, H-6), 6.83 (H-1)	9.89 (Ar-CHO) 11.6 (OH-2)	6.5 (protons of pyran double bond)
40. Clauszoline E	11.30 (DMSO)	8.2 (*s*)	7.52 (*d*, 7.7)	6.97 (*d*, 7.7, H-6) 6.80 (*d*, 7.7, H-7)	10.36 (H, CHO at 3) 9.85 (*br*, 1H, OH)	5.32 (*d*, 5.5, 1H, OH 3.83 (H-3), 3.17 (*dd*, 17, 5.1, H4), 2.83 (H4. 1.38, 1.30 (2Me)
41. Clauszoline G	10.58	8.33 (*s*)	7.61 (*d*, 7.7)	7.04 (*t*, 7.7, H-6) 6.88 (*d*, 7.7, H-7)	10.8 (*s*, CHO at 3) 8.90, 10.36 (H, CHO at 3)	7.18 (0.9; OH) 5.90 (*d*, 9.9, H-3), 1.56 (6H, *s*, 2Me)
42. Clauszoline H	8.02	7.71	7.2 (*d*, 8.1)	6.8 (*d*, 8.1, H-6) 6.85 (H-1)	2.3 (Ar-CH_3) 3.90 (OCH_3-2)	5.56 (H-3), 6.48 (H-4) 1.50 (6H, 2Me)
43. Eustifoline B	7.78	7.92 (*brs*)		7.31 (*d*, 8.3, H-1), 7.14 (*d*, 8.3, H-8), 7.19 (*brd*, H-2), 6.90 (*d*, 8.3, H-7)	2.52 (*s*, Ar-CH_3)	7.28 (*d*, 9.8), 5.79 (*d*, 9.8) 5.11 (*t*, 6.7), 2.17 (2H, *m*), 1.78 (2H, *m*), 1.65, 1.58, 1.5 (3H, *s*, 3Me)
44. Eustifoline D (Furocarbazole)	8.12	7.97 (*brs*)		7.0 (*d*, 8, H-1) 7.26 (*dd*, 8, 1.7, H-2) 7.58 (*dd*, 8.7, 0.7, H-7) 7.35 (*d*, 8.3, H-8)	2.58 (*s*, Ar-CH_3)	7.81 (*d*, 2.0) 7.32 (*dd*, 2.0, 7)
45. Furostifoline (Furocarbazole)	8.28	7.97 (*brs*)		7.25 (*t*, 8.3, H-6), 7.37 (*t*, 8.0; H-7), 7.9 (*d*, 8.0, H-8)	2.67 (Ar-CH_3)	7.30 (*d*, 2.0) 7.00 (*d*, 2.00)
46. Glycomaurine (Eustifoline A)	7.80		7.15 (*d*, 8.5)	5.82 (H-2), 7.28 (H-1) 6.91 (H-6), 7.19–7.26 (H-8 & 9)	2.50 (Ar-CH_3)	1.6 (6H, 2Me)

Table 4. (*continued*)

Alkaloid	NH	H-4	H-5	Other Ar-protons	Protons of substituents on aromatic ring	Other protons
47. Iso-mahanine	7.61 (*s*)		6.62 (*d*, 8.4)	6.50 (*d*, 9.8, H-1), 5.56 (*d*, 9.8, H-2), 7.56 (*d*, 8.4, H-6), 7.56 (*s*, H-7)	2.30 (3H, *s*, Ar-CH_3 4.96 (1H, *brs*, -OH)	1.36 (3H, *s*, 3Me)
48. 7-Methoxy-murrayanine		8.23	7.97	6.83 (*dd*, 8.0, 2.2, H-6) 6.97 (*d*, 2.2, H-8)	3.85 (Ar-OMe) 10.5 (*s*, CHO)	6.90, 5.90 (2H, *d*, 9.80) 1.5 (6H, *s*, 2Me)
49. Mukoenine C	7.76	7.55	7.72 (*d*, 8.1)	6.68 (*dd*, 2.0, 8.1, H-6)	5.69, 6.58 (*d*, 9.7 each) 4.9 (OH-*br*)	1.47 (6H, *s*, 2Me)
50. Murraya-quinone E						
51. Pyrayafoline B	7.70	7.62 (s)	7.9 (*s*)	6.74 (*s*, H-1) 6.71 (*s*, H-8) 4.9 (OH)	2.37 (Ar-CH_3) 6.48 (*d*, 9.8, H-1)	5.58 (*d*, 9.6, H-2) 1.6 (6H, *s*, 2Me)
52. Pyrayafoline C	7.78 (*s*)	7.6 (s)	7.63 (*d*, 8.0)	6.83 (*s*, H-1) 6.69 (*d*, 8.0)	2.38 (*s*, Ar-CH_3) 6.59 (*d*, 9.8, H-1), 5.69 (*d*. 9.8, H-2)	1.7 (6H, *s*, 2CH_3)
53. Pyrayafoline D	7.75	7.64 (s)	7.63 (*d*, 8.4)	6.83 (*s*, H-1) 6.68 (*d*, 8.4, H-6)	2.38 (*s*, Ar-CH_3) 4.47 (OH)	6.62 (*d*, 10.1, H-1′), 5.66 (*d*, 10.1, H-2′), 1.75, 2.15 (2H, *m*, each H-5′, H-6′), 5.10 (7.1, H-7), 1.44, 1.57, 1.65 (3Me)
III. Indolocarbazole						
54. Arcyriaflavin B						
55. Arcyriaflavin C						
56. AT 2433-A1	10.64 (NH-14)		9.18 (H-13)	7.88 (H-10), 9.27 (H-13) 7.73 (H-7, 11) 7.49 (H-12 & 6), 3.70, 3.60 (4′-H)	3.68 (OMe-4′) 1.78 (Hb-2″) 4.81 (OH-3″) 6.91 (H-1′) 5.06 (OH-2′) 4.08 (H-5′) 4.21 (Ha-6′)	6.91 (H-1), 5.06 (OH-2) 5.48 (OH-3′), 2.30 (H-4″) 4.08 (H-5′), 4.21 (Ha-6′), 3.99 (H-6′), 5.13 (H-1″), 1.78 (H-2″), 3.7 (H-3″) 4.81 (OH-3″), 2.22 (N-CH_3-4′)
57. AT 2433-A2	10.64 (NH-16) 2.55 (NH-4″)		9.18	7.88 (H-13), 9.27 (H-13) 7.73 (H-7& 11), 7.49 (H-12 & 6)	1.73 (Hb-2″), 2.32 (Ha-2″) 6.91 (H-1′), 4.81 (OH-3′), 5.06 (OH-2′), 4.08 (H-5′) 3.68 (OMe-4′)	3.72-3.30 (H3′ & 4′) (Ha & Hb5″), 5.1, 5.48 (OH-3′), 4.08 (H-5′), 4.2 (Ha-6′), 3.99 (Hb-6″)
58. AT 2433-B1	10.50 (NH-14)		9.21	7.86 (H-10), 9.32 (H-13), 7.68 (H-7 & 11), 7.48 (H-6 & 12), 8.14	6.44 (H-1′), 3.68 (OMe-4′). 4.32 (H-5′ H6′), 4.6-5.6 (H-1″, OH-2′, 3′, 3″) 1.60 (Hb-2″)	2.03 (H-2), 1.63 (Hb-2″) 2.19 (N-CH_3-4″), 1.3 (H-4″, 5″), 2.03 (H-2″) 3.68 (OMe-3′), 1.60 (H-2′), 2.12 (H-4″)
59. AT 2433-B2	10.50 (NH-14)		9.21	7.86 (H-10), 9.32 (H-13), 7.68 (H-7, H-11), 8.1 (H-8) 7.48 (H-6, H-12) 6.6 (H-1)	3.68 (OMe-4′) 2.03 (Ha-2″) 1.60 (Hb-2″) 6.91 (H-1′)	2.03 (H-2″), 1.60 (Hb-2″) 4.3 (H-2′, 3′, 4′, 5′ Ha-Hb-6′) 1.3 (H-3′, 4″, Ha-Hb-5″), 2.03 (H-2″, 6-5.6, OH-2′, 3′, 3″, H-1″), 3.68-OMe-3′, 1.60 (H-2), 2.12 (H-4″)
60. Protein kinase C inhibitor K-252a	8.64 (*brs*, NH-16)		9.24 (*d*, 7.9)	7.90 (*d*, 8.3, H-13), 8.05 7.49 (*brt*, H-7) (*d*, 7.8) 7.29 (*brt*, H-6) 7.36 (H-12), 7.49 (H-11) 7.95 (H-10)		7.15 (*dd*, 7.4, 4.9, H-1′) 2.04 (*dd*, 14.0, 4.9, H-2′) 3.41 (*dd*, 4.0, 7.4, H-2′) 2.16 (H-5′, 31)

Table 4. (*continued*)

Alkaloid	NH	H-4	H-5	Other Ar-protons	Protons of substituents on aromatic ring	Other protons
61. Protein kinase C inhibitor K-252b	8.62 (*brs*, H-16)		9.22 (*d*, 7.9 H-5)	7.91 (*d*, 8, H-13) 8.05 7.48 (*brt*, H-1) (*d*, 7.8), 7.28 (*br*, *t*; H-15) (H-13) 5.03, 5.04 (AB, 17.3, H-15)		7.12 (*dd*, 7.3, 4.9; H-1), (*dd*, 1.7, 2′), 3.36 (*dd*, 13.8, 4.9, H-2′)
62. Protein kinase C inhibitor K-252c	8.9 (*brs*, H-16) 11.56 (*brs*, H-9) 11.38 (*brs*, H-14)		9.24 (*d*, 7.9, H-5)	7.73 (*d*, 8.1; H-8) 7.1 (*br*, *t*; H-6) 7.4 (*br*, *t*; H-7) 8.05 (*d*, 7.8, H-13) 4.98 (*s*; H-15), 7.4 (*br*, *t*; H-11), 731 (*br*, *t*, H-12) 7.79 (*d*, 8.7; H-10)		
63. Protein kinase C inhibitor K-252d	8.5 (*brs*, H-16) 11.68 (H-9)		9.47 (*d*, 7.7)	7.69 (*d*, 8.3; H-8) 7.49 (*brt*, H-7) 7.27 (*br*, H-6), 8.07 (*d*, 7.8) H-13 5.01 (*brs*, H-15), 7.31 (H-12), 7.50 (*brs*, H-11) 7.60 (*d*, 8.1, H-10)	6.40 (*brd*, 9.5, H-1′) ca.5 (*m*, H-2′) 4.18 (*brq*, H-3′) 4.05 (*brt*, H-4′) 4.48 (*brq*, 7.6, H-5)	1.70 (*d*, 7.2, H-6′) ca. 5.02 (OH)
64. Rebeccamycin	10.30 (H-14), 11.37 (H-16)		9.09 (*d*, H-5)	7.69 (H-7, *d*), 7.45 (*t*, H-6, H-12), 7.5, 7.5 (H-6, H-15) 7.74 (H-11), 9.27 (*d*, H-13)	6.97 (H-1′), 4.45 (3′-OH), 5.36 (6′-OH) 5.03 (2′-OH)	3.48 (4′-OCH_3), 3.53 (4′-H), 3.66 (5′-H), 3.90 (6′-CH_2), 3.45 (3′-H)
65. Staurosporine	8.53 (*s*, H-16)		9.30 (*d*, 7.9)	4.96 (*s*, H-15), 7.27 (*t*, H-6) 7.56 (*d*, H-8), 7.4 (*t*, H-7) 7.98 (*d*, 8.3, H-10), 7.41 (*t*, H-11), 7.96 (*d*, 7.3, H-13)		2.29 (*s*, H-6′), 1.44 (*s*, 3′ -N-CH_3), 6.68 (*s*, H-1′) 3.24 (*m*, H-3′) 4.04 (*d*, 3.2, H-4′)
66. Tan-999	6.28 (*s*, H-16)		9.0 (*d*, 7.8)	7.75 (*d*, 8.5, H-13) 7.35 (*t*, H-6), 6.96 (*dd*, 2.0, 8.5, H-12), 4.96 (ABq, 1.7, H-7), 7.2 (*d*, 8, H-8) 7.46 (*t*, H-7)	3.95 (*s*, Ar-OMe)	3.35 (H-3′), 1.344 3.87 (*d*, 3.5, H-4′-OMe) 3He 2.33 (*s*, H-6′); 1.57 (*s*, 3′ -N-CH_3), 6.53 (*d*, 5.6, H-1′)
67. Tan-1030 A	0.58 (H-16, *s*)		9.31 (*d*, 7.8)	7.98 (*d*, 8.0, H-13) 7.50 (*t*, H-7), 7.32 (*t*, H-6, 12) 4.96 (ABq, 17, H-15) 7.70 (*d*, 8.0, H-8); 8.01 (*d*, 9.1, H-10), 7.44 (*t*, H-11′)		4.73 (*s*, H-4′) 3.43 (*s*, 4′-OMe) 2.47 (*s*, H-6′) 10.45 (*s*, NOH), 7.04 (*d*, 5.2, H-1′), 3.01, 3.63 (*d*, H-2′)
68. UCN-01	8.72 (*d*, 1.1, H-16)					
IV. Biscarbazoles						
69. Bikceniquinone	11.70 10.36		8.24 7.72 (H-5, H-5′)	6.94 (*s*, H-2′), 6.89 (*t*, 8) 7.45 (*d*, 8.3, H-7′) 7.37, 7.23 (H-6, H-6′) 7.65, 7.50 (H-8, H-8′)	1.81, 2.26 (*s*, 2Ar-Me-*s*)	4.01 (*s*-Me)
70. Bis-7-hydroxy-girinimbine A	7.60 (*brs*)	7.61 (*s*, H-4 H-4′)	7.85 (*d*, 8.55 H-5, 5′)	6.98 (*d*, 8.55, H-6)	5.349 (*brs*, OH) 2.32 (Ar-Me)	[6.98 (*d*, 9.7), 5.56 (*d*, 9.7), 1.42, 1.41 (6H, 2Me)]-dimethylpyran ring.
71. Bis-7-hydroxy-girinimbine B	7.90 (*brs*)	7.63 (*s*, H-4 H-4′)	7.83, 7.78 (*s*, H-5′) H-5	7.03 (*s*, H-8), 6.86 (*d*, 8.4 H-6), 3.84, 3.80 (6H, *s*, 2Ar-Me)	2.29, 2.28 (*s*, 2Ar-Me), 9.70, 5.17 (2-OH).	[6.93, 5.79 (*d*, 9.9 each), 6.84, 5.57 (*d*, 9.9 each), 1.48, 1.47, 1.41, 1.40 (12H, *s*, 4Me)] dimethylpyran ring

Table 4. (*continued*)

Alkaloid	NH	H-4	H-5	Other Ar-protons	Protons of substituents on aromatic ring	Other protons
72. Bis-2-hydroxy-3-methyl-carbazole	7.6 (2H, *br*)	7.61	8.03 (2H, 8, 3)	7.22 (*t*, 8.0, H-6; 7, H-6') 7.31 (*t*, 8, H-7, 7')	2.52 (Ar, CH_3-6H) 8.27 (2-OH)	
73. Bismahanine	7.95 7.57 (*brs*)	7.60 7.5	7.8 (*d*, 8.3, H-5) 7.89 (H-5')	6.95 (*d*, 8.3, H-6) 7.16 (H-8)	2.33, 2.31 (2Ar-Me) 5.22, 5.17 (*brs*, 2H) 3.8, 4.02 (2Ar-Me)	6.67, 5.71 (9.8), 5.4 6.8, 5.5 (*d*, 9.8) 2.15 (*m*, H) 1.75 1.71, 1.66, 1.6, 1.5, 1.58, 1.6 (*s*, 3H each)
74. Bismurraya-foline C	7.8	7.8	7.69 (H-5)	6.73 (*d*, 8, H-6)	5.19, 5.05 (2-OH) 2.8 (Ar-Me).	5.16, (*t*, 6.8), 3.2 (H), 1.75 (H), 1.67 (H), 1.60, 1.7, 1.2 (6H)
75. Bismurraya-foline D	7.85	7.85	7.77 (*d*, 6.85)	6.85 (*d*, 8, H-6)	2.8 (6H, Ar-Me) 5.17 (OH) 3.89 (Ar-OMe, 6H)	5.87 (*t*, 6.8), 77 (3.5 *m*, 2H), (3.1 *m*, 1.6-1) (*m*, 8H), 1.56, 1.2 (6H)
76. Bismurraya-quione A	11.79 (2H)		8.23 (*d*, 7.6 -2H)	7.7 (*t*, 7.6, H-7, H-7'), 7.68 (H-8, H-8')	2.08 (6H, *s*, -2Me)	
77. Chrestifoline A	10.16 (*s*) 10.28 (*s*)	7.39	7.78 (*d*, 7.9) 8.08 (*d*, 9, (H-5'')	7.00 (H-2); 7.03 (*t*, 7.9, H-6) 6.9 (H-2), 6.98 (*t*, 7.9, H-6) 7.28 (*t*), 7.27 (*t*), 7.49 (*d*, 7.9, H-8), 7.55 (*d*, 7.9, H-8)	2.47 (Ar-Me) 4.78 (Ar-CH_2)	
78. Chrestifoline B	10.31	7.38 (*s*) 7.81 (*s*)	7.77 7.99	6.90 (*s*, H-2), 7.8 (*t*, 7.9, H-6) 7.13 (*t*, 7.9), 7.28 7.9 (H-7), 7.26 (*t*, 7.9, H-7), 7.39 (*d*, 7.9, H-8) 7.48 (*d*, 7.9, H-8)	3.85 (3H-8), 5.79 (Ar-CH_2)	6.91 (*d*, 9), 5.53, 1.36 (6H)
79. Chrestifoline C	10.31	7.38 7.81 (*s*)	7.77 (*d*, 7.9) 7.98 (*d*, 7.9)	7.12, 7.01 (*t*, 7.9, (H-6 & H-6'), 6.90 (*s*, H-2), 7.26 (*t*, 7.9), 7.28 (*t*, 7.9, H-7 & H-7'), 7.8 (*d*, 7.9, H-8) 738 (*d*, 7.9, H-8)	3.85 (*s*, 3H) 5.80 (*s*, Ar-CH_2)	6.95 (*d*, 9.9), 5.5 (*d*, 5.00) 1.3 (*s*, 3H), 1.68, 2.04 (2H), 1.58, 1.5 (*s*, 3Me)
80. Chrestifoline D	8.20	7.5 7.9 (H-4)	7.92 8.14	6.77 (*d*, 15, H-2) 7.53 (H-2) 7.2 (*d*, 8, H-8) 7.37 (*d*, 8, H-7) 7.2 (H-7)	4.05 (*s*, OCH_3) 10.07 (Ar-CHO)	6.08 (2H, *s*, N-CH_2)
81. Indole dimer	7.8 (*brs*)			7.08–7.21 (*m*, 6H), 7.58 (*m*, 2H)		0.98, 1.19 (*d*, 2Me); 1.85, 2.10, 2.50 (*m*, 6H), 3.33 (*s*, OMe)
82. Murrafoline G	8.04 (*brs*) 10.8 (*br*)	7.77 (*s*)	7.89 (*d*, 8.6) 8.35 (H-5')	6.99 (*d*, 8.6, H-6), 6.66 (*s*, H-2), 7.00 (*t*, 8.6, H-7) 7.21 (*t*, 8.6, H-6), 5.79 (*dd*, 6.7, 12.2), 6.89 (*d*, 8.6, H-8), 7.43 (*t*, 8.6, (H-7'), 7.69 (*d*, 8.6, H-8)	1.58, 2.39, 1.6 (3H, *s* each.) 3.96 (3H, *s*)	2.2 (*dd*, 21.0, 14.0 2.46 (*dd*, 6.7, 14) 1.73 (*s*, 3H)
83. Murrafoline H	8.26 (*brs*) 10.25 (*br*)	7.78 (*s*) 7.6 (*s*)	7.89 (*d*, 8.1) 7.93 (*d*, 1.8)	6.97-7, 0 (H-6), 7.09 (*dd*, 1.8, 8.1) 7.30 (*d*, 8.1)	2.38 (3H, *s*, C-Me) 1.6 (*s*, 6H), 1.1 (*s*, OMe), 1.7 (*s*, OMe)	2.00 (*m*), 2.38 (*m*) 2.26 (*s*, H.60) 5.77, 6.91 (*d*, 9.5 each)
84. Murranimbine		7.65 (H-) 8.00 (H-)	7.84 (*d*, 7.7) 8.08 (*d*, 7.3)	6.87 (*t*, 6.7, H-6), 6.17 (*d*, 8, H-8) 7.19 (*t*, 7.0, H-6) 6.61 (*t*, 7.0, H-7), 7.30 (*t*, 7.0, H-7), 7 (*d*, 7.7, H-8')	2.32 (Ar-CH_3) 2.30 (Ar-CH_3)	4.07 (*m*), 6.17 (*d*, 2.9) 1.60 (6H), 1.54, 006 2.08, 2.21, 2.74 (1H, each)

Table 4. (*continued*)

Alkaloid	NH	H-4	H-5	Other Ar-protons	Protons of substituents on aromatic ring	Other protons
85. Murrastifoline A	10.5 (*s*)	8.13 (*d*, 7.9) 8.10 (*d*, 2.0)	8.13 (*d*, 7.9) 8.1 (*d*, 2.0)	6.85, 6.88 (H-2 & H-2′) 7.21 (H-6), 7.16, 7.67 (H-8 & H-8′)	3.57, 4.03 (3H, -OMe), 2.52, 2.8 ($Ar-CH_3$)	
86. Murrastifoline B	10.59	7.63, 7.81 (*d*, 1.5, H-4)	8.13 (*d*, 7.9) (H-5)	6.86, 7.0 (H-2 & H-2′) 7.20, 7.21 (H-6 & H-6) 7.34, 7.44 (H-7 & H-7′) 7.22, 7.65 (H-8 & H-8′)	3.61, 4.03 (3H, OCH_3) 2.52 (*s*, $Ar-CH_3$)	
87. Murrastifoline C	10.33	7.58 7.36 (H-4)	8.07 (*d*, 7.9) (*d*, 7.9, H-5)	6.90 (*s*, H-2) 7.12 (*t*, 7.9, H-6), 7.99 (*t*, 7.9, H-6′), 7.29 (*t*, 7.9, H-7) 7.20 (*t*, 7.4, H-9′), 7.46, 7.35 (*d*, 7.9 each, H-8 & H-8′)	3.95 (*s*, OMe), 2.5 ($Ar-CH_3$), 6.02 (*s*, CH_2), 6.96 (*d*, 9.9) 5.77 (*d*)	1.38 (*m*, 7, 1.70, 3H) 2.08 (*m*, 2H), 5.19 (*t*, 7) 1.67, 1.59 (*s*, 3H each)
88. Murrastifoline D	8.53	7.56 7.87 (*s*, H-4)	7.88 (*d*, 7.9) 7.98 (*d*, 7.4, H-5)	7.00 (*s*, H-2′), 6.92 (*t*, 7.9, H-6), 6.94 (*t*, 7.9, H-6′) 6.75 (*d*, 7.9, H-8) 6.95–7.05 (*t*), 7.21 (*d*, 9.2)	4.11 (3H), 2.56, 2.43 (*s*, 3H each. $Ar-CH_3$)	1.42 (3H), 4.33 (*dd*, 5.9, 9.2), (3H)
89. Murrastifoline E	8.94	7.58 (*s*)	7.98 (*d*)	7.02 (*s*, H-2), 6.92 (*d*, 8-1), 7.01 (*t*, 8.1, H), 6.89 (*d*, 8.1), 6.97 (*t*, 8.1), 6.73 (*d*, 8.1)	4.11 (OCH_3) 2.56 ($Ar-CH_3$) 2.42 ($Ar-CH_3$)	1.50, 1.8 (2′, $2CH_3$) 2.42 (3H, 3′-CH_3) 4.48 (2″ CH_3)
90. Murrastifoline F	10.7	7.71 (H-4)	7.50 8.22 (H-5 & H-5′)	6.57, 7.19 (H-6 & H-6′) 6.75, 7.05 (H-2 & H-2′) 6.17, 6.74 (H-8 & H-8′) 7.12, 7.21 (H-7 & H-7′)	2.52, 2.05 (*s*, $Ar-CH_3$)	3.37, 4.11 (*s*, 3H each)

substituted carbazolequinones exhibited interesting variations. The methyl signals of **1**a–g appeared at δ15.3 to δ15.7 while those of **6**a–g appeared at δ14.8 to 15.00. The olefinic tertiary carbons (C-2) of **1**a–g resonated at δ131.5–131.7 whereas the C-3 signals in **2**a–g appeared in the region δ134.1–134.9. In murrayaquinone B and C the allylic methyl signals appeared at δ16.1 and δ18.0 and both C-2 signals at δ131.5.

4. *Mass Spectra*

The characteristic mass spectral peak at m/z 248 of the carbazolopyrilium ion (**7**) is consistently observed in the mass spectra of alkaloids containing a unit capable of generating such fragments like eustifoline A (**187**), eustifoline B (**275**) and several other alkaloids. Alkaloids containing the dihydrogirinimbine unit (**8**) have been found to yield a mass spectral peak at m/z 210 (**9**) like that obtained from dihydrogirinimbine itself. Substitution or modification of functional groups in the carbazole skeleton changes the m/z value **11** as shown in the case of hydroxydihydromurrayacine (**10**).

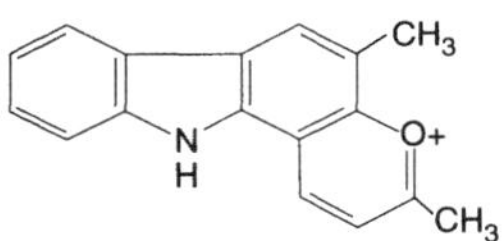

(**7**) carbazolopyrilium ion

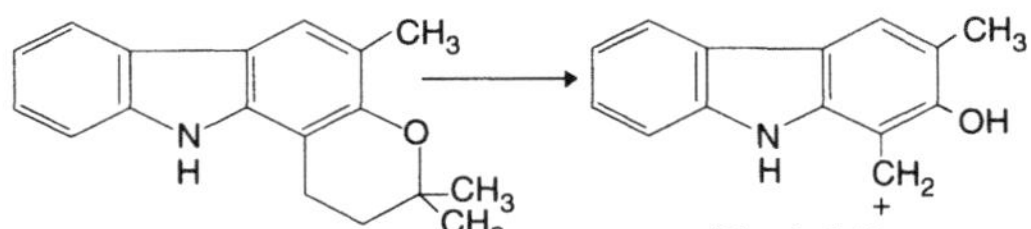

(**8**) (**9**) m/z 210

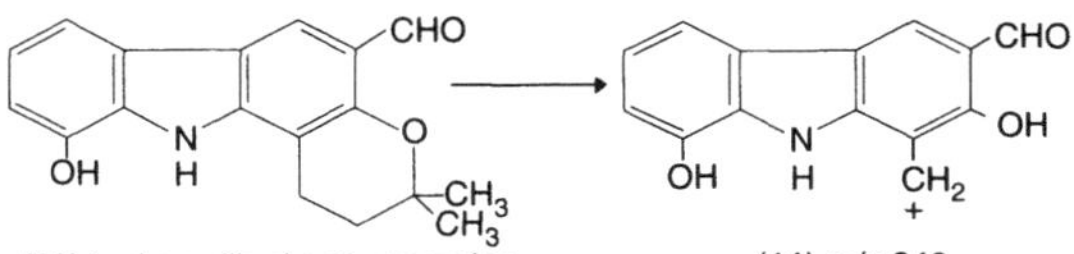

(**10**) hydroxydihydro murrayacine (**11**) m/z 240

The mass spectral peak at m/z 210 has been found to arise from several structures other than the dihydrogirinimbine unit as in the case of (**12**) and (**13**).

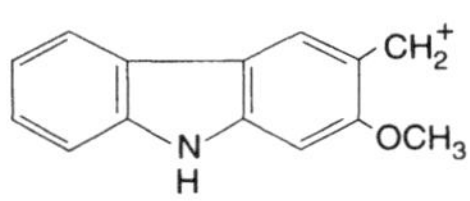

From 2-methoxy-3-methyl carbazole
(**12**) m/z 210

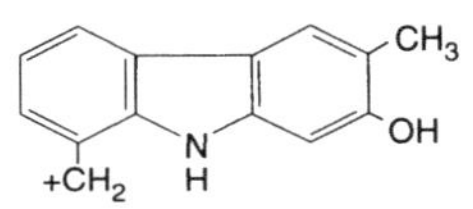

From 8-(2':2' dimethyl)allyl-2-hydroxy-3-methyl carbazole
(**13**) m/z 210

Alkaloids bearing prenyl and geranyl chains substituted in the aromatic ring undergo benzylic cleavage giving rise to fragments C_4H_7 or C_9H_{15} respectively. Thus clauszoline A (**269**) (M^+361) has a base peak at m/z 306 (M^+-55) as a result of cleavage of a C_4H_7 unit from the C_5 side chain while eustifoline D (**276**) (M^+333) showed the base peak at m/z 210 (M^+-123) due to the loss of a C_9H_{15}-unit. From a 2′:2′ substituted pyran system bearing a methyl and an additional C_6-residue as in mahanimbine, the formation of the carbazolopyrilium ion at m/z 248 due to the loss of a C_6-residue (M^+-83) has also been observed.

5. X-ray Crystallography

X-ray crystallographic methods have been used to confirm the structures of several alkaloids. In some instances, the structures were primarily arrived at from X-ray data as in the case of staurosporine.

References, pp. 224–230

B. Chemical Methods

Degradative reactions have been very little used for structure elucidation during this period. Some reactions and transformations utilised for structural studies will, however, be described in discussing the structures of various alkaloids.

C. Synthesis

There has been wider interest in the synthesis of carbazoles and carbazole alkaloids since the last review (*13*). The Borsche method (see below), the Diels-Alder reaction, cycloaddition and electrocyclic reactions, transition metal catalysed reactions, photolytic and thermal methods have been utilised successfully for carbazole alkaloid synthesis.

1. Synthesis from Monocyclic Systems

A. Borsche Method

1) Synthesis of Clausenal

2-Hydroxymethylene-5-methylcyclohexanone (**14**) on condensation with diazotised O-methoxytoludine (**15**) (Chart I) under Japp-Klingemann condition furnished **16** (*9*) which on cyclisation with a mixture of acetic acid and hydrochloric acid (3:1) furnished 1-methoxy-3-methyl-8-oxo-5, 6,7,8-tetrahydrocarbazole (**17**). The latter on dehydrogenation with 10% Pd/C in boiling decalin furnished **18** which on methylation with diazomethane gave 1,8-dimethoxy-3-methylcarbazole (**19**). This on DDQ oxidation furnished 1:8 dimethoxy-3-formyl-carbazole or clausenal (**20**).

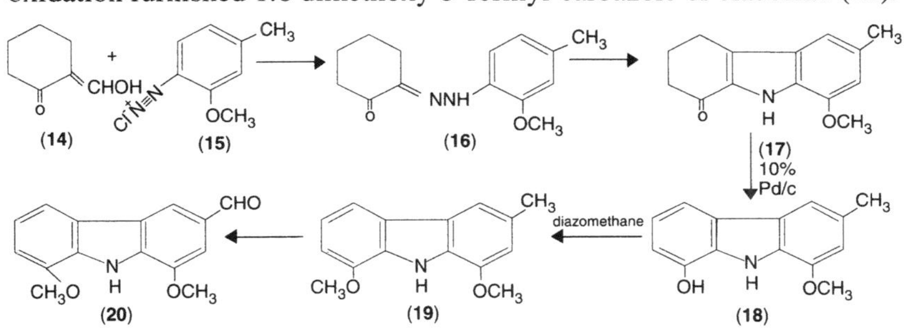

Chart 1. Synthesis of clausenal

2) Synthesis of Clausenol

Starting with **21** and **22**, 1-hydroxy-8-methoxy-3-methylcarbazole identical with clausenol (**25**) was obtained (*8*) through the steps detailed below (Chart 2).

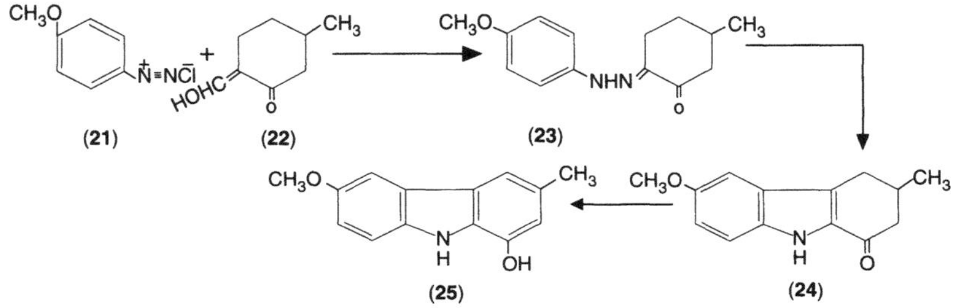

Chart 2. Synthesis of clausenol

B. Synthesis of Murrayaquinone B

Chart 3 illustrates an interesting method utilising a benzene derivative as starting material (*74*). Thus condensation of 4-(1′,1′-dimethylallyloxy)-benzaldehyde (**26**) with methylazidoacetate (**27**) furnished the azidocinnamate **28** which on heating in toluene followed by methylation with methyl iodide gave methyl ether **30**, R=CH_3. The latter on Claisen condensation with 4-methylbutyrolactone gave lactone **31**, converted to alcohol **32** on heating in aqueous dioxane which on oxidation furnished aldehyde **33**. Cyclisation with boron trifluoride etherate and methylation gave 1,7-dimethoxy-3-methyl-8-(3-methylbut-2-enyl)-9-H-carbazole (**34**) which on photoxidation furnished murraya-quinone B (**35**).

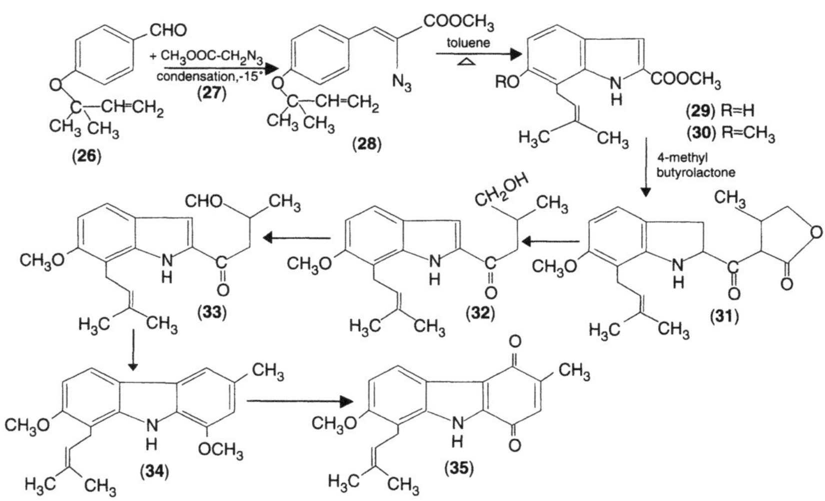

Chart 3. Synthesis of murrayaquinone B

C. Aryl Derivatives

Aryl derivatives have been utilised for the preparation of diphenylamine derivatives which have been used as substrates for carbazole synthesis.

i) Synthesis of Murrayaquinone A

Diphenylamine derivative **37** synthesised by reaction of aniline with *O*-methyl-*p*-quinone (**36**) (Chart 4) on cyclisation with $Pd(OAc)_2$ afforded the carbazoloquinone murrayaquinone A (**38**) (*120*).

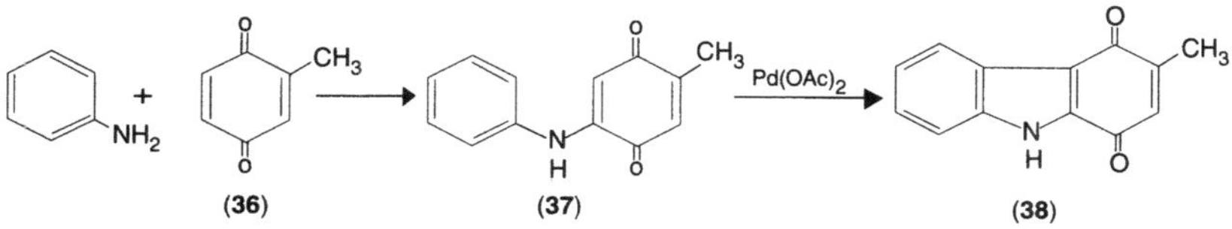

Chart 4. Synthesis of murrayaquinone A

ii) Synthesis of Pyrayaquinone A, B

Both linear (**41**) and angular (**42**) pyranoacetanilides were obtained (*120*) by treatment of **39** with acetylene **40** (Chart 5). Hydrolysis to **43** or **44** and subsequent reactions with *O*-methyl-*p*-benzoquinone (**36**) gave the respective pyranoquinonoid amines **45** or **46** which on cyclisation afforded pyrayaquinone A (**47**) and pyrayaquinone B (**48**).

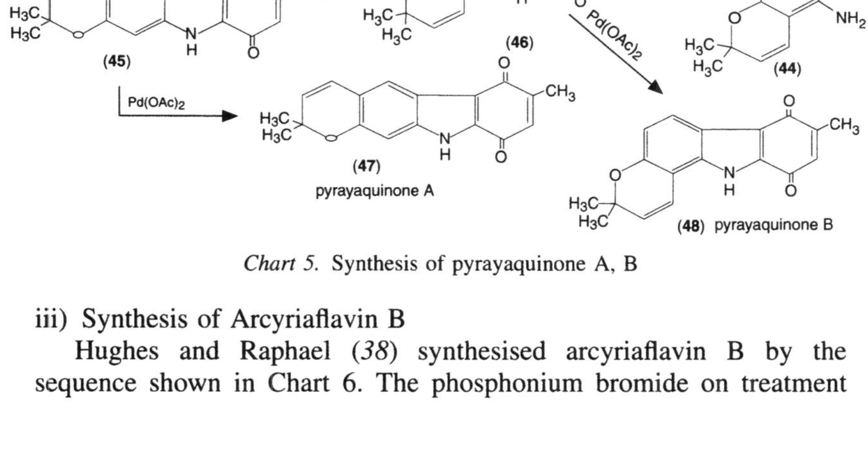

Chart 5. Synthesis of pyrayaquinone A, B

iii) Synthesis of Arcyriaflavin B

Hughes and Raphael (*38*) synthesised arcyriaflavin B by the sequence shown in Chart 6. The phosphonium bromide on treatment with O-nitrosocinnamaldehyde gave a mixture of stereoisomeric dienes which was converted to an E, E-diene **50** on treatment with iodine in toluene at room temperature. Diels-Alder reaction with maleimide furnished **51** which on dehydrogenation with DDQ yielded terphenyl derivative **52**. Reaction of the latter with triphenyl phosphine in refluxing collidine yielded **53** which on demethylation furnished arcyriaflavin B (**54**).

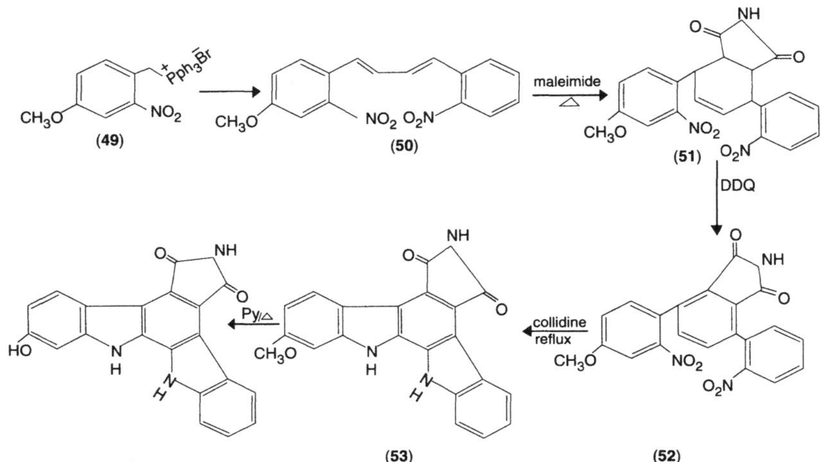

Chart 6. Synthesis of arcyriaflavin B

iv) Transition Metal Catalysed Synthesis

Transition metal mediated syntheses of carbazoles constitute an interesting new area. Some of these syntheses beginning mainly from aryl derivative are presented here. A review of some of this work has been published by Knolker (*63*) whose group has made interesting contributions in this area. For detailed references this review may be consulted.

a) Palladium-Promoted Cyclisation

Diphenylamines have been cyclised to carbazoles using palladium acetate $Pd(OAc)_2$. Generally one equivalent of $Pd(OAc)_2$ is necessary but substrates with electron withdrawing groups like NO_2 or OAc groups require a higher proportion of $Pd(OAc)_2$. The reactions may sometimes be acid catalysed which suggests that electrophilic attack of palladium(II) on the aromatic rings is the rate determining step. Murrayastine (**56**) and related alkaloids have been synthesised by this method (*30*) from diphenyl amines like (**55**).

(55) (56) murrayastine

An iodo-substituted arene (**57**) in the presence of $Pd(OAc)_2$ and triethylamine at 150°C in a stainless steel trough has been (*63*) utilised for the synthesis of carbazole-1-carboxylic acid (73%) (**58**).

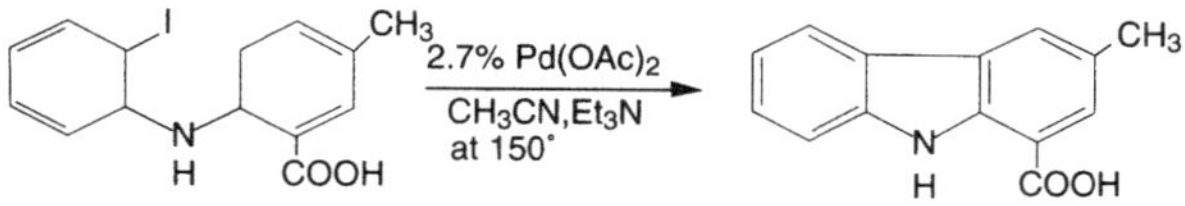

(57) (58)

Cyclic β-diketones such as **60** have been condensed with arylamines **59** to obtain N-aryl-enaminones **61** which with equivalent amount of palladium acetate in acetonitrile (*63*) provide the corresponding carbazole derivative (**62**).

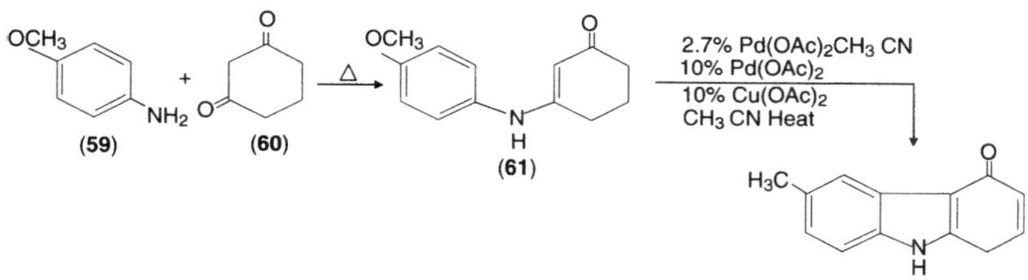

(59) (60) (61) (62)

Transmetallation of arylmetal derivative **63** has been utilised for the synthesis of carbazole **65**. The generation of σ-aryl palladium(II) by oxidative addition is superior as stoichiometric amounts of mercury and palladium are required (*63*).

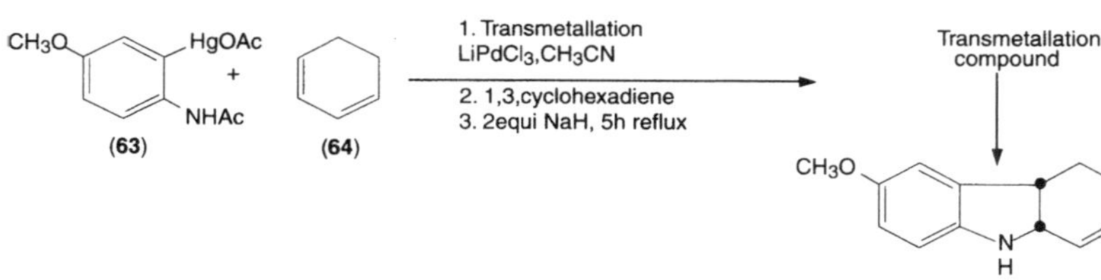

(63) (64) (65)

Heteroannulation of 1,3-dienes by Aryl Halides

Intramolecular nucleophilic attack on an π-allylpalladium complex formed by an aniline leads to heteroannulation of 1,3-dienes, an example being the synthesis of 3,4,4a,9-tetrahydro-9a-H-carbazole by palladium acetate catalysed heteroannulation of 1,3-cyclohexadiene (**64**). Thus the reaction of 2-iodoaniline (**66**) with (**64**) in trimethylamine as solvent in the presence of bis-(triphenylphosphine) palladium-(II)-acetate as catalyst yielded tetrahydrocarbazole (**67**) in (70%) yield. The mechanism of the reaction has been discussed by Knolker.

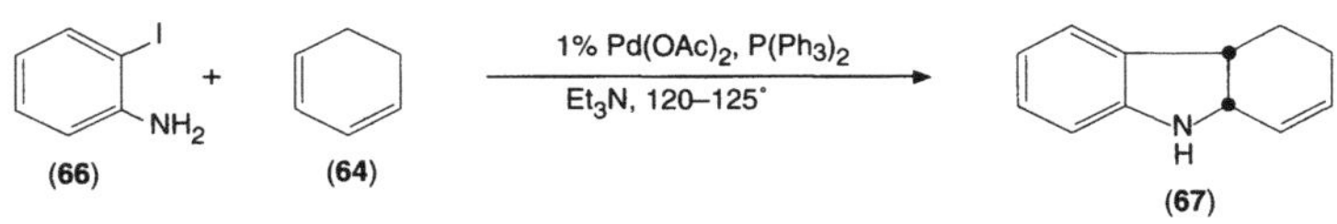

b) Iron Mediated Oxidations

The electrophilic aromatic substitution of arylamines by an iron-complexed cation and subsequent oxidative cyclisation of the iron-diene complex to 9-H-carbazole derivative has been utilised for the synthesis of several alkaloids (*2*). Syntheses of carbazomycin A and B, were described in the last review (*13*). The method has been extended to syntheses of other alkaloids like deoxycarbazomycin, carbazomycin E, mukonine, mukoeic acid and koeniline. Thus iron-complexed cation **68** reacts with 4-methoxy-2,3-dimethylaniline (**69**) to give the ortho-substituted arylamine **70** in a regio- and stereoselective manner (Chart 7). On oxidation with Fatiad's very active MnO_2 in toluene at 25°C the metal-complexed arylamine cyclises directly to 4-deoxycarbazomycin B (**72**).

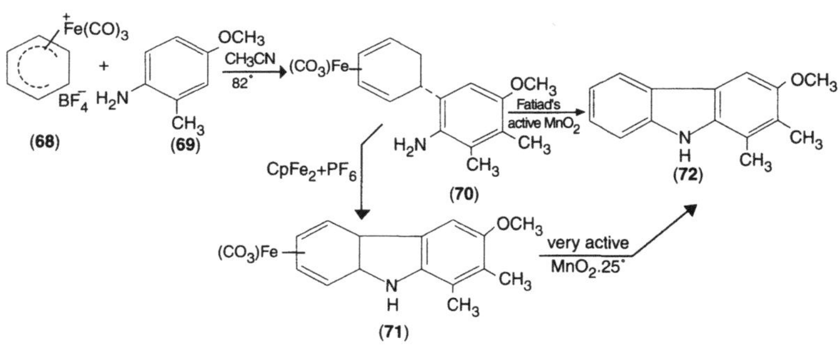

Chart 7. Synthesis of 4-deoxycarbazomycin B through iron-mediated oxidation

The iron mediated oxidation is considered to take place by cyclodehydrogenation to an intermediate 4a,9a-dihydro-9H-carbazole probably via a single electron (SET) oxidation. A twenty electron complex which demetalates instantaneously would be generated by a second dehydrogenation. Isolation of tricarbonyliron complexed 4a,9a-dihydrodihydroxy-9H-carbazole was accomplished by SET oxidation with ferricenium hexafluorophosphate (*63*). The iron complexed carbazole on oxidation with very active MnO_2 at 25°C in toluene gave deoxycarbazomycin B (Chart 7). The method has also been utilised for syntheses of carbazomycin A, B and E (carbazomycinal) (*64*) as well as for syntheses of mukonine, mukoeic acid, murrayanine and koeniline (*65*) using appropriately substituted anilines (Chart 8).

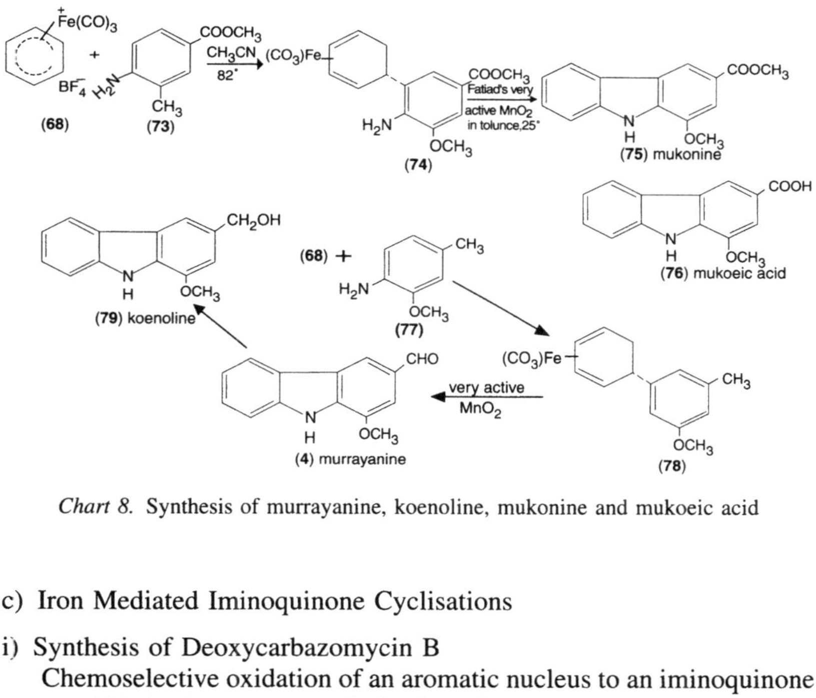

Chart 8. Synthesis of murrayanine, koenoline, mukonine and mukoeic acid

c) Iron Mediated Iminoquinone Cyclisations

i) Synthesis of Deoxycarbazomycin B

Chemoselective oxidation of an aromatic nucleus to an iminoquinone system (*66*) in the presence of tricarbonyliron-cyclohexadiene unit and further cyclization of the iminoquinone iron complex (**70**) is illustrated in Chart 9. The tricarbonyliron complexed 4b,8a-dihydrocarbazole-3-one (**80**) on demetallation with triethylamine N-oxide afforded 3-hydroxy-1,2-dimethylcarbazole (**83**) and on further methylation 4-deoxycarbazomycin (**72**). The method has also been utilised for syntheses of isocarbazomycin B, carbazomycin A (Chart 10), murrafoline A and murrayaquinone A (Chart 11) (*66*).

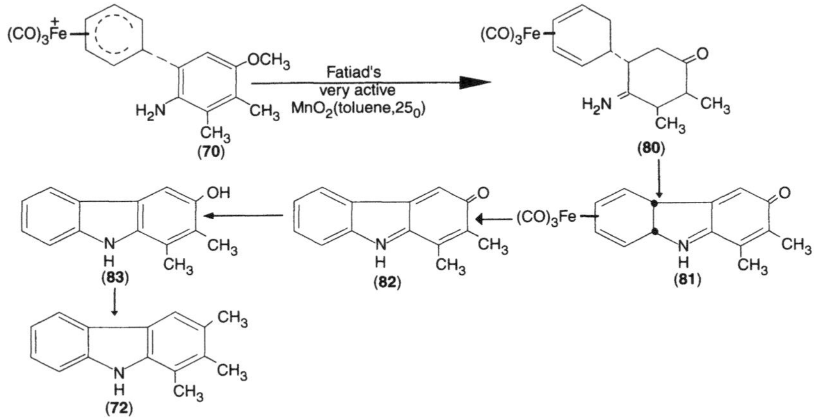

Chart 9. Synthesis of deoxycarbazomycin B

(ii) Synthesis of Isocarbazomycin B (**87**) and Carbazomycin A (**88**)

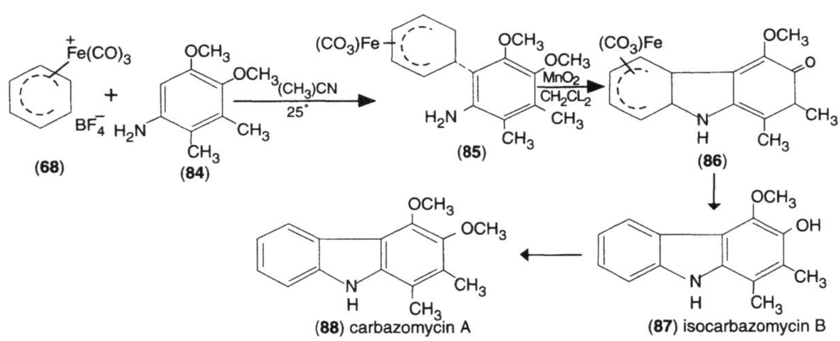

Chart 10. Synthesis of carbazomycin A and isocarbazomycin B

(iii) Synthesis of Murrafoline A (**93**) and Murrayaquinone A (**38**)

The carbonyl reactivity of tricarbonyliron complexed 4b,8a-dihydrocarbazole-3-one for nucleophilic addition has been used to synthesise murrayafoline A and murrayaquinone A. The iron complexed cation reacted with 2,4-dimethoxyaniline (**89**) to yield the complex **90** which after oxidation with MnO_2 and cyclisation with thalium trifluoroacetate yielded 4b,8a-dihydrocarbazole-3-one **91**. 1,2-Addition of methyl lithium at low temperature then gave **92** which on dehydration with *p*-toluenesulphonic acid yielded murrayafoline A (**93**). Demethylation and oxidation with Fermi's salt then gave murrayaquinone A (**38**) (*66*, *28*).

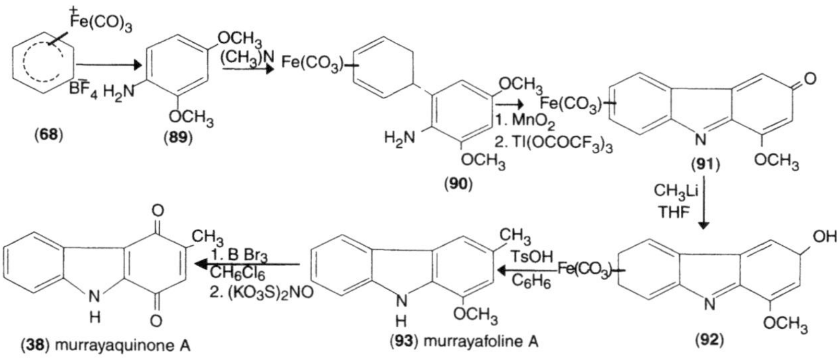

Chart 11. Synthesis of murrafoline A and murrayaquinone A

iv) Synthesis of Mukonine and Mukonidine

Tricarbonyliron mediated syntheses of mukonine and mukonidine were effected by Knolker and Wolpert (*66*). Reaction of the tricarbonyliron complex with arylamine **94** gave complex **95** (Chart 12) which on stirring in toluene and THF in air cyclised to dihydrocarbazole iron complex **96**. Aromatization and concomitant demetallation with ferricenium hexafluorophosphate (**97**) in the presence of sodium bicarbonate gave mukonine (**75**). On the other hand, complex **98** on cyclisation with THF at 25°C yielded dihydrocarbazole which was dehydrogenated to mukonidine (**99**).

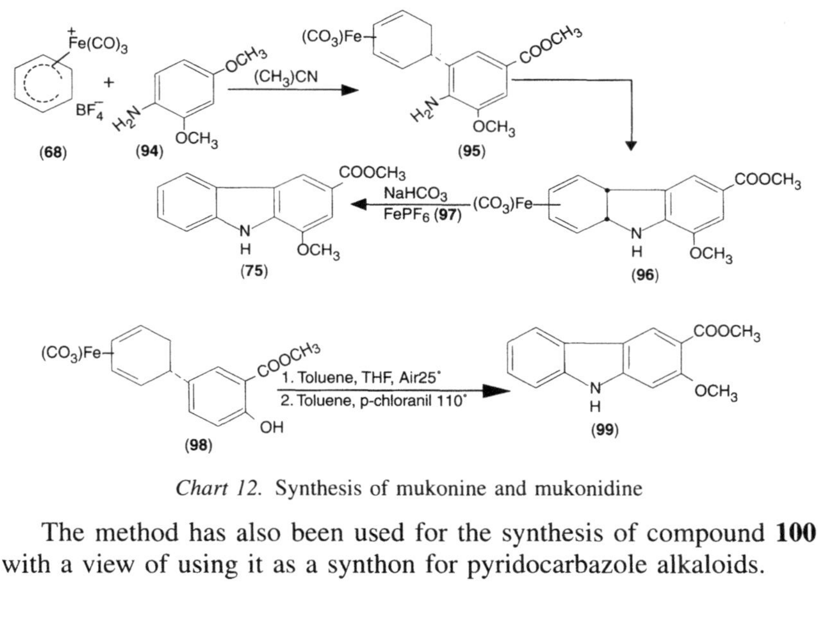

Chart 12. Synthesis of mukonine and mukonidine

The method has also been used for the synthesis of compound **100** with a view of using it as a synthon for pyridocarbazole alkaloids.

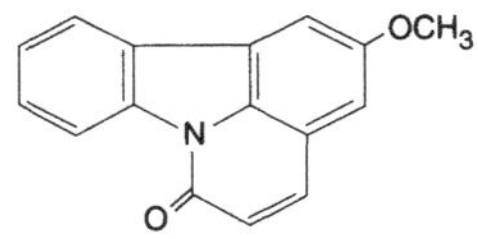

(100)

d) Mercury (II) mediated heterocyclisation of 2(2′-cyclohexenyl)-N-methylaniline (*63*)

2-(2′-Cyclohexenyl)-N-methylaniline (**101**) in alcohol on exposure to mercury(II) acetate yields 1-alkoxy-1,2,3,4-tetrahydrocarbazoles (**102**). The yields depend on the conditions and the reagents used. Thus carbazole was obtained when **102** was dehydrogenated with Pd/C but **103** was formed on dehydrogenation with DDQ.

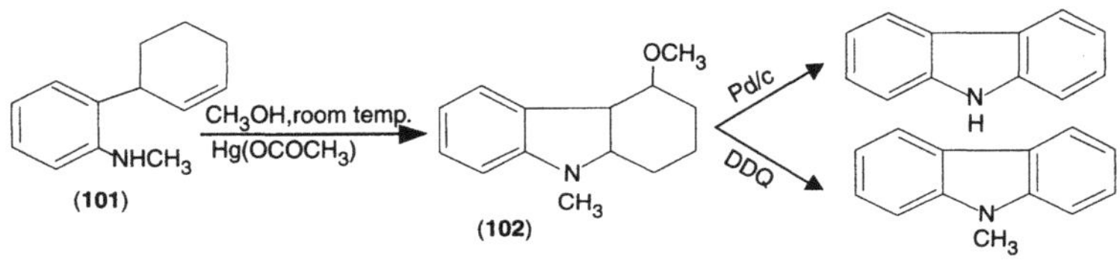

D. Radical Cyclisation: Synthesis of Carbazomycin B

Clive *et al.* (*19*) synthesised carbazomycin B (**105**) by radical cyclisation of the tosylated cyclohexene derivative **104** by deprotonation and dehydrogenation.

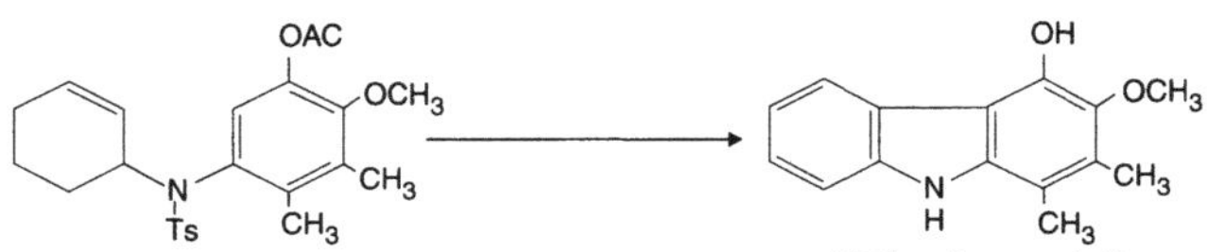

2. Synthesis from Bicyclic Systems

A. From Indole Derivatives

i) Synthesis of Carazostatin

Reaction of indoleacetic acid (**106**) with octanoic anhydride (*50*) furnished 1-heptylpyrano-(3,4-b)-indole-3-one (**107**). Reaction of the latter (Chart 13) with 3-trimethylsilylpropyonate (**109**) yielded carbazole derivative **108** which on reduction furnished carazostatin (**110**).

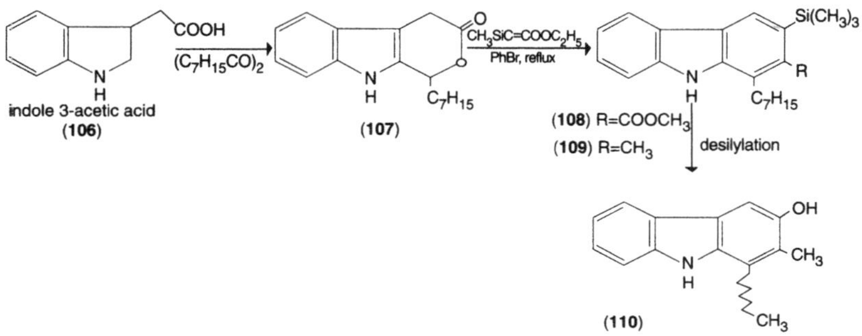

Chart 13. Synthesis of carazostatin

1,2-dialkyl-3-trimethylsilylcarbazole (**108**) also furnished (**110**) on mercurodesilylation followed by hydroboration and oxidation in a reaction sequence used in the case of carbazomycin B and hyellazole (*79*).

ii) Synthesis of Rebeccamycin

a) 7-Chloroindole (**111**) with CH_3MgI and 1-benzyloxymethyl-2,3-dibromomaleimide (**112**) in benzene and a small amount of hexamethylphosphoramide (HMPA) gave the 2:1 adduct **113** and a 1:1 by-product

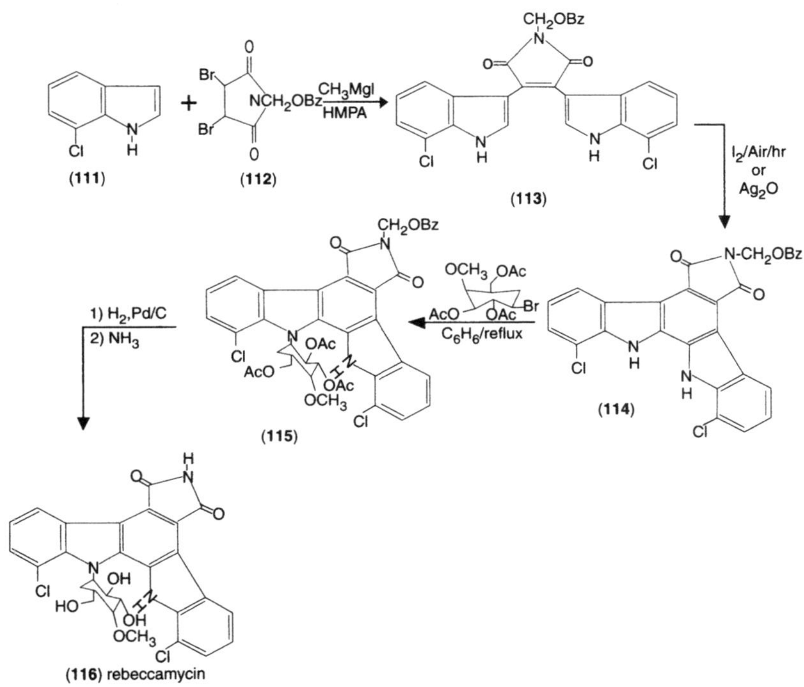

Chart 14. Synthesis of rebeccamycin

(Chart 14). The former on photocyclisation in the presence of iodine gave **114** which on glycosidation with 1-bromo-2,3-tri-O-acetyl-4-O-methyl-glucose in benzene gave **115**. On removal of the benzoyloxy and acetyl groups by hydrogenation and aminolysis respectively, rebeccamycin (**116**) was obtained in 95% yield. The absolute configuration of (**116**) was that of D-glucose (*53*).

In a one pot method, cyclisation and glycosidation in benzene were effected by reacting **111** and **112** in the presence of Ag_2O, probably by thermal cyclisation to **115** of the triene system **113** (*53*).

b) The synthesis of the aglycone of rebeccamycin has been reported through a shorter route as depicted below (Chart 15). 7,7-Dichloro-*bis*-indigo (**117**) on Wolff-Kishner reduction and acetylation furnished the monoacetyl-*bis*-indole **118**. This on heating in a sealed tube with N-benzyloxymethylmaleimide (**119**) at 105°C for 8 days furnished the aglycone of rebeccamycin (**120**) probably through a Diels-Alder reaction followed by the loss of acetic acid and subsequent dehydration (*53*).

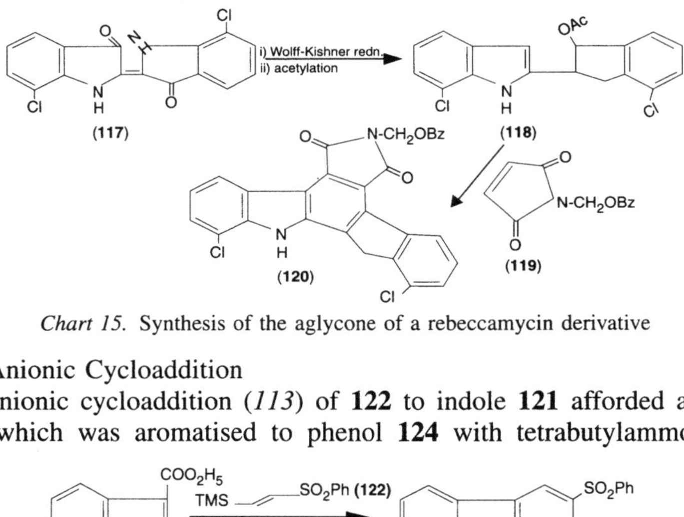

Chart 15. Synthesis of the aglycone of a rebeccamycin derivative

iii) Anionic Cycloaddition

Anionic cycloaddition (*113*) of **122** to indole **121** afforded adduct **123** which was aromatised to phenol **124** with tetrabutylammonium

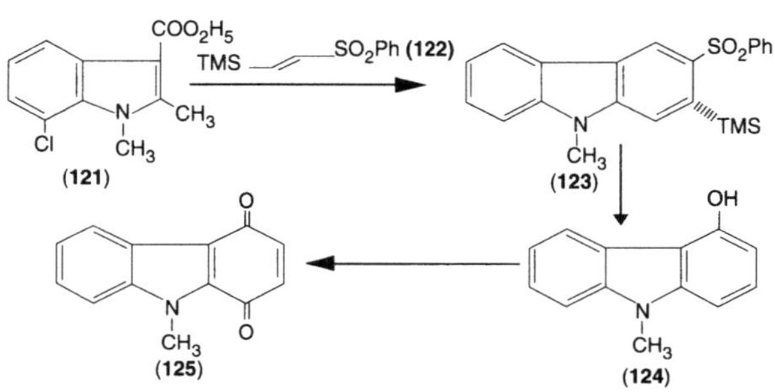

Chart 16. Anionic cycloaddition for synthesis of carbazolequinones

fluoride (TBAF) from which carbazolequinone **125** was obtained by oxidation (Chart 16).

The above method has also been utilised in the synthesis of murrayaquinone A (*113*). Condensation of **126** with **122** in the presence of LDA in THF gave **127** as a single isomer which on methylation with CH_3I in the presence of potassium tertiary butoxide gave two diastereoisomers (**128a**, **128b**) the stereochemistry of which was determined using the nuclear Overhauser data. The mixture on treatment with TBAF in THF afforded a phenol **129** which on treatment with bis-(trifluoroacetoxy)-iodo-benzene yielded the carbazolequinone **130** (Chart 17). Removal of the methoxymethyl group from the indole nitrogen with hydrochloric acid in methanol, furnished murrayaquinone A (**38**).

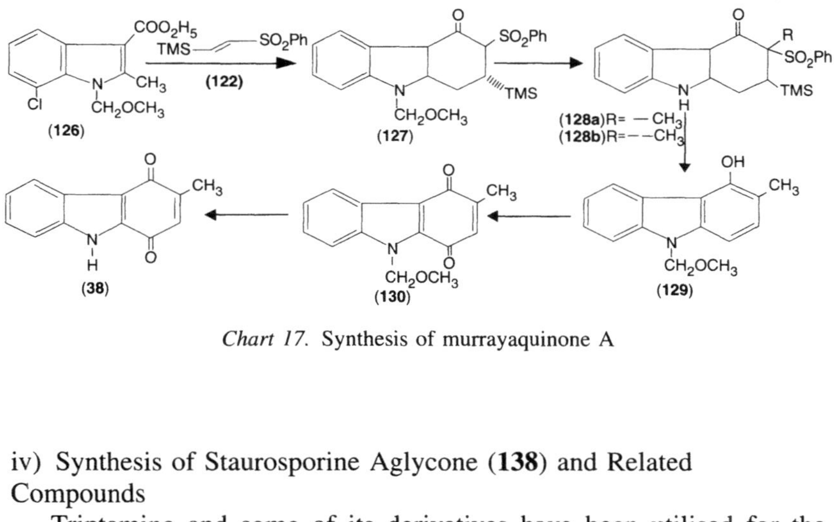

Chart 17. Synthesis of murrayaquinone A

iv) Synthesis of Staurosporine Aglycone (**138**) and Related Compounds

Triptamine and some of its derivatives have been utilised for the synthesis of staurosporine and some related alkaloids.

a) DDQ oxidation of the bis-indole from tryptamine and indole acetylchloride provided the diketocompound **134** which on reduction with $NaBH_4$ and acetylation in the presence of 4-dimethylaminopyridine (DMAP) cyclised to the pentacetate **136** (Chart 18). Deacetylation with $NaHCO_3$ in aqueous methanol and subsequent irradiation provided the aglycone of staurosporine (**138**) (*98*).

b) Substance **138** was also synthesised by Magnus and Sear (*73*) from tryptamine through phthalamide derivative **139** (Chart 19). Formylation of sulphonyl derivative **140** with 2,2-dichloromethylmethyl

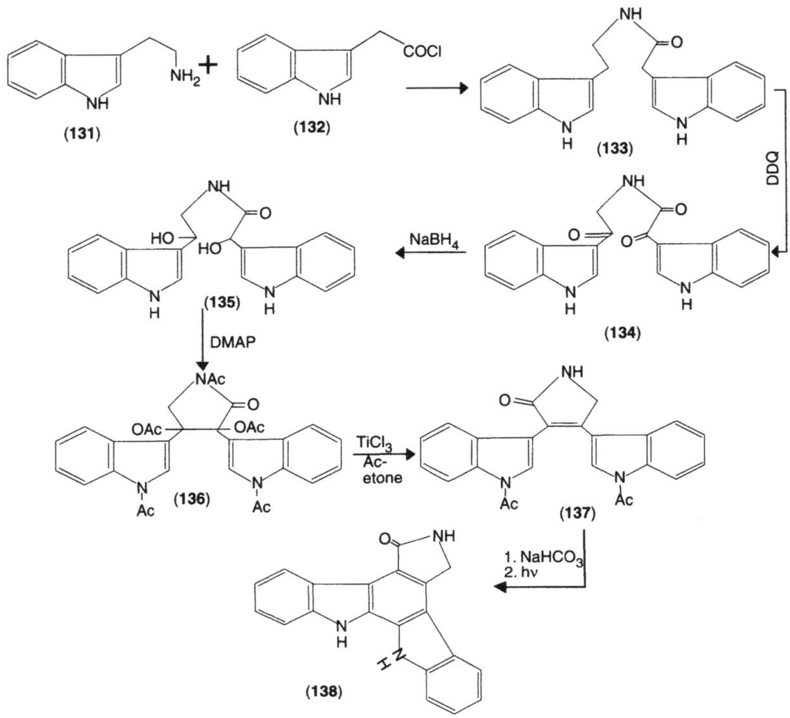

Chart 18. Synthesis of aglycone of staurosporine

ether-titanium tetrachloride at 35° yielded **141** which gave imine **142** on treatment with 2-aminostyrene. The 2,3-quinodimethane intermediate generated in the usual way yielded the pentacyclic carbazole **143** which on dehydrogenation with DDQ in toluene furnished the indolecarbazole **144**. Removal of the phthalimide group with hydrazine hydrate in THF at 20°C gave **145** which on treatment with phosgene in dichloromethane followed by $TiCl_3$ (0°C) furnished the hexacyclic carbazole **146**. On selective removal of the 1-$MeOC_6H_4SO_2$ group using $LiNH_3$-THF **146** gave **147** while treatment with KOH/Glyme, **146** gave **148** thereby offering the potential to attach a carbohydrate substrate regiospecifically.

c) In another sequence (*114*) staurosporine was synthesised as shown in Chart 20. Indolylmagnesium bromide (**149**) on reaction with N-benzylated dibromomaleimide (*112*) gave the bis-indole (**150**) which on DDQ oxidation in the presence of p-toluene-sulphonic acid gave (**151**).

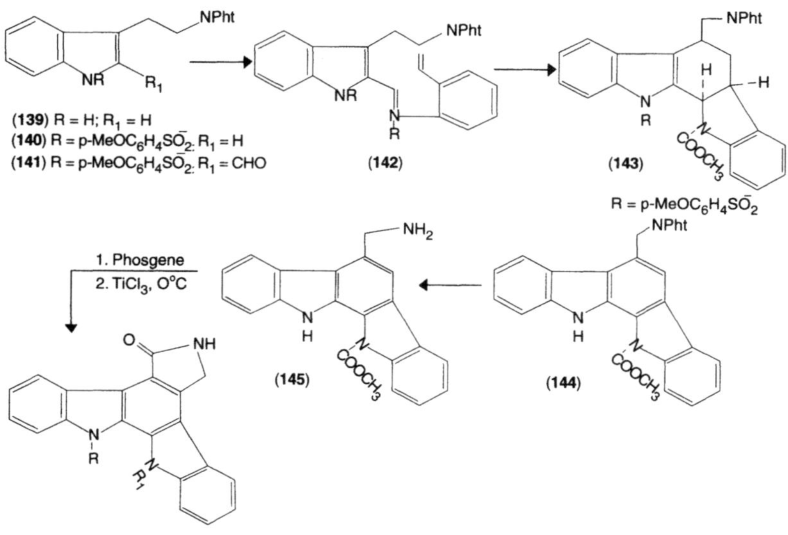

(146) R = p-MeOC$_6$H$_4$SO$_2$; R$_1$ = COOCH$_3$;
(147) R = R=H. R$_1$ = COOCH$_3$
(148) R = p-MeOC$_6$H$_4$SO$_2$; R$_1$ = H

Chart 19. Synthesis of staurosporine aglycone (Magnus and Sear)

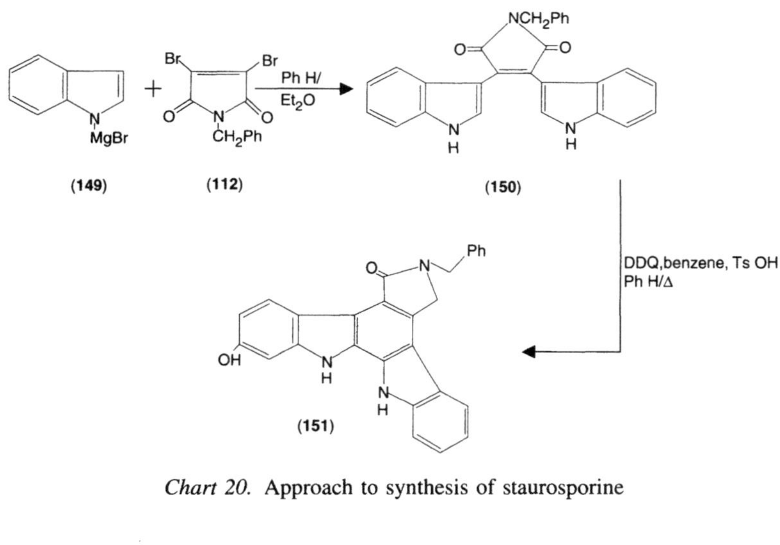

Chart 20. Approach to synthesis of staurosporine

d) Intermediate **151** was also prepared by Hughes and Raphael (*37*) starting with 1,4-Di-(O-nitrophenyl)-butadiene and N-benzylmaleimide to give **152**. Following the same series of reactions as in the case of arcyriaflavin B (see page 147–148) the N-benzylimide **153** was obtained which on Clemmensen reduction gave the lactam as the synthon for staurosporine.

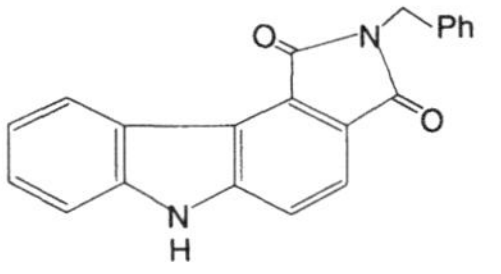

(**152**)

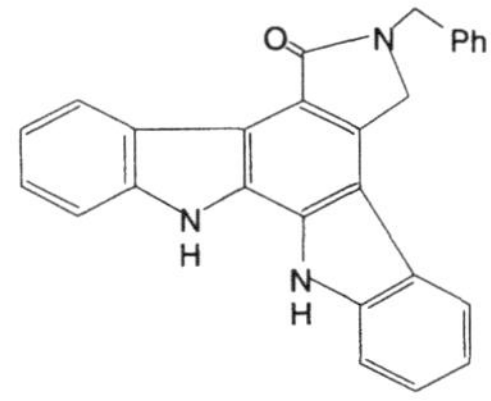

(**153**)

v) Syntheses of Arcyriaflavin B and C

Arcyriaflavin B (**54**) was synthesised (*103*) from the bisindole arcyriarubin B (**154**) by treatment with H_2SO_4 (Chart 21) while arcyriarubin C (**155**) on similar treatment yielded arcyriaflavin C (**156**).

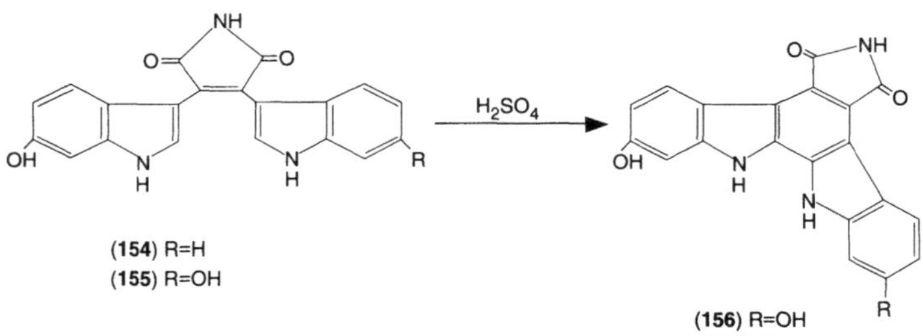

Chart 21. Synthesis of arcyriaflavin B and C

vi) Synthesis of Pyrrolocarbazole

a) Synthesis using the Diels-Alder Reaction

N-Sulphonylvinylindole (**157**) in a thermal Diels-Alder reaction with N-phenylmaleimide (**158**) in xylene furnished the pyrrolocarbazole derivative **159** (Chart 22). The reaction involves (4 + 2) cycloaddition followed by a thermally induced (1,3,-H) shift (*99*).

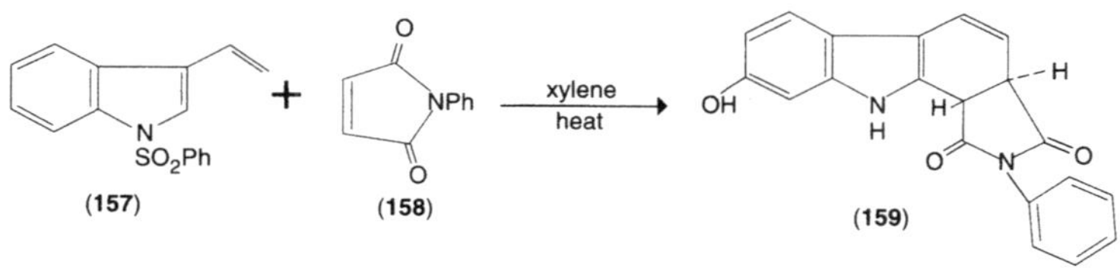

Chart 22. Synthesis of (3,4a)-pyrrolo-1,2-tetrahydro carbazole

b) A useful synthesis of thiosubstituted pyrrolocarbazole (**161**) has been effected by reacting 1-aminomethyl-2-[2′-(3H)-thionylvinyl]-indole (**160**) with maleimide (*3*).

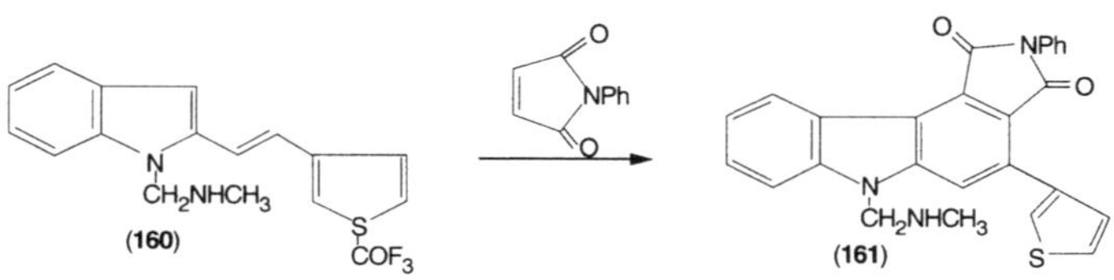

Some pyrrolocarbazoles are promising as antitumor agents and were effective against human breast cancer cells, epidermal carcinoma cells and melanoma cells with little toxicity against normal cells. Thus analogues **162–165** of 8,10-dimethoxyellipticine have been synthesised by this method as shown below.

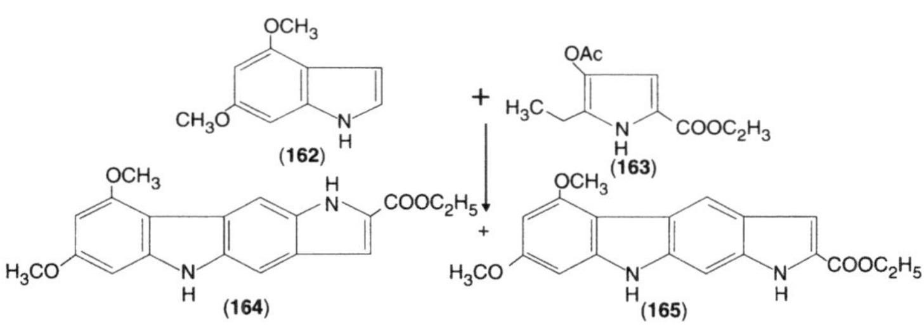

c) Shanon *et al.* (*15*, *16*) reported a novel synthesis of pyrrolo-(3,2-b)-carbazole from **166** and **167** using montmorillonite MK 10 clay (Chart 23).

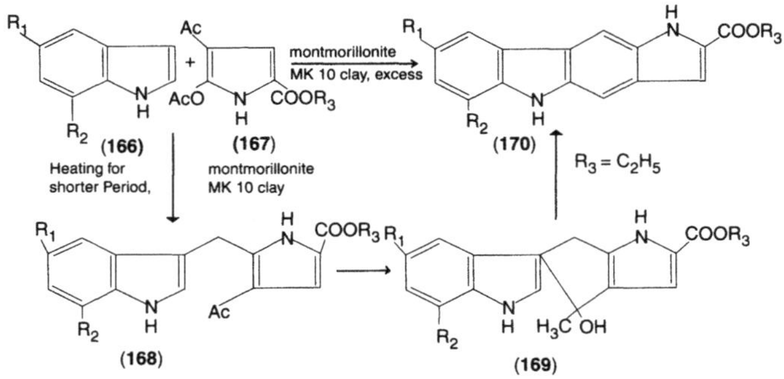

Chart 23. Shanon's synthesis of pyrrolocarbazole

Heating of the starting materials for a shorter period furnished intermediates **168** and **169**. Presence of an electron donating group in the indole as in **170A** led to formation of increasing proportions of [2, 3b]-type pyrrolocarbazoles, since a 4/6-methyl or methoxy group enhances electrophilic substitution at C-2, thus increasing the relative proportion of (2,3b)-isomers while N-methyl substitution in the indole induces electrophilic substitution at C-3. With halogen substitution at 5 or 6, (3,2b)-pyrrolocarbazoles are formed.

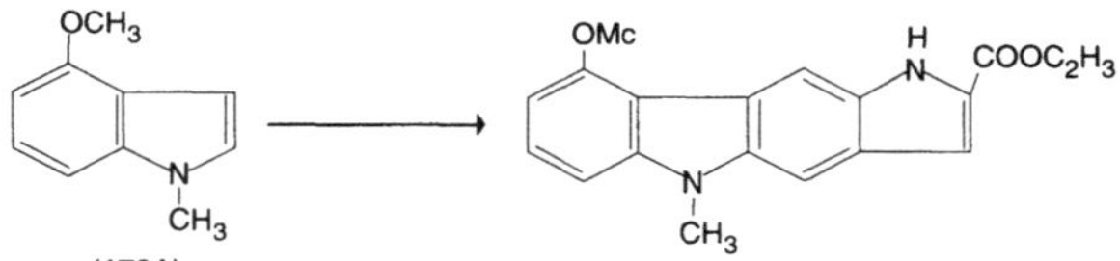

B. From Bicyclic System other than Indole

Pyrayafoline B was synthesised (*31*) starting from 7-acetylamine-chromene **171** and 4-bromo-2-methoxytoluene **172** as illustrated in Chart 24. On refluxing these two constituents in the presence of copper, pyridine and K_2CO_3 for 43 hrs, the N-acetyldiphenylamine derivative **173** was obtained. Cyclisation of the deacetylated diphenyl-amine (**174**) obtained after hydrolysis with $Pd(OAc)_2$ in DMF furnished O-methylpyrayafoline B **175** whose demethylation afforded pyrayafoline B (**176**).

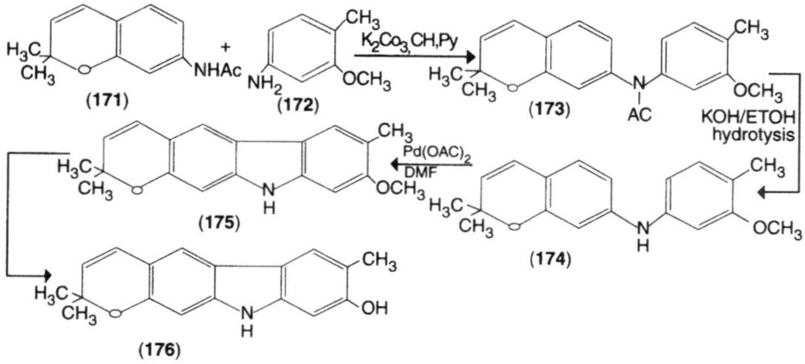

Chart 24. Synthesis of pyrayafoline B

3. Synthesis from Tricyclic Systems

i) Synthesis of Murrayaquinone B

Kapil and Ramesh (*92*) synthesised murrayaquinone B (**35**) starting from 7-hydroxy-3-methyl-1-oxo-1,2,3,4-tetrahydrocarbazole (**177**) (Chart 25). Its acetoxy derivative **178** on aromatisation with Pd/C furnished 1-acetoxy-3-methyl-7-hydroxycarbazole (**179**) as one of the products. Condensation of the latter with 2-methyl-3-butene-2-ol gave 1-acetoxy-7-hydroxy-3-methyl-8-(3′-methylbut-2-enyl)-carbazole (**180**) and a further methylated product **181** which on hydrolysis gave murrafoline B (**34**). Murrayafoline B on oxidation with pyridinium dichlorodichromate (PCC) furnished murrayaquinone B (**35**).

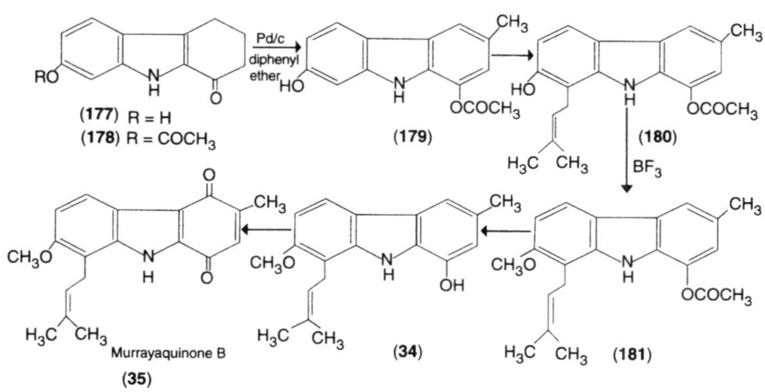

Chart 25. Synthesis of murrayaquinone B

ii) Synthesis of Bisalkaloids, Murrafolines D, G, H

Furukawa *et al.* (*29*) reported synthesis of the bisalkaloids (Chart 26) murrafoline D, G and H by refluxing girinimbine (**182**) and murrafoline A or 1-methoxy-3-methylcarbazole (**93**) with Nafion 117 which resulted in the formation of murrafolines D, G and H **183–185** (Chart 26).

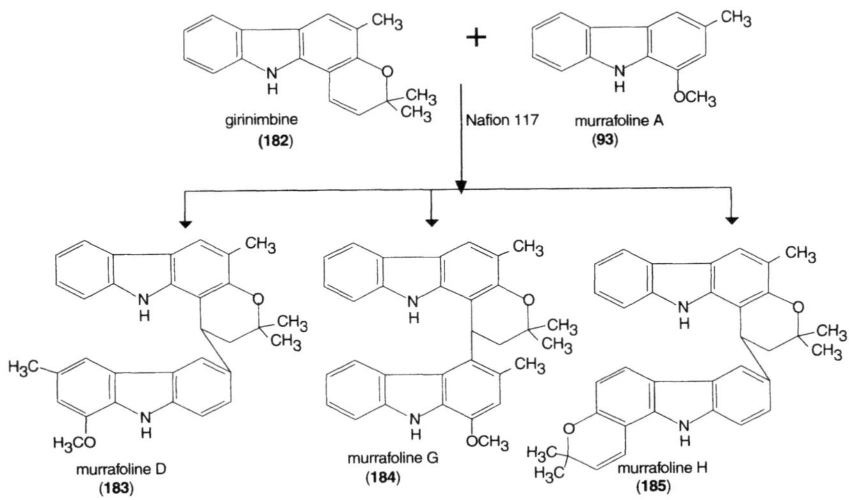

Chart 26. Synthesis of murrafoline D, G, H

iii) Synthesis of Glycomaurine

Glycozoline (6-hydroxy-3-methylcarbazole) (**186**) when condensed with 3-chloro-3-methylbutyne in the presence of the phase transfer catalyst tetrabutylammonium bromide (Chart 27) furnished glycomaurine (**187**) and its isomer **188** (*68*).

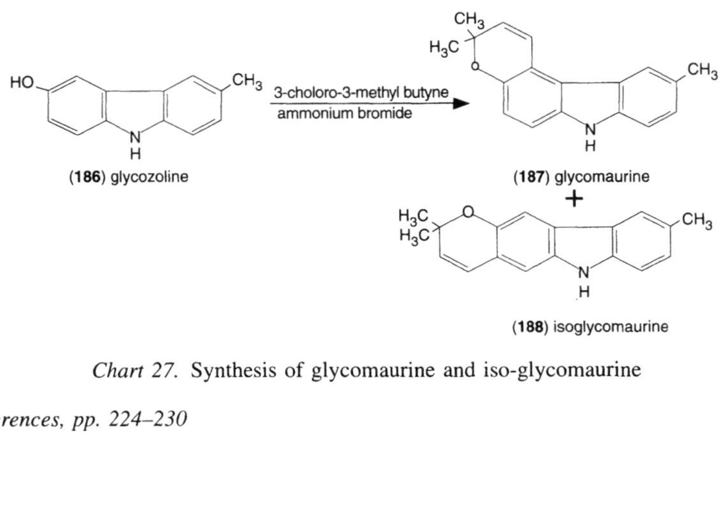

Chart 27. Synthesis of glycomaurine and iso-glycomaurine

4. Synthesis of Carbazoles by Electrocyclisation Methods

Kawasaki and Sakamoto (*59*) have presented a comprehensive review of electrocyclisation methods for carbazole synthesis. Such syntheses have been effected by use of either thermal or photoinduced 6π-electrocyclisation of indolylhexatrienes and subsequent aromatisation. Some of these syntheses are described below.

a) Hyellazole (**191**) and its chloroanalog **192** have been synthesised (*55*) by heating 2,3-divinylindoles **189** and **190** with Pd/C in decalin as shown below. On the other hand 2,3-vinylindole (**193**) (Chart 28) on heating in O-dichlorobenzene in the presence of Pd/C or DDQ afforded carbazoles **195** and **196** the second of which on reduction gave deoxycarbazomycin (**72**) (*36*).

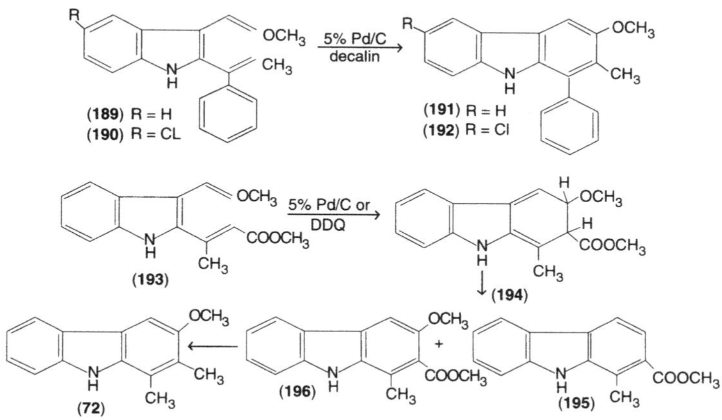

Chart 28. Synthesis of hyellazole, chlorohyellazole and deoxycarbazomycin

b) Starting with a 2,3-divinyl-indole containing an electron withdrawing group in the vinyl fragment, Pindur and Adam (*90*) obtained 1,3-dihydrocarbazoles by cyclisation and a 1,5-hydrogen shift (Chart 29). The dihydrocarbazoles on dehydrogenation with Pd/C furnished the appropriate carbazoles.

Thermolysis of **197** in the presence of Pd/C produced carbazole **199** directly. Electron withdrawing groups in the vinylindole promote the cyclisation as compared with vinylindoles containing electron donating groups.

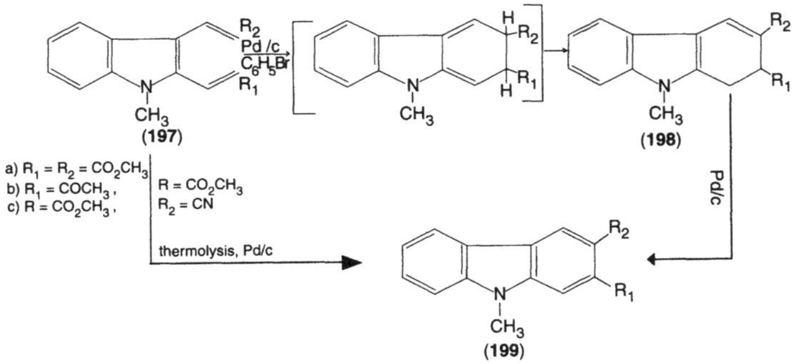

Chart 29. Synthesis of 1,2-dihydrocarbazole and substituted carbazoles

c) The benzocarbazoles **202** were obtained (Chart 30) through cyclisation of indole 2,3-quinodimethanes (**201**) prepared by flash pyrolysis (400–410°) of 3-ethyl-2-vinylindole (**200**) in five minutes (*54*).

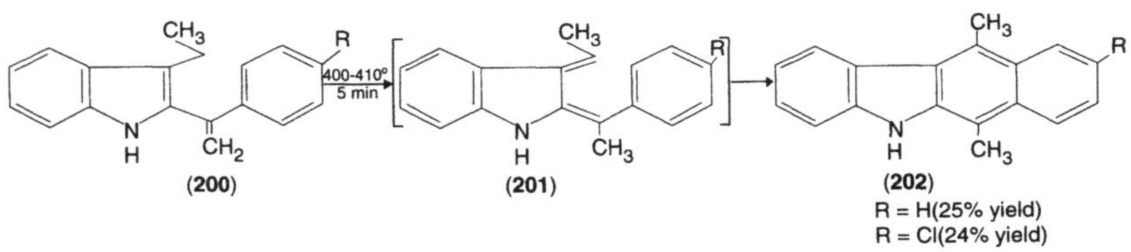

Chart 30. Synthesis of carbazoles involving flash pyrolysis

d) Acid-catalysed electrocyclisation of bis-indoles has been utilised for the synthesis of arcyriaflavin C (**156**) from arcyriarubin C (**155**) (see page 160). Several other indolocarbazoles have been synthesised by this method (*98*).

5. Synthesis by Photolytic Methods

Several syntheses of carbazoles have been effected using photolytic methods, some of which are presented here.

a) An annulation method for regioselective synthesis of aromatic compounds has been applied by Danheiser (*20*) to the synthesis of hyellazoles (Chart 31). Irradiation of indolyl-diazoketone **204**, obtained from N-substituted indolylketone **203**, in the presence of acetylenes **205**, at room temperature and subsequent heating gave carbazole **209** through the intermediates **206**, **207** and **208**. Carbazole **209** was converted into hyellazole **191** by the cross coupling reaction of the triflate of **209** with phenyltrimethylstannane.

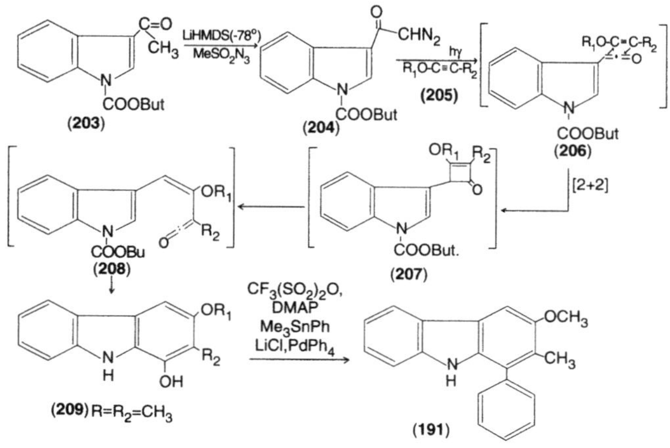

Chart 31. Photochemical synthesis of hyellazole and related compounds

b) Synthesis of (a)-annulated Carbazoles by Photolytic Method Benzo-(a)-carbazoles of type **211** have been synthesised from 3-styrylindoles **210** (a–h) by irradiation in the presence of iodine when isomerisation, 6π electrocyclisation and dehydrogenation led to the formation of the products (Chart 32). For this type of photocyclisation reaction the presence of an electron withdrawing group in the olefinic fragment is desirable (*5*, *21*).

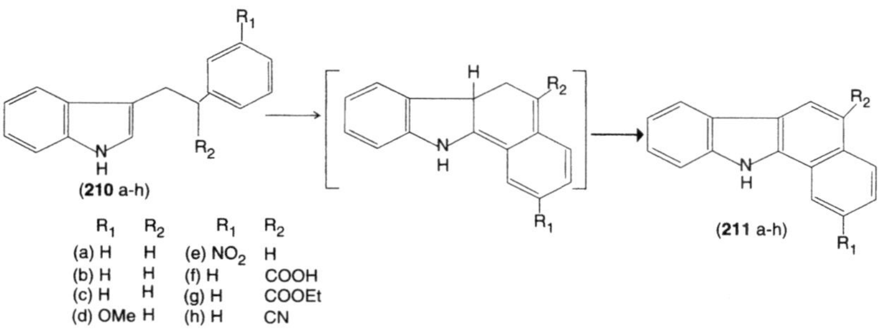

Chart 32. Synthesis of benzo-(a)-carbazoles

c) Bioactive carbazole alkaloids like staurosporine, rebeccamycin and arcyriaflavin which contain the indole (2, 3a)-carbazole skeleton have been synthesised by thermal and photoinduced electrocyclic reactions (*38*, *53*, *98*).

d) Oxidative irradiation of 2-styrylindole **212** (Chart 33) yielded benzo-(c)-carbazole (**214**) (*21*).

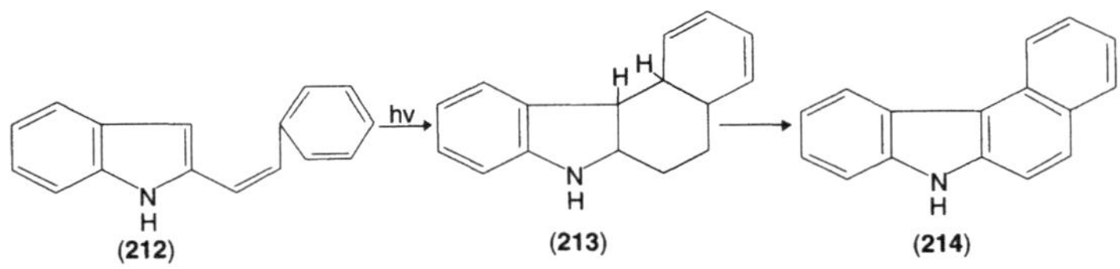

Chart 33. Benzo-(c)-carbazole from 2-styrylindole

Photocyclisation (*39*) of pyridoethylenes with a cyano group at the alpha position as in **215** resulted in the formation of a cyanocarbazole **216** due to opening of the pyridine ring in initially formed product (Chart 34).

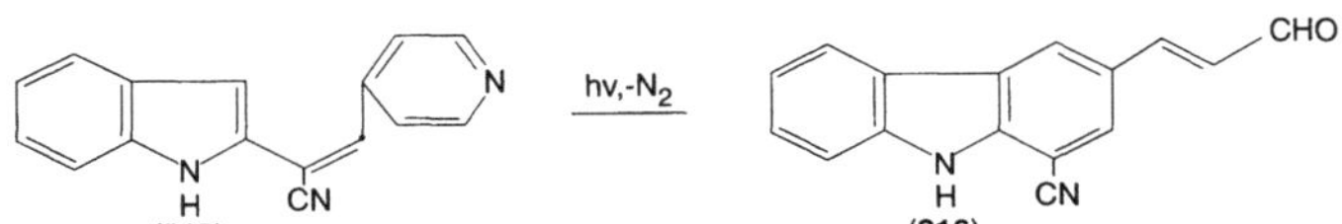

Chart 34

III. Biogenesis of Carbazole Alkaloids

Previously it was suggested (*56*) that 3-methylcarbazole is the key compound from which carbazole alkaloids originate in higher plants. On the other hand, 2-methylcarbazole has been considered to be the precursor of carbazoles found in some lower plants (*13*). Both 2-prenylated and 3-prenylated indoles were considered to be precursors of 2-methyl or 3-methylcarbazoles (*13*). As early as 1943, after structure elucidation of echinulin (*37*), it was thought that the primary electrophilic site for prenylation of indoles was the 3-position of indole nucleus and that the prenyl unit may subsequently migrate to some other positions. In fact, Kapil (*56*) as well as Erdtman (*24*) suggested that 3-methylcarbazole arises from 3-prenylated indole via 2-prenylated indole. In 1990, isolation from *Murraya gleni* by Kumar *et al.* (*67*), a dimer (**217**) containing both a 2-methyltetrahydrocarbazole unit and a 3-prehylated unit, provided the first circumstantial evidence for the formation of a 2-methylcarbazole from a 3-prenylated indole while somewhat earlier paniculadin A (**218**) was isolated in 1985 from the carbazole alkaloid bearing plant *Murraya paniculate* (*61*). These facts reinforced to the concept that 2-methyl- or 3-methylcarbazole are precursors of carbazole alkaloids of some lower or higher plants. The occurrence of a 3-polyprenylated indole like nominine (**219**) in *Aspergillus tubingensis* which also contains the carbazole alkaloids tubingensin A, B (**220**, **221**) and aflavazole (**222**) points out further that prenylated or polyprenylated indoles are precursors of carbazole alkaloids (*109*, *110*, *111*).

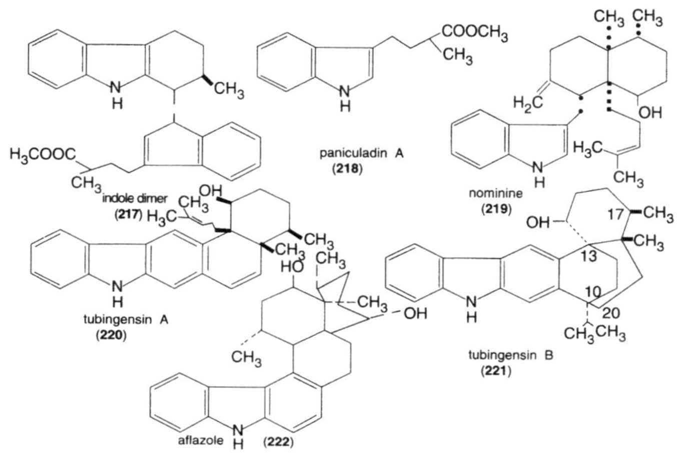

Biosynthetic studies on carbazole alkaloids of lower plants have led to a different view (*13*) about the genesis of carbazole alkaloids in these. In such studies on the carbazomycins, Nakamura *et al.* (*52*) showed that tryptophan (**223**) is the precursor of the carbazole skeleton of carbazomycin B (**105**) while the methyl groups at C-1 and C-2 derive from a pyruvate unit (*62*). Biosynthetic studies on staurosporine (**224**) and on rebeccamycin (**116**) have revealed that tryptophan is the precursor of these alkaloids. In the case of rebeccamycin (**116**) two molecules of tryptophan, one of glucose and one of methionins have been shown to be involved (*88*) while Meksuriyen and Cordell (*76*) have shown that the aglycone of staurosporine (*138*) arises from two molecules of tryptophan. Tryptophan however, does not contribute to formation of the phthalimide fragment of staurosporine (**224**).

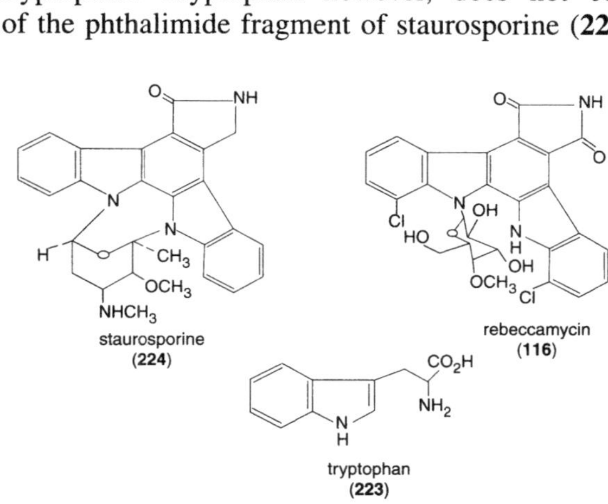

It therefore seems that the biosyntheses of carbazole alkaloids isolated from different types of sources follows different paths.

IV. Biological and Therapeutic Properties of Carbazoles and Carbazole Alkaloids

Various biological properties and therapeutic applications of some carbazoles and carbazole alkaloids have been reported in previous reviews (*1*, *10*, *11*, *13*). Biological and therapeutic properties reported most recently are described below:

a) Antibiotic Activity

Clausenol has been found to be highly active against both gram positive and gram negative bacteria as well as against fungi. Clausenine, on the other hand, had poor activity against gram negative bacteria like *E. coli*, *B. subtilis*, *Salmonella typhimurium*, *Pseudomonas aeruginosa*, *Staphylococcus aureus* and fungi like *Candida albicans* and *Trichophyton rubrum* (*7*, *8*). Clausenal was found to be active against both gram positive and gram negative bacteria as well as against the fungi mentioned above (*9*). AT 2433-A1, AT 2433-A2, AT 2433-B1, and AT 2433-B2 have rebeccamycin-like activity (*4*). They were also active (*75*) against organisms like *Micrococcus luteus* (ATCC 9341), *B. subtilis* (ATCC 6633), *Staphylococcus aureus* (A 9537), *Streptococcus faecalis* (A 20688) and *Streptococcus facium* (ATCC). AT 2433-A1 is at least 8 times more active than AT 2433-B1. AT 2433-B1 is also active against the gram-negative *E. coli* (SS 1431).

The above compounds were also active against the fungi *C. albicans*, *C. krusai*, *C. tropicalis*, *Trichophyton tonsuram*, *T. rubrum*, *Microsporum gypseum and M. catis*. The indolocarbazole staurosporine had substantial activity against *Candida albicans* (6.25), *Candida pseudotropicalis* (3.13), *Saccharomyces sake* (3.13), *Aspergillus brevipes* (3.13), *Trichophyton rubrum* (6.25), *Sclerotinia cinerea* (0.78) and *Piricularia oryzae* (0.78) at the MIC (mg/ml) listed in brackets against each microorganism (*106*). Alkaloid SF 2370 (identical with K-252a) was found active against *Micrococcus luteus*, *Micrococcus flavus* FDA 16, *Corynebacterium bovis* 1810 at a concentration of 6.25 mg/ml (*101*).

b) Antitumor and Tumor-Promoting Activities

9-Formyl-3-methylcarbazole and carbazole showed weak inhibitory activity against the growth of mouse melanoma (*7*) B-16 cells and adriamycine resistent P388 mouse leukemia cell lines at a dose of 25 µg/ml. In the brine shrimp assay 7-methoxymukonal (**244**) was more active than 7-methoxyheptaphylline (**260**) whereas against the NT-29 cell lines, 7-methoxyheptaphylline was more active than 7-methoxymukonal (*6*).

N-Methylcarbazole (**103**) is a carcinogen and is a mutagenic constituent of tobacco smoke. The alkaloid is converted by the primary culture of rat hepatocytes to N-hydroxymethylcarbazole (**225**). The hydroxylation of N-methylcarbazole is likely to be considered a toxicant step (*117*) whereas its dealkylation to carbazole (**1**) is likely to be a detoxication process. 3-Hydroxymethyl-1-methoxycarbazole (koeniline) (**79**) has been shown to cytotoxic (*25*).

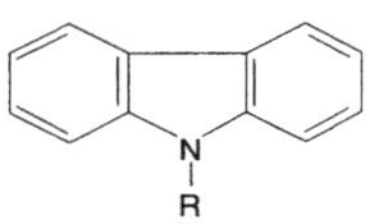

(**103**) R=CH_3
(**225**)

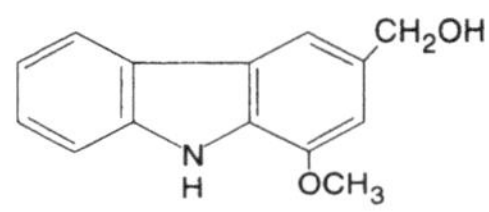

(**79**)

Antibiotics AT 2433-A1 and AT 2433-B1 were found to be effective in prolonging the life spans of mice transplanted with leukemia P 238 tumor. Rebeccamycin (**116**) has been found to have antitumor activity. Growth inhibition of mammary carcinoma by **226** has been observed at a concentration of 1 µg/ml (*93*).

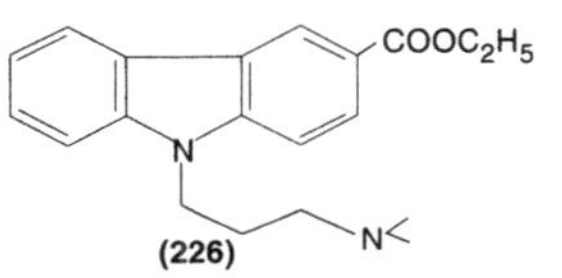

(**226**)

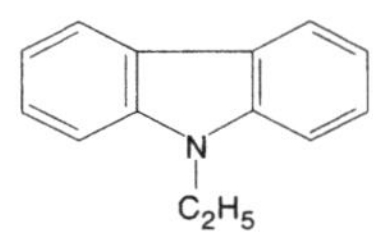

(**227**)

Some pyrrolocarbazoles are promising as antitumor agents. They are effective against human breast cancer cells, epidermal carcinoma cells and melanoma cells with little toxicity against normal cells. Staurosporine (**224**) showed in vitro neoplastic activities against cancer cells (*81*) and a strong cytotoxic effect on the growth of Hela S3 cells (*106*) with an IC_{50} value × 10^{-12} M. Staurosporine is also cytotoxic against NB-1 cells. It induces elongation of neurites and cell enlargement (*80*). 9-Methyl- and 9-ethylcarbazoles are mutagenic towards several strains of *S. typhimurium* though carbazole is not mutagenic (*13*). 9-Ethylcarbazole (**227**) induced UDS in rat hepatocytes while the 9-methyl analog did not. None of these carbazoles produced a tumerogenic response in new born CD-1 mice at a total dose level of 1.75 µg/ml. The potential bifunctional DNA intercalating property of (**228**) has been reported (*89*).

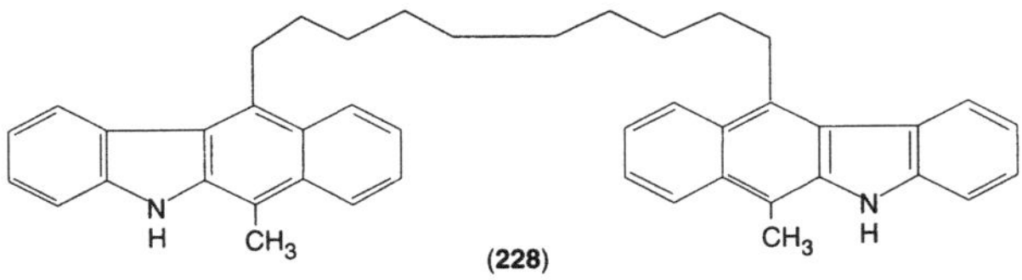

(228)

c) Cardiovascular Modulating Activity

Murrayaquinone A (*38*) has been found to produce a triphasic inotropic response (*105*). This triphasic pattern of inotropism was unaffected by reserpine, metoprolol or cimitidine treatment and is mediated through a mechanism involving ATP production. Arachidonic acid induced sudden death in rabits was counteracted after Bay U 3405 (**229**) treatment. Evidently it was a thromboxane A2 antagonist with antithrombolic activity. Murrayazolinine (**230**) caused sudden lowering of blood pressure in experimental subjects (*77*). The action of murrayazolinine is probably not mediated through mascurine H_1 or H_2 receptors. Staurosporine has been found to have hypotensive activity. Protein kinase inhibitors K-252b and K-252c seriously affect the functions of platelete, most cells and several other cells and tissues (*84*).

Bay U 3405 may represent a novel therapeutic approach to the treatment of ischemia reperfusion injury.

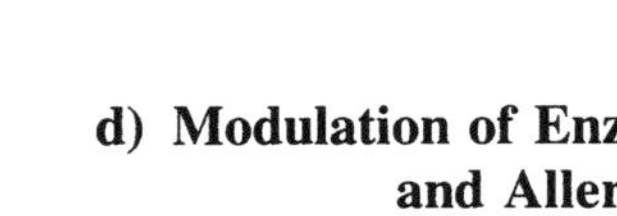

(229)

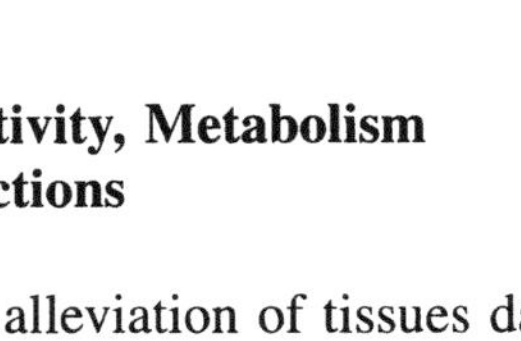

(230)

d) Modulation of Enzyme Activity, Metabolism and Allergic Reactions

Carazostatin (**110**) may be helpful for alleviation of tissues damage due to the action of superoxide radicals and subsequent peroxidative disintegration of cell membranes. The inhibition of lipid peroxidation induced by free radicals generated by carazostatin is higher than that with the brain protective agent lunarizine which has free radical scavenging activity (*50*, *58*). As an oxidant it is better than butylated hydroxytoluene. Carbazomycine B and C were found to inhibit lipoxygenase activity (*118*) which could be related to their free radical scavenging effects (*118*).

Alkaloids of the indolocarbazole group have the general property of inhibiting the activity of protein kinase C and their structural feature is considered to be the requirement for such activity. The carbazole alkaloid K-252-a, -b, -c, -d, UCN-01, Tan-999, Tan-1030 A, and staurosporine inhibit protein kinase-C activity. K-252-b appeared to lose the calmodulin selectivity. Not only Ca^{2+}/calmodulin dependent activity but also the basal activity of Ca^{2+}/calmodulin-dependent phosphodiesterase activity was affected. Tan-999, Tan-1030 A, and staurosporine were found to be macrophage activating substances. They augment the phagocytic activity of murine macrophage cells. The activity of a lysosomal enzyme, β-glucouronidase, is a marker for macrophages related to bacterial infection. Tan-999, increased the activity of this enzyme at a concentration of 5 µg/mal. Tan-1030 A was to some extent active but staurosporine was practically inactive. It has been pointed out that the specific inhibitors of protein kinase C are expected to act as macrophage activators and could provide protection against tumor development and microbial infection through phagocytic activity (*57*, *84*, *104*, *106*, *107*).

The autoinflammatory activity of carprofen (**231**) is probably dependent on the inhibition of neutrophil macrophage function. It stimulates acid secretion without effecting basal acid secretion. This enhancement of secretagogus stimulated acid secretion was dependent upon extracellular calcium. It has been suggested that the compound acts at a post-receptor site between adenylate-cyclase and a protein pump. The drug probably increases calcium efflux through the plasma membrane and decreases the endogenous prostaglandin E_2 content. Studies on photohaemolysis of red blood cells with carprofen showed that the dehalogenation took place *via* the intermediate (2-carbazolyl)-propionic acid in the presence of hydrogen donors. Irradiation of carprofen resulted in decarboxylation *via* free radicals and formation of 2-ethylcarbazole (*11*, *13*).

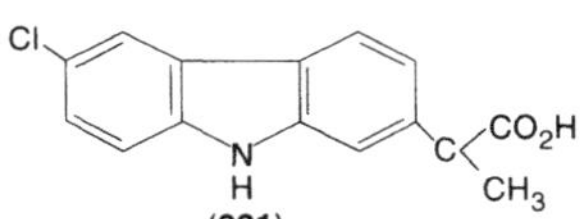

(231)

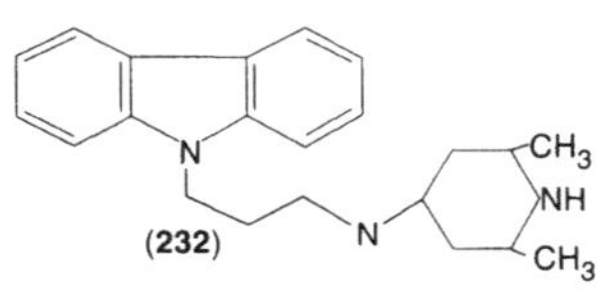

(232)

BMY 1402 or rincazole (**232**) is a novel neuroleptic and antipyretic agent (*26*). The effects induced by phencyclidine, a δ-opioid antagonist, are counteracted by **232**. It indirectly affects dopamine neurons, displaying selectivity for A10 dopamine cells. It increases the concentration of neurotensin in the caudate nucleus (*91*).

Pyrrolocarbazoles from a marine sponge, *Dictyodentrilla* sp., have been shown to be inhibitors of aldose reductase (*100*) catalysing the reduction of aldoses to polyols which on accumulation in cells may result in diabetes. Aldose reductase inhibitors provide prevention and cure of such ailments. Some of the pyrrolocarbazoles are being given trials as antidiabetic agents.

Studies on the mechanism of stimulation of NADPH oxidation by N-methylcarbazole (**103**) based on inactivation of cytochrome 450 2 BI by carbazole showed that the stimulation of NADPH oxidation is due to 1-hydroxy-N-methylcarbazole, a major metabolite of N-methylcarbazole. This metabolite probably causes NADPH, DNA damage.

e) Other Properties

i) SF-2370, identical with K-252a (**317**), has been found to have effects against rice plant diseases and greenhouse blights caused by *Rhizocotonia solani*, *Xanthomonas campestris* pv *oryzae* and *Piricularia oryzae* (*101*).

ii) Aflavazole displays significant antifeedant activity against the fungivorous beetle, *Carpophilus hemipterus* (*109*).

iii) Some synthetic 3,6-diaminocarbazole derivatives were found to bind to adenosine derivatives.

iv) Antioxidant properties. Carazostatin (**110**) is a strong antioxidant as tested by oxidation of methyl linoleate and soybean phosphatidyl choline (PC) liposomes in aqueous dispersion induced by free radicals (*50*, *58*). Carbazomycin B (**105**) was found to have moderate activity. Carbazoles without a free hydroxyl group did not suppress the oxidation appreciably. The antioxidant activity of carazostatin in the oxidation of soybean P.C. liposome was stronger than that of α-tocopherol. Vitamin C acted as synergist in combination with carazostatin in the oxidation of soybean P.C. liposomes. Carazostatin is the most active of all carbazoles in inhibiting lipid peroxidation of rat brain homogenate. It has also been suggested that natural carbazoles and their effective derivatives can protect tissues from peroxidative damage due to the generations of free radicals (*50*).

References, pp. 224–230

V. Chemistry of Carbazole Alkaloids

A. Alkaloids from Higher Plants

i) C_{13}-Alkaloids

1. 9-Carboethoxy-3-methylcarbazole

9-Carboethoxy-3-methylcarbazole (**233**), $C_{16}H_{15}NO_2$, (M^+253.1103) m.p. 78–80°, from the roots of *M. koenigii* (*7*) was considered to be a carbazole derivative because of the UV spetrum [λ_{max} 250–52, 262, 278 (sh), 304, 305 nm] and ir data (λ_{max} 1717, 1482, 1442, 1373, 1340, 1320, 1300, 1250, 1224, 1044, 800, 700 cm^{-1}). An ethoxycarbonyl group (δ1.56, 3H, *t* and 4.59, $J = 2$Hz) and an aromatic methyl group (δ2.53, 3H, *s*) were evident from the ^{1}H NMR data. These assignments were also supported by the ^{13}C NMR data (δ152.4, *s*, 62.9, *t*, and 14.5, COOET, 21.2 *q*, Ar-CH_3). There were also seven distinct aromatic proton signals at 7.8–8.3. The methyl substituted carbazole skeleton of **233** was supported by the mass spectral base peak at m/z 180. The ^{1}H NMR signals at δ8.18 and 8.30 (1H each, $J = 8$ Hz) were attributed to the H-1 and H-8 protons of a carbazole nucleus. From all these data the alkaloid was assigned the structure 9-carboethoxy-3-methylcarbazole which has been confirmed by synthesis.

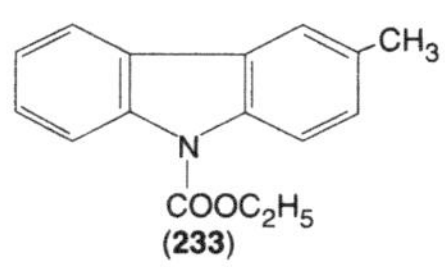

(233)

2. Clausenol

Clausenol (**25**), $C_{14}H_{13}NO_2$ (M^+227), m.p. 139°, UV (λ_{max} 228, 253, 303, 356 with logϵ 4.30, 4.36, 4.02, 3.50) and ir (λ_{max} 3395, 1620, 1580, 1200, 975, 850, 820 cm^{-1}) indicated that it was a carbazole derivative (*8*). Its reaction with $FeCl_3$ showed that it was phenolic. The ^{1}H NMR spectrum showed signals for OH (δ10.56), -NH- (δ9.50), aromatic protons H-4 (7.50,*d*), H-5 (δ7.40*d*) H-8 (δ8.31, *d*) H-7 (δ6.80 *dd*), H-2 (δ6.5, *d*) and an aromatic methoxy group (δ3.91). Protons H-2 and H-4 were not ortho coupled suggesting that C-3 and C-1 were substituted. H-5 was not ortho coupled and was slightly shielded indicating that the methoxyl was at C-6. The similarity of the hydroxyl signal to that of

1-hydroxycarbazole and the upfield shift of the C-2 and C-4 carbons in the ^{13}C NMR spectrum suggested that the hydroxyl group in the alkaloid was located at C-1. The tosyl derivative on reduction furnished 3-methyl-6-methoxycarbazole (glycozoline). Hence clausenol was formulated as 1-hydroxy-3-methyl-6-methoxycarbazole (**25**). The structure was confirmed by synthesis.

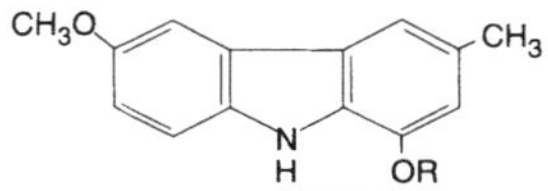

(**25**) R=H
(**234**) R=CH_3

3. Clausenine

Clausenine (**234**), $C_{15}H_{15}NO_2$ (M^+241), m.p. 151°, from the stem bark of *C. anisata* (*8*) was a carbazole derivative from UV (λ_{max} 226, 242, 299, 340, 354 mm with logε 4.34, 4.32, 4.00, 3.44, 3.42, respectively) and ir data (λ_{max} 3400, 1580, 1208, 940, 840, 820 cm^{-1}). Its ^1H NMR spectrum exhibited signals to those of clausenol except for replacement of the hydroxyl proton by that of a methoxy group (δ3.80). The molecular formula and the physical data suggested that it was the methyl ether of clausenol which was confirmed by methylation of clausenol (**25**) to clausenine (**234**).

4. 9-Formyl-3-methylcarbazole

9-Formyl-3-methylcarbazole (**235**), $C_{14}H_{11}NO$ (M^+209), m.p. 58–60°, from the roots of *Murraya koenigii* Spreng. (*7*) was a carbazole derivative from the UV (λ_{max} 222, 267, 291sh, 315 nm) and ir (λ_{max} 1498, 1588, 1488, 1452, 1359, 1317, 1148, 808, 747 cm^{-1}) spectra. The ^1H NMR (400 MHz) and ^{13}C (100 MHz) NMR data together with HMBC and ^1H-^1H COSY spectra led to its formulation as 9-formyl-3-methylcarbazole. This was confirmed by synthesis.

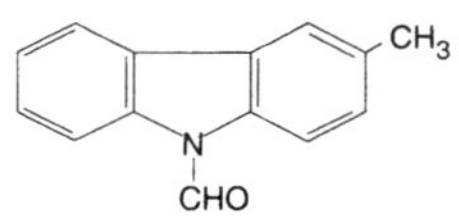

(**235**)

5. Murrayaline B

Murrayaline B (**236**), $C_{15}H_{13}NO_3$ (M^+255), m.p. 240–47°, from *M. euchrestifolia* (*45*) had a UV spectrum (λ_{max} 223, 259 (sh) 303, 380 nm) resembling that of murrayaline A (**237**). The ^{1}H NMR spectrum exhibited signals for an aryl methyl (δ2.34), an aromatic methoxy group (δ4.02), an aldehydic proton (δ10.58) and aromatic protons, H-4 (δ7.79) and H-5 (δ8.16) besides two deuterium exchangeable protons at δ8.4 and δ10.19. Irradiation of the methoxy at δ4.02 and the aromatic methyl caused enhancements of the signals of H-6 (δ6.89) resp. H-4 (δ7.79), showing that the methoxyl was at C-7 and the methyl at C-3. Methylation with methyl iodide of **236** furnished murrayaline A showing that murrayaline B was demethylated murrayaline A (**237**) or 2-hydroxy 3-methyl-7-methoxy-8-formylcarbazole.

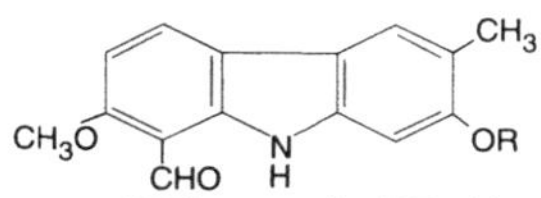

(**236**) murrayaline B,R = H
(**237**) murrayaline A,R = CH_3

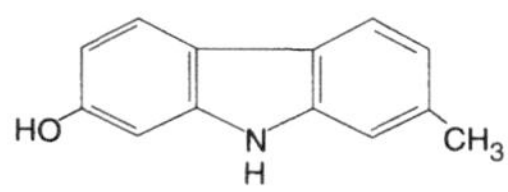

(**238**)

6. 2-Methyl-7-hydroxycarbazole or 2-Hydroxy-7-methylcarbazole

The alkaloid was isolated from aerial parts of *Cimicifuga simplex* (Ranunculaceae) by Kusano *et al.* (*69*). It was assigned the structure 2-hydroxy-7-methylcarbazole (or 2-methyl-7-hydroxycarbazole) (**238**) from physical data.

7. N-Methoxy-3-hydroxymethylcarbazole

N-Methoxy-3-hydroxymethylcarbazole (**239**), $C_{14}H_{13}NO_2$ (M^{+}227.0827), was obtained as a colorless oil from *M. euchrestifolia* (*46*). From its physical data (UV, ir, ^{1}H NMR and ^{13}C NMR) the N-methylcarbazole formulation of the alkaloid was suggestive. Its identity with the borohydride or catalytic reduction product of N-methoxy-3-formylcarbazole (**240**) confirmed its structure as **239**.

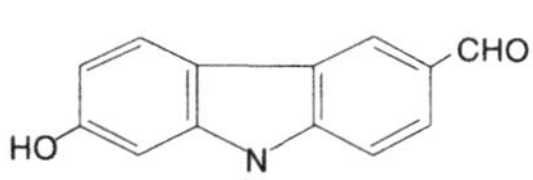

(**241**)

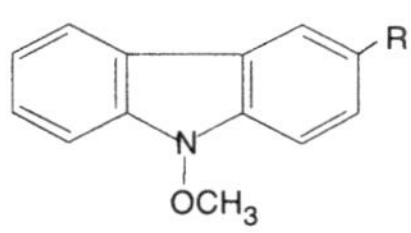

(**239**) R = CH_2OH
(**240**) R = CHO

8. 3-Formyl-7-hydroxycarbazole

3-Formyl-7-hydroxycarbazole (**241**), $C_{13}H_9NO_2$ (M^{+}211), was obtained from *Murraya euchrestifolia* (*46*) as a colorless powder. The

UV spectrum showed that it was a 3-formylcarbazole derivative. The presence of hydroxyl and NH functions was apparent from the ir spectrum while signals for the formyl group (δ10.10 *s*) the H-5, H-4 (δ8.62), H-1 (δ7.53) and H-2 (δ7.98) protons appeared in the ^{1}H NMR spectrum. Hence the substance was 3-formyl-7-hydroxycarbazole (**241**).

9. O-Methylmukonal

O-Methylmukonal (**242**), $C_{14}H_{11}NO_2$ (M^+225) m.p. 189-189°, from the roots of *M. siamensis* (*97*) showed UV and ir signals suggesting that it was a 3-formylcarbazole derivative. The ^{1}H NMR data indicated the presence of an aromatic methoxy group (δ4.50), an aldehyde function (δ10.45), signals for H-4 (δ8.50) and H-5 (δ7.12) protons and a singlet for H-1. Hence the alkaloid was O-methylmukonal.

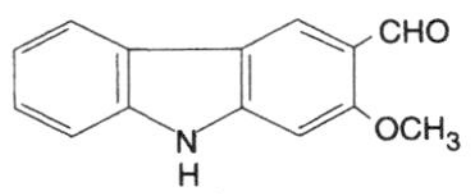

(242)

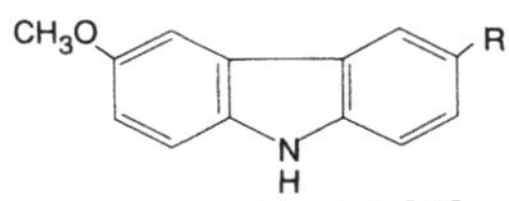

(243) R = CHO
(186) R = CH3

10. 3-Formyl-6-methoxycarbazole

3-Formyl-6-methoxycarbazole (**243**), $C_{14}H_{11}NO_2$ (M^+225), m.p. 135–36°, from *Clausena lansium* (*71*) had a UV spectrum characteristic of a 3-formylcarbazole (λ_{max} 229, 245, 281, 297, 334 nm with logε 4.38, 4.17, 4.35, 4.47, 4.80). The ^{1}H NMR spectrum was similar to that of glycozoline (**186**) whose aromatic methyl signal was however replaced by the signal of a formyl group. Hence, the structure of the alkaloid was 3-formyl-6-methoxycarbazole, a structure confirmed by synthesis.

11. 7-Methoxymukonal

7-Methoxymukonal (**244**), $C_{14}H_{11}NO_3$ (M^+241) from the root bark of *Clausena harmandiana* (*6*) had a UV spectrum (δ_{max} 222, 238, 244, 254, 288, 302, 340 nm with logε 4.27, 4.30, 4.30, 4.20, 4.33, 4.54 and 3.8 respectively) showing it to be a 3-formylcarbazole derivative. The ir spectrum (λ_{max} 3350 and 1619 cm^{-1}) suggested that the formyl group was chelated to the hydroxyl group while the ^{1}H NMR spectrum was similar to that of 7-methoxy-O-methylmukonal (**245**) with the methoxy signal of the latter being replaced by that of a hydroxyl group, chelated to the 3-formyl at C-3. This together with the ^{13}C NMR data led to formulation of the alkaloid as 2-hydroxy-3-formyl-7-methoxycarbazole or 7-methoxymukonal (**244**).

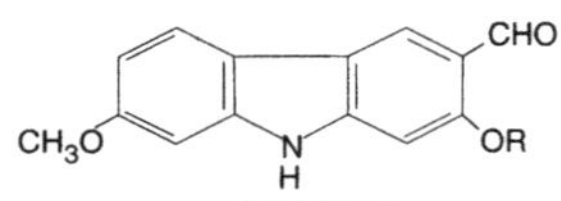

(**244**) R = H
(**245**) R = CH_3

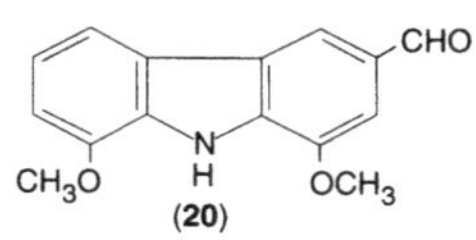

(**20**)

12. Clausenal

Clausenal, $C_{15}H_{13}NO_3$ (**20**), (M^+255), m.p. 198, from *Clausena heptaphylla* (*9*), was a derivative of 3-formylcarbazole with UV (λ_{max} 231, 257, 261, 327 m, with logε 4.45, 4.09, 4.10, 3.7) and ir (λ_{max} 3400, 1660, 1610, 1576, 1208 cm^{-1}) spectral data. The ^{1}H NMR spectrum showed the presence of -NH- (δ8.81, *br*s) H-4 (δ8.60, 1H, $J = 2.8$ Hz), H-5 (δ7.78 *dd*, $J = 8$ Hz and 1.8 Hz), H-2 (δ7.44 *d*, $J = 1.82$ Hz), H-6 (δ7.25 *m*), H-7 (δ6.98 *dd*, $J = 8$ Hz) as well as two singlets of aromatic methoxy groups (δ4.02, 4.07). In consideration of the NMR data, the two methoxyls were placed at C-1 and C-8. A mass spectral fragment at m/z 227 represented a formylcarbazole residue while the peaks at m/z 240 (M-15) and 212 (227–15) suggested the presence of two aromatic methoxyls. Hence structure 1,8-dimethoxy-3-formylcarbazole was assigned to it. This was confirmed by synthesis.

13. 6-Methoxymurrayanine

6-Methoxymurrayanine (**246**), $C_{15}H_{13}NO_3$ (M^+255), m.p. 231.33°, from *Clausena lansium* (*71*) exhibited UV (λ_{max} 239, 251, 294, 335, 349 nm, logε 4.35, 4.20, 4.40, 4.36, 3.85, 3.30, respectively) and ir data which showed the presence of a 3-formylcarbazole. The ^{1}H NMR signals were similar to those of murrayanine with an additional methoxyl replacing an aromatic proton and the signal of H-5 not ortho coupled. Hence the additional methoxyl was located at C-6. Thus, together with the ^{13}C NMR data, the alkaloid was formulated as 6-methoxymurrayanine (**246**).

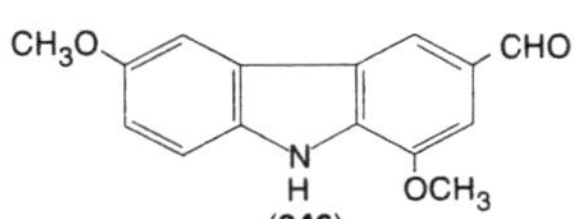

(**246**)

14. 7-Methoxy-O-methylmukonal

7-Methoxy-O-methylmukonal (**245**), $C_{15}H_{13}NO_3$ (M^+255.0875), m.p. 119–220°, was isolated from *Murraya siamensis* (*97*). The UV and ir spectrum showed that it was a 3-formylcarbazole, The ^{1}H NMR spectrum was similar to that of O-methylmukonal (**242**) with an aromatic proton being replaced by that of a methoxy group at δ3.85. The ^{1}H NMR data showed that H-5 was ortho coupled (δ7.97, *d*, *J*-8.3 Hz) and H-6 was both ortho and meta coupled (δ6.83, *dJ*-8.3, 2.2 Hz) showing that the additional methoxy group was at C-7. Hence the alkaloid was **245**, a structure further supported by the ^{13}C NMR spectrum.

15. Murrayaline C

Murrayaline C (**247**), $C_{15}H_{11}NO_4$ (M^+279), was obtained as a pale yellow powder from *M. euchrestifolia* (*45*). Its UV spectrum resembled that of murrayaline A (**237**) and its ^{1}H NMR spectrum had signals similar to those of murrayaline B (**236**) with the signal of the methyl group at C-3 being replaced by the signal of a formyl group (δ 9.55) which was evidently chelated to the hydroxyl at C-2. Therefore, murrayaline C was 2-hydroxy-3,8-diformyl-7-methoxycarbazole.

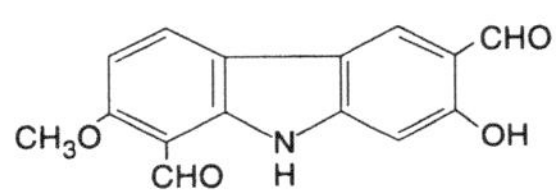

(247)

16. Carbazole-3-carboxylic Acid

Carbazole-3-carboxylic acid (**248**), $C_{13}H_9NO_2$, m.p. 270–72° isolated from *Clausena lansium* (*71*) had UV absorption band, at λ_{max} 230, 237, 269 nm with logε 4.78, 4.7, 4.22. The ^{1}H NMR spectrum and the identity of its methyl ester with carbazole-3-methyl carboxylate (**249**) led to its formulation as carbazole-3-carboxylic acid (**248**).

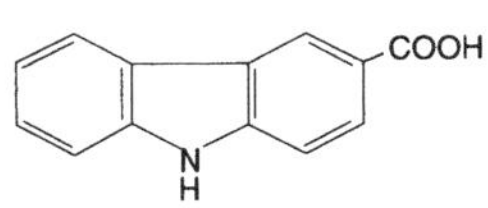

(248)

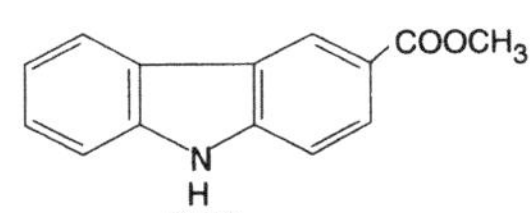

(249)

17. 3-Carbomethoxycarbazole

3-Carbomethoxycarbazole (**249**), $C_{14}H_{11}NO_2$ (M^+225), m.p. 168–170, from *Clausena lansium* (*71*) exhibited UV absorption maxima at 241,

251, 273 (sh), 287, 320, 332, 348 nm with logε 4.57, 4.50, 4.57, 4.68, 3.82, 3.73, 3.30 characteristic of a carbazole derivative. From its ^{1}H NMR spectrum the presence of a carbomethoxy group and other signals for a carbazole derivative were readily deduced. From these data the alkaloid was formulated as 3-carbomethoxycarbazole.

18. Clauszoline C

Clauszoline C (**250**), $C_{16}H_{15}NO_4$ (M^+285), was obtained from *Clausena excavata* (*43*) as a yellow powder. Its UV spectrum (λ_{max} 205, 222, 246, 282, 309, 320, 336 (sh) nm and ir bands (λ_{max} 3467, 1710, 1618, 1577 cm^{-1}) suggested that it was a carbazole derivative with a carbomethoxy group. Presence of three methoxy signals in the ^{1}H NMR spectrum at δ3.80, 3.82, and 3.89 showed it to have two aromatic methoxyls besides the carbomethoxy group. In NOE experiments irradiation of the methoxyl at δ3.89 and 3.80 caused enhancements of H-6 (δ6.82) and H-8 (δ7.02) signals showing the presence of a methoxy group at C-7. Irradiation of the NH-signal (δ10.36) caused enhancement of the signals of H-1 and H-8. Mass spectral peaks m/z 254 (M-31) and 226 (M-59) further indicated the presence of methoxyl and carbomethoxy groups, inferences also supported by ^{13}C NMR signals at δ167.4 and at δ51.6. Hence clauszoline C was assigned structure (**250**).

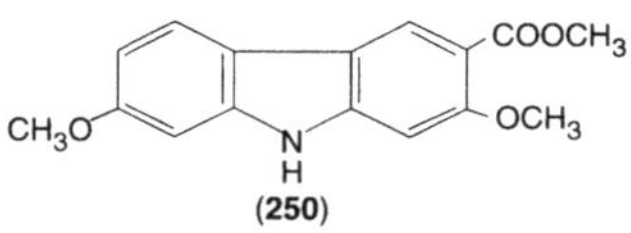

(250)

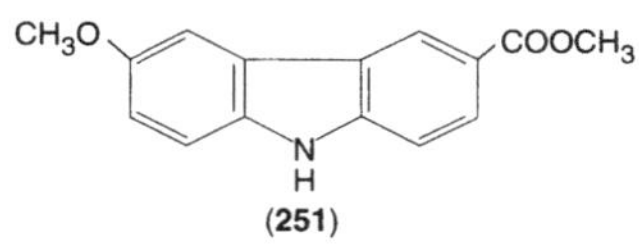

(251)

19. 3-Carbomethoxy-6-methoxycarbazole

6-Methoxycarbazole-3-methyl carboxylate (**251**), $C_{15}H_{13}NO_3$ (M^+255), m.p. 147–148°, from *Clausena lansium* (*71*) had a UV spectrum resembling that of a carbazole derivative. The presence of a carbomethoxy function and an aromatic methoxy group (δ3.9) was apparent from the ir and ^{1}H NMR spectra. On the basis of the spectral data the substance was formulated as 6-methoxycarbazole-3-methyl carboxylate.

20. Clauszoline I

Clauszoline I (**252A**), $C_{14}H_{11}NO_3$ (M^+241), was isolated from *Clausena excavata* (*47*) as a pale yellow oil. Its UV λ 223 (sh) 238, 247 (sh), 269, 275, 312, 322, 336 nm and ir spectra were similar to those of mukoeic acid. The ^{1}H NMR spectrum contained a set of four signals characteristic of an unsubstituted carbazole ring A. The ^{1}H NMR data together with biosynthetic considerations suggested the presence of a 1-oxygenated-3-carbomethoxycarbazole unit. Mass spectral peaks at M^+-59 and M-31 further confirmed the presence of -COOMe and -OMe groups. Hence the structure of 1-hydroxy-3-carbomethoxycarbazole was assigned to clauszoline I.

21. Clauszoline J

Clauszoline J (**252B**), $C_{15}H_{13}NO_4$ (M^+271), m.p. 252–256°, was isolated from *Clausena excavata* (*47*) as yellow prisms. Its UV spectrum (λ_{max} 240, 276, 308, 318 nm) and the ir data were suggestive of the presence of a carbazole skeleton with a -COOH group. The ^{1}H NMR spectrum exhibited ABC type signals at δ7.93, 6.76 and 6.96 and two one proton singlets in the aromatic region as well as signals of two methoxy groups (δ3.88 and 3.82). Presence of a hydroxyl group (m/z 254, M^+-OH) and a carboxy group was also inferred from the mass spectrum (m/z 226, M^+-COOH). Irradiation of the -NH-signal caused enhancement of H-1 at δ7.01 and H-8 at δ6.96. Irradiation of the H-4 signal at δ8.33 caused enhancement of the H-5 signal (δ8.93) while irradiation of the methoxy signal at δ3.82 caused enhancements of signals of H-5 and H-6 and irradiation of the methoxy signal at δ3.88 caused enhancement of the H-1 signal. Hence the methoxy groups were at C-7 and C-2 respectively and the structure of clauszoline J was assigned as **253B** or 2,7-dimethoxycarbazole-3-carboxylic acid.

22. Clauszoline K

Clauszoline K (**252C**), (M^+225) from *Clausena excavata* (*47*) had UV (λ_{max} 226 (sh), 232, 252 (sh), 293, 324 nm) and ir spectral data consistent with the presence of a formyl-(λ_{max} 1680 cm^{-1}) carbazole skeleton corroborated by a singlet at δ10.08 and an OMe group at 3.92 in the ^{1}H NMR spectrum. Further ^{1}H NMR signals were those at δ7.47, δ7.91 and 8.51 of H-1, H-2 and H-4, the latter deshielded by a -CHO group at C-3 and of H-6, H-8 and H-5 at δ6.94, δ6.96 and δ8.06. From the enhancement of the signals of H-6 (δ6.94) and at H-8 (δ6.96) on irradiation of the OMe signal at δ3.92 the position of the methoxy group was located at C-7. Hence clauszoline K was 3-formyl-7-methoxycarbazole.

References, pp. 224–230

23. Clauszoline L

Clauszoline L (**252D**), $C_{15}H_{13}NO_3$ (M^+255.1873), with a CH_2 unit more than clauszoline K was obtained as a colorless oil from *Clausena excavata* (*47*). The presence of methoxy and carbomethoxy groups was indicated by mass spectral peaks at m/z 224 (M^+-CH_3) and at 196 (M^+-59). The ^{1}H NMR spectrum was similar to that of clauszoline K with the formyl signal of the latter being replaced by the signal of the carbomethoxy group which suggested that the carbomethoxy was at C-3. These results and NOE experiments led to formulation of clauszoline L as (**252D**).

24. Clauszoline M

Clauszoline M (**252E**), $C_{13}H_9NO_3$ (M^+227.0587), was isolated as a pale yellow powder from *Clausena excavata* (*47*). Its UV (λ_{max} 217, 240, 275, 290, 356 nm) and ir spectra (λ_{max} 3342, 1651 cm^{-1}) showed it to be a phenolic carbazole derivative containing a formyl group. Proton peaks at δ6.90, 7.05 and 7.58 in the ^{1}H NMR spectrum were attributed to H-7, H-6, H-5 while singlets at δ6.94 and δ8.4 were assigned to H-1 and H-4 respectively. The signal of the formyl proton was at δ9.9 and that of the hydroxyl at δ11.4 both being mutually hydrogen bonded. Hence the structure was 2,8-dihydroxy-3-formylcarbazole.

(**252A**) clauszoline I : R_1 = OH; R_3 = $COOCH_3$; R_2 = R_4 = R_5 = H
(**252B**) clauszoline J : R_2 = R_4 = OCH_3; R_3 = COOH, R_1 = R_5 = H
(**252C**) clauszoline K : R_3 = CHO; R_4 = OCH_3 ; R_1 = R_2 = R_5 = H
(**252D**) clauszoline L : R_3 = $COOCH_3$; R_4 = OCH_3; R_1 = R_2 = R_5 = H
(**252E**) clauszoline M : R_2 = R_5 =OH; R_3 = CHO; R_1 = R_2 = H

ii) C_{18}-Alkaloids

1. Clauszoline B

Clauszoline B (**253**), $C_{18}H_{15}NO_3$ (M^+293), isolated as a yellow oil from *Clausena excavata* (*43*) had UV [λ_{max} 204, 242, 252 (sh), 264, 291 (sh), 304, 376 (nm)] and ir maxima (λ_{max} 3450, 3300, 1650, 1630 cm^{-1}) resembling those of a pyranocarbazole derivative, clauszoline A (see p. 190). The structure was presumably similar to that of clauszoline A. The high intensity base peak at m/z 278 was suggestive of the carbazolopyrilium ion **253A**, typical of many pyranocarbazoles. The ^{1}H NMR spectrum lacked the signals for the prenyl unit found in clauszoline A. In the NOE spectra, irradiation of H-4 (δ6.45) enhanced the doublets at δ7.43 (H-5), the aldehyde (δ9.89) and the doublet of H-6 (δ6.90). Thus the alkaloid was formulated as **253**.

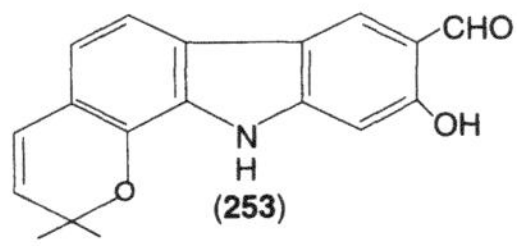

(253)

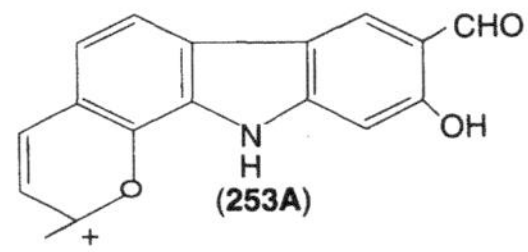

(253A)

2. Clauszoline D

Clauszoline D (**254**), $C_{18}H_{19}NO_5$ (M^+329), $[\alpha]_D \pm O$ a yellow oil obtained from *C. excavata* (*43*), had UV maxima [λ_{max}, 206, 217, 376, 292, 304 (sh), 356 nm] similar to those of a 2,8-dioxygenated 3-formylcarbazole derivative. This conclusion was also supported by the ^{1}H NMR spectrum which contained deshielded signals for H-4 (δ8.30, s) and H-5 (δ7.57, *d*, $J = 7.7$) and a formyl proton singlet (δ9.97, *s*). In NOE experiments, irradiation of the formyl proton caused enhancement of the H-4 signal which further supported position of the formyl group at C-3. The ^{1}H NMR data also showed that the methyls (signals at δ1.33 and 1.31) were attached to a carbon carrying oxygen and the presence of ABC type signals having geminal and viscinal couplings at δ3.46 (1H, *dd*, $J = 1.43, 1.8$ Hz), δ2.89 (1H, *dd*, $J = 9.5, 14.3$ Hz) and δ3.73 (1H, broad *d*, *J*-9.5 Hz). Two other ^{1}H NMR signals at δ4.28 and 3.72 were assigned to two hydroxyl protons. In the mass spectrum the loss of a $-C(OH)(CH_3)_2$ group was evident from the peak at m/z 240. Evidently the substituent at the 1-position could be expanded to $-CH_2CH(OH)-C(OH)(CH_3)_2$ and thus the structure of clauszoline D was assigned as **254**.

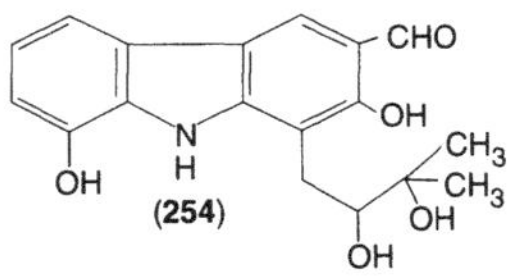

(254)

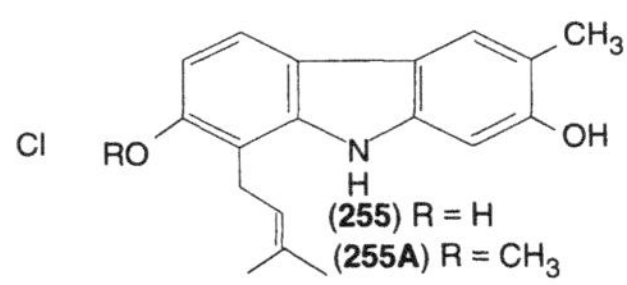

(255) R = H
(255A) R = CH_3

3. Euchrestine A

Euchrestine A (**255**), $C_{18}H_{19}NO_2$ (M^+281, 1414), was obtained as a colorless oil from *M. euchrestifolia* (*44*). The presence of a 9,7-dioxygenated carbazole skeleton was evident from the UV spectrum, while the ^{1}H NMR spectrum exhibited signals of an aromatic C-methyl group (δ2.38), a dimethylallyl chain (δ1.78, 1.89, 3H each, *s*), δ3.59 (2H, *d*), δ5.36 (*t*, $J = 6$, 7 Hz), signals for H-4 and H-1 (1, H, *s* each) and ortho coupled H-5 and H-6. Comparison of the ^{1}H NMR data with those of isomurrayafoline B (**255A**) led to structure **255**.

4. Eustifoline D

Eustifoline D (**256**), $C_{15}H_{11}NO$ (M^+221.0841), a furocarbazole was obtained as a colorless oil from *M. euchrestifolia* (*42*). It showed UV (λ_{max} 206, 224, 350, 268, 298, 310, 340, 354 nm), ir (ν_{max} 3470 cm^{-1}) and ^{1}H NMR signals for the α- and β-furan protons (δ7.81, 7.32, 1 proton each *d*, *J*-2.6 Hz). Irradiation of H-4 caused enhancement of the signal at the β-position of the furan nucleus showing that the furan was fused to the 5:6 position with the furan oxygen at C-6 of the carbazole skeleton. These observations and ^{1}H NMR signals for an aromatic C-methyl and other protons of a carbazole system led to its formulation as **256**.

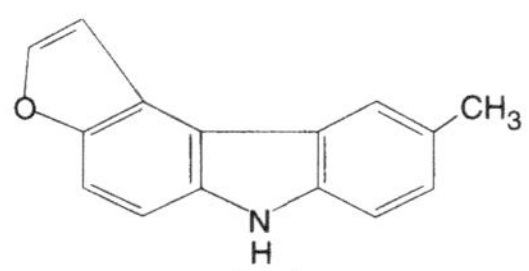

(256)

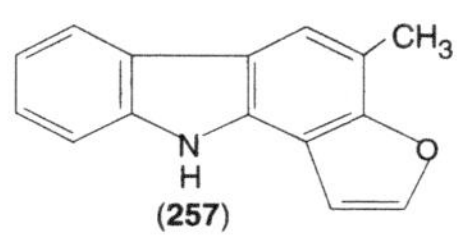

(257)

5. Furostifoline

Furostifoline (**257**), $C_{15}H_{11}NO$ (M^+221.0840), was obtained as an oil from *Murraya euchrestifolia* (*42*). Its UV and ir data were characteristic of a 2-oxygenated carbazole. The ^{1}H NMR spectrum contained signals of α and β furan protons as two doublets (δ7.73, 7.00, 1 each $J = 2.0$ Hz) and also showed that ring A of the carbazole skeleton was unsubstituted (δ8.06, 7.25, 7.37, 7.49). Enhancement of the H-4 signal after irradiation of the methyl signal at C-3 led to structure **257** for furostifoline.

6. Glycomaurine

Glycomaurine (**187**), $C_{18}H_{17}NO$ (M^+263), m.p. 195–196°C, was isolated in 1989 from *Glycosmis mauritiana* (*68*). The UV spectrum [λ_{max} 215, 231, 262, 270 (sh), 284 (sh), 298 (sh), 312, 322, 378 nm with logε 4.18, 4.23, 3.90, 3.87, 3.65, 3.57, 3.82, 3.85, 3.22 respectively] was suggestive of a pyranocarbazole supported also by the mass spectral base peak at m/z 248 (M^+-15) which was due to generation of the carbazolepyrilium ion. Presence of a 2,2-dimethyl-Δ^3-pyran system was evident from the ^{1}H NMR data of δ1.46 (6H, *s*) 5.82 and 7.28 (two *d*,

J-9 Hz each). The H-4 signal appeared at δ7.92. Hydrogenation afforded a dihydro derivative (**258**). From the ^{1}H NMR and mass spectral data structure **187** was advanced for glycomaurine. Structures of glycomaurine and isoglycomaurine (**188**) were confirmed by syntheses (see p. 164). Eustifoline A reported from *Murraya euchrestifolia* by Ito and Furukawa in 1990 (*42*) is identical with glycomaurine.

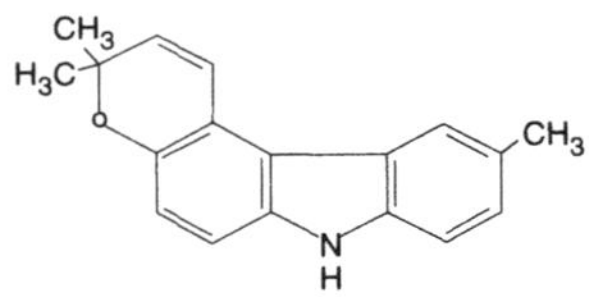
(187)

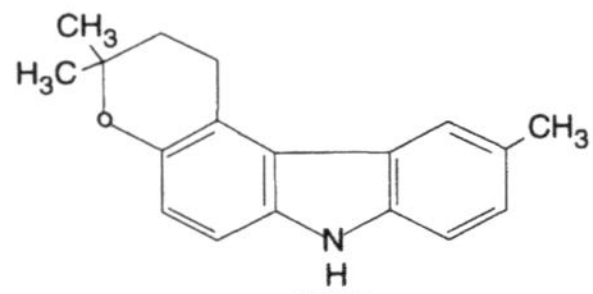
(258)

7. Glycomaurol

Glycomaurol (**259**), $C_{18}H_{19}NO$ (M^+265.1466), m.p. 149–50°, was isolated from the stem bark of *Glycosmis mauritiana* (*68*). Its UV [λ_{max} 230, 243 sh, 255 sh, 268 sh, 289, 298, 327 nm, (logε 4.15, 4.08, 3.97, 3.89, 3.79, 3.88, 3.05)] and ir spectra showed it to be a phenolic carbazole. The ^{1}H NMR spectrum established the presence of a dimethylallyl side chain [δ1.76, 1.95 (3H, *s*, methyl), 3.99 (*d*, *J*-6.5 Hz), 5.40 (*d*, *J*-6.5 Hz)], a phenolic hydroxyl (δ4.68) and an aromatic C-methyl group (δ2.52) with the signals of the aromatic ring similar to those of glycomaurine (**187**). Hence glycomaurol was 3-methyl-5-(3′,3′-dimethylallyl)-6-hydroxycarbazole (**259**) which was confirmed by its conversion to dihydroglycomaurine (**258**) by acid treatment.

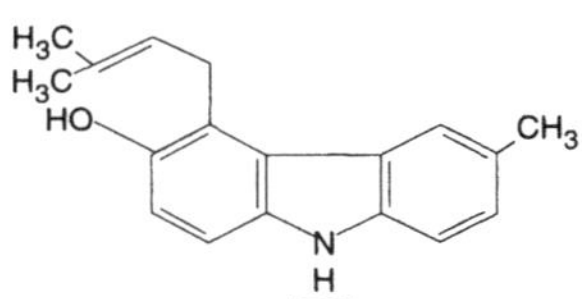
(259)

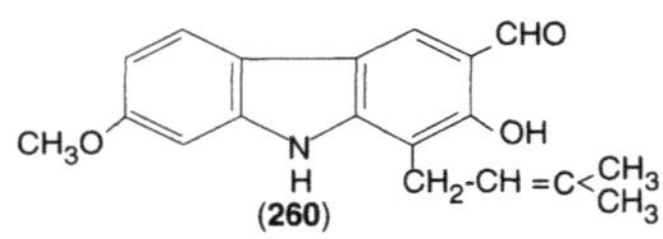
(260)

8. 7-Methoxyheptaphylline

7-Methoxyheptaphylline (**260**), $C_{19}H_{19}NO_3$ (M^+309), was isolated from *Clausena harmandiana* (*6*). The UV, ir and ^{1}H NMR spectra exhibited presence of a formyl group at C-3, of -OH, -NH and -OMe functions and of a dimethylallyl chain. Comparison with the ^{1}H NMR data of 7-methoxymukonal (**244**) allowed location of the dimethylallyl chain at C-1. The methoxy group was placed at C-7 because of the ^{13}C NMR spectrum (C-8, δ95.57). Hence the alkaloid possessed structure **260**.

9. 7-Methoxymurrayacine

7-Methoxymurrayacine (**261**), $C_{19}H_{17}O_3$ (M^+307.1208), m.p. 211–213°, was isolated from *Murraya siamensis* (*97*). The UV (λ_{max} 306, 354 nm with logε 4.58, 4.19 respectively) and ir (ν_{max} 3460, 1628, 1156 cm^{-1}) spectra showed it to be a carbazole alkaloid. The ^{1}H NMR spectrum exhibited signals for H-4 (δ8.23), H-5 (δ7.97, $J = 8$, 4 Hz), H-6 (δ6.83, *dd*, $J = 8.4$, 2.2 Hz) and H-8 (δ6.97, 1H, *d*, $J = 2.2$ Hz). Additional ^{1}H NMR signals at δ1.54 (6H, *s*) 6.90, 5.90 (1H each, $J = 9.8$ Hz each) established the presence of a 2,2-dimethyl-Δ^3-pyran system, an aldehydic proton (δ10.45, *s*) and an aromatic methoxy group (δ3.85). The formyl group was placed at C-3 and the 2,2-dimethylpyran system was at the 2:1 position while the methoxy group was placed at C-7 as H-8 was only meta coupled. A carbazolopyrilium ion at m/z 292 (M^+-15) was evident in the mass spectrum. Hence the alkaloid was formulated as 7-methoxymurrayacine which was supported by the ^{13}C data.

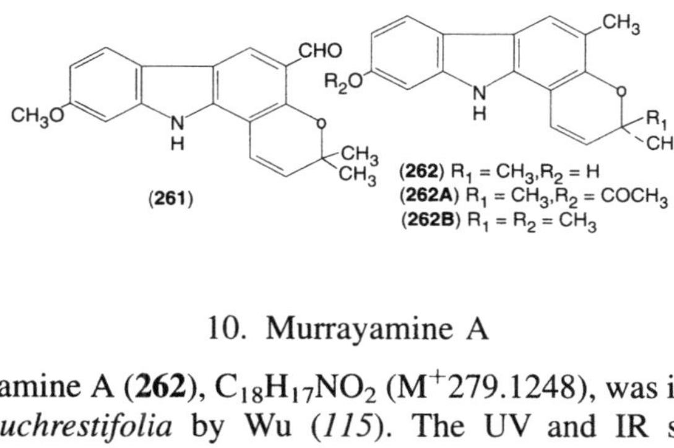

(**261**)

(**262**) $R_1 = CH_3, R_2 = H$
(**262A**) $R_1 = CH_3, R_2 = COCH_3$
(**262B**) $R_1 = R_2 = CH_3$

10. Murrayamine A

Murrayamine A (**262**), $C_{18}H_{17}NO_2$ (M^+279.1248), was isolated from *Murraya euchrestifolia* by Wu (*115*). The UV and IR spectra were consonant with a carbazole and it furnished a monoacetyl derivative **262A** and a monomethyl ether **262B**. The ^{1}H NMR spectrum showed the presence of -NH- (δ7.76), and phenolic hydroxyl groups (δ4.91) while an arylmethyl signal (δ2.31) and signals of aromatic protons at δ7.72 (1H, *d*, $J = 8.3$ Hz), 6.68 (1H, *dd*, $J = 8.3$, 2.2 Hz) and 7.81 (1H, *d*, $J = 2$ Hz) indicated a 3-methylcarbazole skeleton. On acetylation H-6 and H-8 experienced dimagnetic shielding showing that the hydroxyl was at C-7. The six proton singlet at δ1.47 and AB type signals at δ5.69 and δ6.58 (1H each, *d*, $J = 8.7$ Hz) showed the presence of a 2,2-dimethylpyran ring which was considered to be angularly fused to the carbazole skeleton as 4′-H of the chromene ring was shifted down-field by δ0.28 ppm on changing the solvent from chloroform to DMSO-d_6. Hence murrayamine A was assigned structure **262**.

11. Pyrayafoline B

Pyrayafoline B (**176**), $C_{18}H_{17}NO_2$ (M^+279.1266), from the same stem bark of *M. euchrestifolia* (*44*) had an UV spectrum (λ_{max} 228, 252, 286, 296 (sh), 329, 353 nm) of a carbazole derivative. In the MS a high intensity peak at m/z 264 suggested that it might be a 2,2-dimethyl-Δ^3-pyranocarbazole. The ^{1}H NMR spectrum substantiated these inferences as it contained signals of an aromatic C-Me group (δ2.37), H-4 and H-5 protons (δ7.62 and 7.49), those of a 2,2-dimethyl-Δ^3 pyran system [(δ1.46), two doublets at 6.48, 5.58 (1H, $J = 6$ Hz each)] and signals of aromatic protons at 6.74 and 6.71. Methylation with diazomethane gave an O-methyl derivative (**175**). The NMR signals of H-1 and H-4 were enhanced when the methoxy and aromatic C-methyl groups were irradiated which indicated that the methoxy group was at C-2 and the methyl attached to C-3. Structure 2-hydroxy-3-methyl-6,7-(2,2′-dimethyl-Δ^3-pyrano)-carbazole of pyrayafoline B (**176**) was confirmed by synthesis of O-methylpyrayafoline B (**175**).

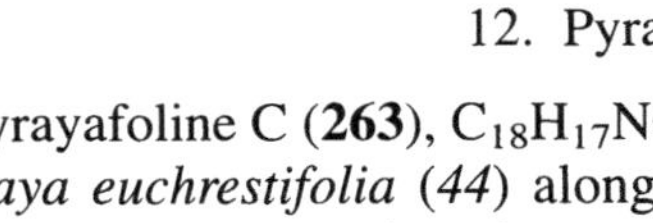

(**175**) R = CH_3
(**176**) R = H

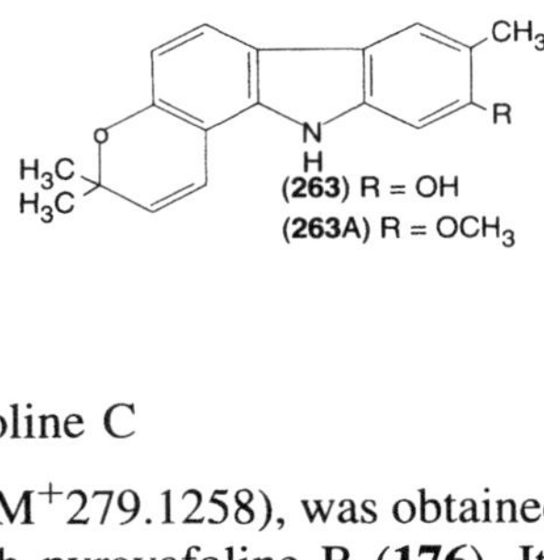

(**263**) R = OH
(**263A**) R = OCH_3

12. Pyrayafoline C

Pyrayafoline C (**263**), $C_{18}H_{17}NO_2$ (M^+279.1258), was obtained from *Murraya euchrestifolia* (*44*) along with pyrayafoline B (**176**). Its high intensity mass spectral peak at m/z 264 (M^+-15) suggested it to be a pyranocarbazole while the UV and IR spectra indicated that it was phenolic. The presence of a 2,2-dimethyl-Δ^3-pyran system (DMP) was readily discernible from the ^{1}H NMR spectrum. Treatment with diazomethane furnished an O-methyl derivative identical with pyrayafoline A (**263A**).

13. Mukoenine A

Mukoenine A (**264**), $C_{18}H_{19}NO$ (M^+265), m.p. 104–107°, was obtained as colorless prisms from *Murraya koenigii* (L.) Spreng by the

Furukawa group (*49*). The UV [λ_{max} 217, 238, 260 (sh), 288, 300, 321, 333, 356] and ir spectrum (δ_{max} 3600, 3460) showed it to possess a 2-hydroxycarbazole skeleton. The ^{1}H NMR spectrum had signals of an aryl methyl (δ2.40), an imine group (δ7.84) and a hydroxyl function (δ5.18). Signals typical of a prenyl chain [δ3.62 (2H), 5.37 (1H), 1.91 (3H), 1.79 (3H)] and the protons at C-5 (δ7.93) and C-4 (δ7.68) were apparent. Irradiation of the C-methyl group caused an enhancement of the H-4 signal. Evidently mukoenine A had an unsubstituted ring A and an aryl methyl group at C-3 while the hydroxyl and the prenyl chain were at C-2 and C-1 respectively. Hence the structure of mukoenine A was **264**.

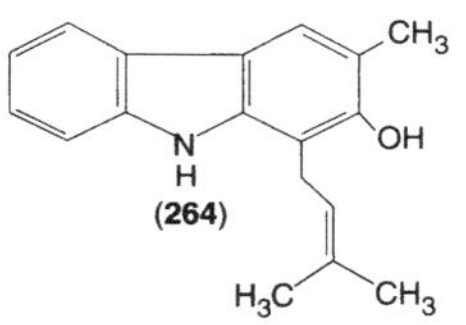

(264)

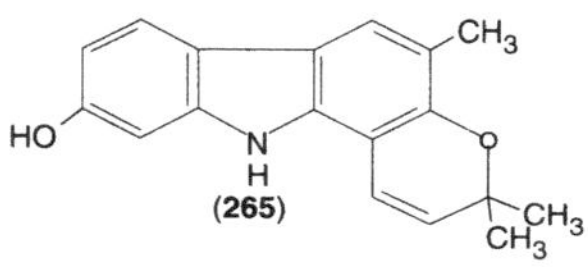

(265)

14. Mukoenine C

Mukoenine C (**265**), $C_{18}H_{17}NO_2$ (M^+279), was obtained as a colorless oil from *M. koenigii* (L.) Spreng (*49*). The UV [λ_{max} 204, 234, 248 (sh), 278, 288, 298, 325 (sh), 343] and ir spectrum (ν_{max} 3600, 3460, 1625, 1610) together with the mass spectral peak at m/z 264 (M^+-15) suggested the presence of a 2′,2′-dimethyl-4-Δ^3-pyranocarbazole. Irradiation of the aromatic C-Me enhanced the H-4 signal and the *ortho*-coupled aromatic protons at C-5 and C-6 which were also visible. The hydroxyl group was placed at C-7, leading to the structure **265** for mukoenine C.

15. Murrayaquinone E

Murrayaquinone E (**266**), $C_{18}H_{17}NO_3$ (M^+295.1205) was obtained from *M. euchrestifolia* as a brown oil (*45*). The UV spectrum was very similar to that 3-methyl-7-methoxy-8-prenyl-1,4-quinone (**267**) while the ir spectrum (ν_{max} 1620, 1650, and 3400, 3450 cm^{-1}) also indicated that it was a phenolic 1,4-quinonoid alkaloid. The ^{1}H-NMR spectrum indicated the presence of an aryl methyl group (δ2.14) which was long range coupled to an olefinic proton at δ6.45 (*d*, *J*-1.5 Hz) and *ortho*-coupled protons at δ7.95 (H-5) and δ6.88 (H-6), as well as a prenyl chain [6.32 (1H), 3.56 (2H), 1.88 (3H, *s*) and 1.80 (3H, *s*)]. The ^{1}H-NMR pattern was very similar to that of murrayaquinone B (**267**), a 3-methyl-

7-methoxy-8-prenyl-1,4-quinone except for the extra signal of the methoxy group in the latter. Hence structure **266** was assigned to murrayaquinone E.

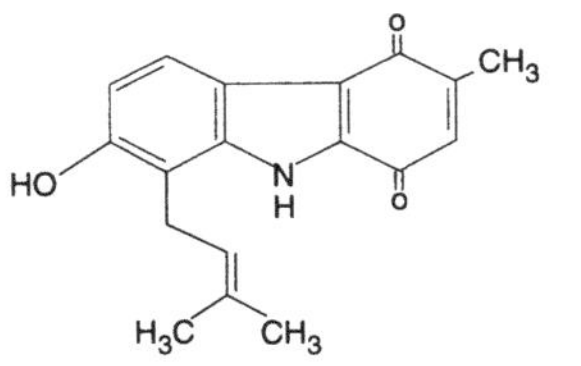

(266)

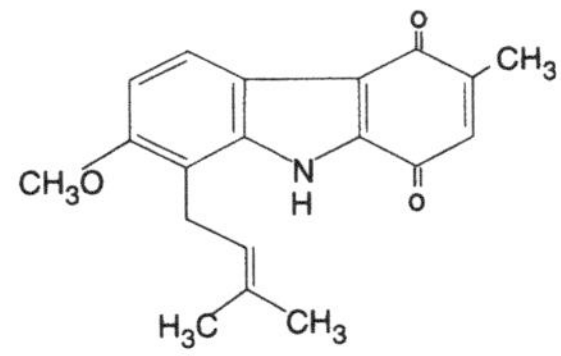

(267)

16. Clauszoline H

Clauszoline H (**268**), $C_{19}H_{19}NO_2$ (M^+293.1416), was obtained as a colorless powder from *Clausena excavata* by Ito *et al.* (*47*). The UV (λ_{max} 230, 258, 288, 310, 323, 348, 363 nm) and ir (ν_{max} 3471) spectra showed it to be a phenolic carbazole while the high intensity mass spectral peak at m/z 278 (M^+15) showed it to be a 2′,2′-dimethyl-Δ3′ pyranocarbazole. The ^{1}H NMR data showed the presence of aryl methyl (δ2.34) and aryl methoxy (δ3.90) groups, a 2′,2′-dimethyl-Δ3 pyran ring (δ6.4, 5.56), each a 1H *d*, $J = 9.8$ Hz, 1.50 (6H, *s*). Irradiation of H-4 at δ7.71 caused enhancement of the signals of H-5 (δ7.12) and the aryl methyl (δ2.34) showing that the methyl was at C-3, while enhancements of H-1 (δ6.8) and H-4′ (δ6.48) on irradiation of OMe and H-5 placed the methoxy group at C-2 and the 2′,2′-dimethyl-Δ^3-pyran ring at C-7 with the oxygen at C-8. Hence structure **268** was assigned to clauszoline H.

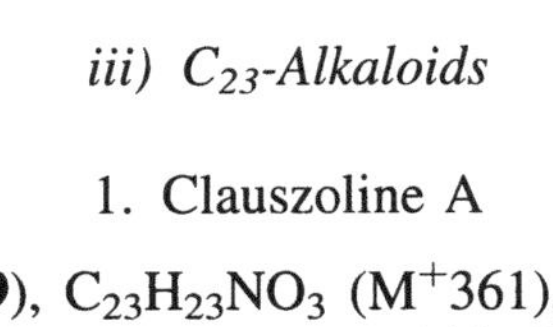

(268)

iii) C_{23}-Alkaloids

1. Clauszoline A

Clauszoline A (**269**), $C_{23}H_{23}NO_3$ (M^+361), was obtaind as a pale yellow powder from *Clausena excavata* (*43*). The UV (λ_{max} 204, 240, 265, 306, 375 nm) and ir (ν_{max} 3450, 3440, 1630 cm^{-1}) spectra indicated that it was a hydroxycarbazole. The ^{1}H NMR spectrum contained signals of a hydrogen bonded hydroxyl (δ11.64), a formyl group (δ9.89, *s*), a typical deshielded H-4 (δ7.98) and further showed the presence of a 2,2 dimethyl-Δ^3-pyran ring (δ1.53, 6H, *s*, δ6.48, 5.63, 1H each, $J = 9.9$ Hz each), H-5 (δ 7.4 *d*, $J = 7.7$ Hz), H-6 (δ6.90, 1H, $J = 7.7$ Hz). A carbazole pyrilium ion at m/z 346 confirmed the presence of a 2′,2′-dimethyl-Δ^3-pyran ring which was evidently at positions 7 and 8 and supported by enhancement of the signals of H-6 (δ6.90) and H-4′ (δ6.48) on irradiation of the H-5 proton. Irradiation of H-4 also enhanced the aldehyde signal at C-3 and H-5. Signals of the prenyl chain appeared at δ5.36 (1H, *t*, *J*-7.3 Hz), 3.65 (2H, *d*, *J*-3 Hz), 1.94 (3H, *s*) and 1.79 (3H, *s*), loss of the latter was evident from the mass spectral peak at m/z 306 (M^+-55). Hence **269** was the structure of the alkaloid.

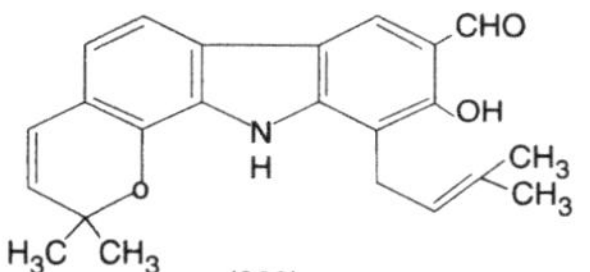

(269)

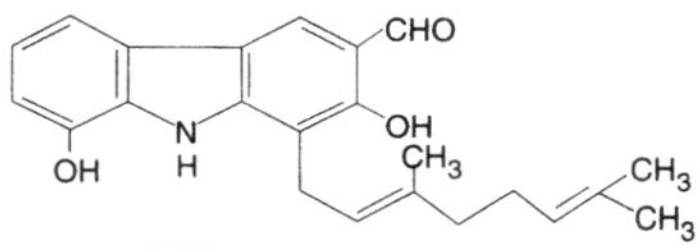

(270)

2. Clauszoline F

Clauszoline F (**270**), $C_{23}H_{25}NO_3$ (M^+363.1840), was obtained as a pale yellow powder from the stem bark of *Clausena excavata* (*43*). Its UV [λ_{max} 204, 216, 242, 276, 291 300 (sh), 354 nm] and ir (ν_{max} 3590, 3450, 3300 (br) 1630, 1585 cm^{-1}) spectra showed it to be a 3-formylcarbazole with a hydroxyl group. The ^{1}H NMR spectrum also indicated presence of a chelated formyl group (δ9.90, *s*) and a geranyl chain with signals at δ1.56, 1.60, 1.92 (three methyls), 2.10 (4H), 3.67 (2H, *d*, $J = 7.0$ Hz), 5.07 (1H, *m*), 5.35 (1H, *t*, $J = 7.0$ Hz) the latter being supported by mass spectral peaks at m/z 294 (M^+-C_5H_9) and at m/z 240 (M^+-C_9H_{15}). In NOE experiment enhancements of H-1′ (δ3.67) and 3′-Me (δ1.92) suggested location of the side chain at C-1, thus leading to structure **270** for clauszoline F.

3. Euchrestine B

Euchrestine B (**271**), $C_{24}H_{29}NO_2$ (M^+363.2173), was obtained as a pale yellow oil from *M. euchrestifolia* (*44*). The UV spectrum was very similar to that of euchrestine A thus suggesting the presence of a 2,7-dioxygenated carbazole while the ^{1}H NMR spectrum indicated the presence of H-4, H-1 and ortho coupled H-5 and H-6 protons. Signals of an aromatic methoxyl (δ3.90), an aromatic C-methyl (δ2.38) and the protons of a geranyl side chain (δ1.57, 1.62, 1.88 (three methyls, *s*), 2.06 (4H, *m*), 3.62 (2H, *d*, $J = 6$–7 Hz), 5.32 (1H, *t*, $J = 6.7$ Hz), 5.07 (1H, *m*) were also evident. Irradiation of the methoxy signal enhanced the H-6 signal showing that the methoxy group was attached to C-7, so that euchrestine B could be formulated as **271**.

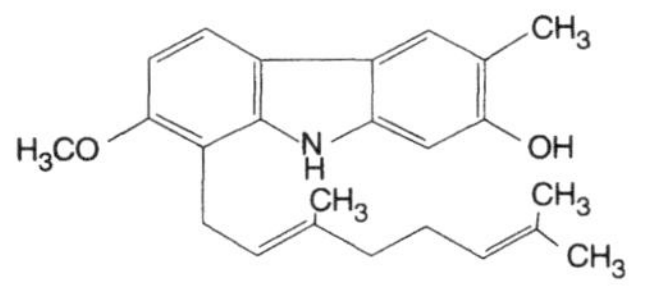

(271)

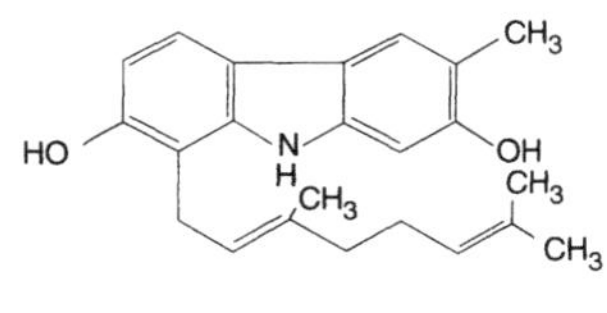

(272)

4. Euchrestine C

Euchrestine C (**272**), $C_{23}H_{27}NO_2$ (M^+349), was obtained as a brown powder from *M. euchrestifolia* (*44*). Like euchrestines A and B, its UV spectrum showed the presence of a 2,7-dioxygenated carbazole system. The mass spectral peak at m/z 266 (M^+-83) and the ^{1}H NMR signal of H-1 (δ6.81) and other aromatic proton signals similar to those of euchrestine B, but without the methoxy signal of the latter, permitted placement of the geranyl side chain at C-8 and formulation of euchrestine C as **272**.

5. Euchrestine D

Euchrestine D (**273**), $C_{23}H_{29}NO_2$ (M^+351), was obtained as a pale yellow oil from *M. euchrestifolia* (*44*). The UV data were similar to those of euchrestine A, B and C, so evidently it possessed a 2,7-dioxygenated carbazole skeleton. Signals for H-5 (δ7.74, *dd*, $J =$ 14.1 Hz), H-6 (δ6.88, *dd*, $J = 2.4$, 8.4 Hz), H-8 (δ6.82, *d*, $J = 2.4$ Hz) and H-4 (δ7.56) and those of a geranyl side chain appeared in the ^{1}H NMR spectrum. The O-dimethyl-ether **273A** exhibited analogous signals for the aromatic protons and those of the geranyl side chain in addition to the signals for two aromatic methoxy groups. The geranyl side chain could thus be placed at C-1 with the structure being formulated as **273**. This structural assignment was also supported by the NOE data.

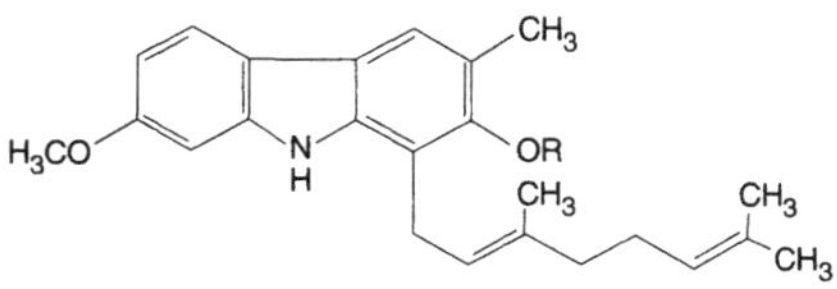

(**273**) R = H (**273A**) R = CH_3

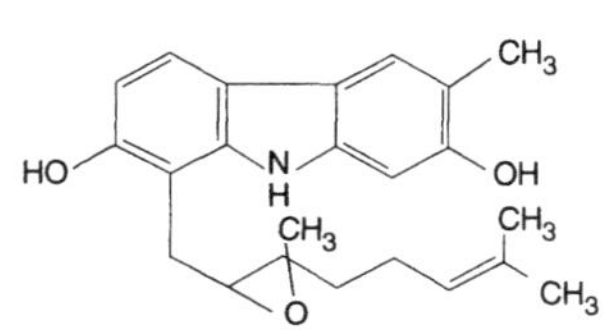

(**274**)

6. Euchrestine E

Euchrestine E (**274**), $C_{23}H_{27}NO_3$ (M^+365.1993), was isolated from *M. euchrestifolia* (*45*) as a pale brown racemate. A mass spectral peak at m/z 226 (M^+-139) was suggestive of the loss of a geranyloxy chain. The ^{1}H NMR spectrum had signals of an epoxide proton (δ2.38), H-4 (δ7.65) and H-5 (δ7.66, *d*), the latter being coupled to H-6 (δ6.73, *d*, $J = 8.4$ Hz) while COSY showed the presence of three mutually coupled protons (δ2.84, 3.10, 3.99) arising from a benzylic methylene and three adjacent methines on an oxirane ring in the geranyloxy chain. Further, a methylene signal at δ2.17 coupled to a vinylic proton at (δ5.07) and other oxymethylene protons at δ1.70 was found to overlap with two allylic methyl signals and a methyl attached to an oxygenated carbon. Similarity of the ^{1}H NMR spectrum between euchrestine C and euchrestine E except for some signals of the oxogeranyl chain in the latter led to structure **274** for euchrestine E.

7. Eustifoline B

Eustifoline B (**275**), $C_{23}H_{25}NO$ (M^+331), from *M. euchrestifolia* (*42*) showed a high intensity mass spectral peak at m/z 248 showing it to be a pyranocarbazole alkaloid. The UV, IR and ^{1}H NMR spectra were very similar to those of eustifoline A (**275A**) except for signals at δ5.11 (1H, *t*, $J = 6.7$ Hz), 2.17 (2H, *m*), 1.78 (2H, *m*), 1.65 (3H, *s*) and 1.58 (3H, *s*) but no methyl signal at δ1.49 like that in **275A**. However the C_{10} unit appeared to be similar to that in mahanimbine. Hence structure **275** was assigned to eustifoline B which was consistent with the UV, ir, and ^{1}H NMR data.

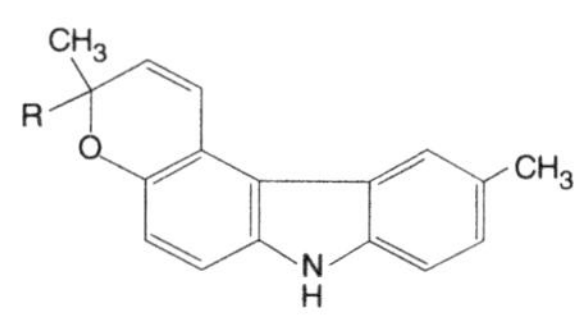

(275) R = $-CH_2CH_2CH=C(CH_3)_2$

(275A) R = CH_3

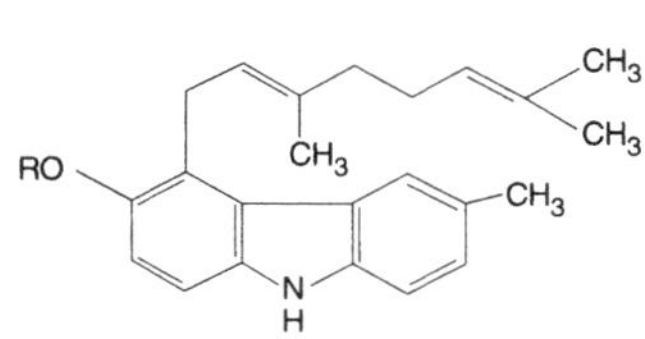

(276) R = H　　**(276A)** R = CH_3

8. Eustifoline C

Eustifoline C (**276**), $C_{23}H_{27}NO$ (M^+333), was isolated from *M. euchrestifolia* (*42*) as a brown powder. Its UV and ir spectra suggested that it was a 3- or 6-oxygenated carbazole alkaloid. The ^{1}H NMR spectrum indicated the presence of an aromatic methyl group (δ2.51), a hydroxyl group (δ4.90) and a geranyl side chain [δ1.56, 1.63, 1.94 (3H, *s*, each), 2.09 (4H, *m*), 4.0 (2H, *d*, $J = 6.4$ Hz), 5.05 (1H, *m*) and 5.41 (1H, $J = 6.4$ Hz)], and a mass spectral peak at m/z 210 (M^+-123) arising from loss of a C_9H_{15} fragment representing the geranyl unit was also observed. Enhancement of the H-4 signal on irradiation of the benzylic methylene of the geranyl unit suggested that the geranyl unit was substituted at C-5. In the methyl ether **276A** of eustifoline C the H-7 signal (δ7.08) was enhanced by irradiation of the methoxy group. Hence the hydroxyl group attached to C-6 and geranyl chain to C-5. Therefore, eustifoline C was formulated as **276**.

9. Isomahanine

Isomahanine (**277**), $C_{23}H_{25}NO_2$ (M^+347.1885), m.p. 184°, isolated from *M. koenigii* (*94*) had UV [λ_{max} 220, 238, 286, 294, 324, 338 nm, logε 4.51, 4.48 (sh), 4.29, 5.52, 3.68, 3.74] and ir [ν_{max} 3580 (-OH), 3410 (-NH) cm^{-1}] spectra characteristic of a phenolic pyranocarbazole further supported by the base peak at m/z 264 (M^+-C_6H_{11}) of a carbazole pyrilium ion as in mahanine (*10*). The presence of singlet at δ7.56 (H-5), ortho-coupled H-3 and H-4 signals at δ6.62 and 7.56 ($J = 8.4$ Hz), a methyl singlet at δ2.38 and an OH signal at δ4.96 in the ^{1}H NMR spectrum and comparison of the latter with that of mahanine placed the OH group at C-7 and the methyl at C-6 with a pyran unit fused at the 2:1 position thus leading to structure **277**.

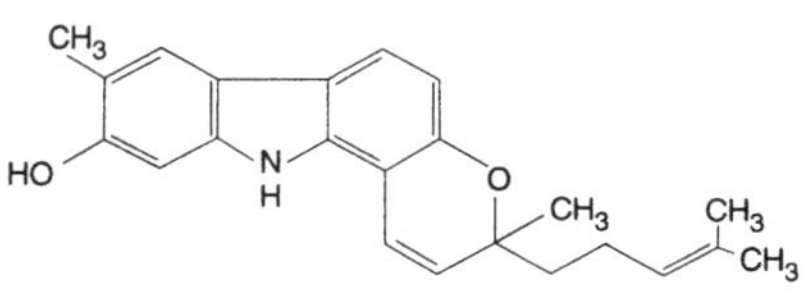

(277)

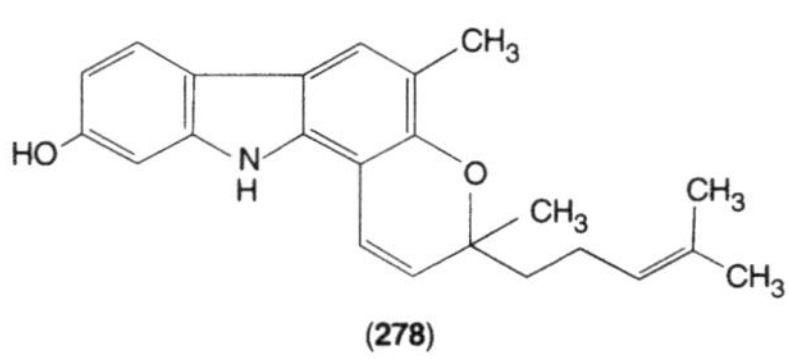

(278)

10. (+)-Mahanine

(+)-Mahanine (**278**), $C_{23}H_{25}NO_2$ (M^+347), [α]+34°, from *Murraya euchrestifolia* (*115*) had an UV spectrum typical of carbazoles. Its ir band at 3550–3200 cm^{-1} showed the presence of NH and phenolic hydroxyl groups while its positive response to ferric chloride and the formation of an O-acetate confirmed the phenolic character. The ^{1}H NMR spectrum and physical properties were identical with those of (−)-mahanine except for the optical rotation (*10*). The CD spectrum of (+)-

mahanine displayed a positive Cotton effect in the 265–276 region due to the styrene chromophore. The absolute configuration of (+)-mahanine at C-3′ was *S*. Conversely (−)-mahanine has the *R*-configuration at the chiral centre.

11. Murrayaline D

Murrayaline D (**279**), $C_{24}H_{27}NO_3$ (M^+377.1984), obtained as a brown oil from *Murraya euchrestifolia* (*45*) had UV and ir data typical of a phenolic carbazole. The mass spectral peak at m/z 254 (M^+-123) (**279A**) could arise by scission of a C_9H_{15}-fragment representing a geranyl side chain. Signals of a methoxy group at δ3.95, an aldehyde function at C-3 (δ9.98), ortho-coupled H-5 and H-6 protons (δ7.88, 6.95, *d*, $J = 8.8$ Hz each) singlets for H-1 and H-4 at δ7.4 and 8.08 and the signals of a geranyl chain [δ1.56, 1.60, 1.91 (3 methyl signals), 2.10 (4H, *m*), 3.68 (2H), 5.08, 1H and 5.36 (1H, *t*)] appeared in the ^{1}H NMR spectrum. NOE experiments suggested the E configuration for the side chain double bond. Enhancement of the proton signals at C-6 on irradiation of the methoxy signal showed that the methoxy group was at C-7 and the geranyl chain was at C-8. Hence structure **279** was assigned to murrayaline D.

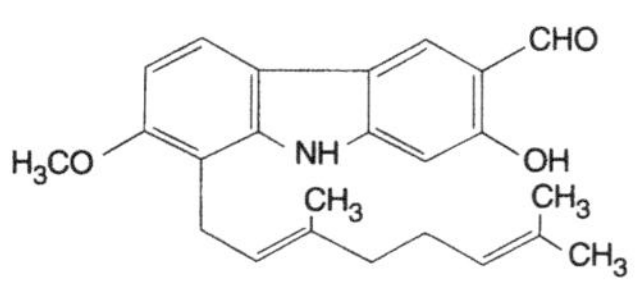

(279)

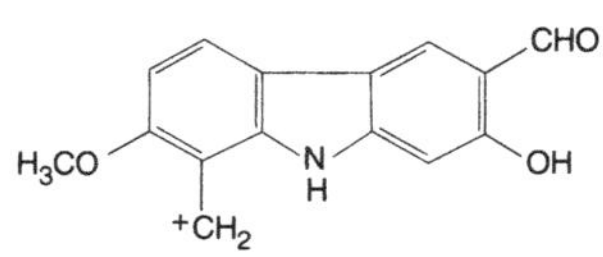

(279A) m/z 254

12. Murrayamine B

Murrayamine B (**280**), $C_{24}H_{27}NO_2$ (M^+361), was isolated from *M. euchrestifolia* (*115*). The UV spectrum indicated presence of a 1-methoxypyranocarbazole skeleton. Mass spectral fragments at m/z 346, 292 and 278 showed scission of methyl, $-CH_2CH{=}C(Me)_2$, and $-CH_2CH_2CH{=}C(Me)_2$ fragments. The ^{1}H NMR spectrum showed the presence of mutually coupled deshielded protons at C-4 and C-5 (δ7.52, 7.66), an aromatic methyl at δ2.31, *gem*-dimethyls at δ1.57 and 1.65 and a tertiary methyl adjacent to an oxygen atom at δ1.45 together with signals for two methylene groups and an olefinic proton at δ1.76 (2H, *t*) 2.17 (2H, *m*) and 5.11 (2H, *t*), respectively, characteristic of the signals of a pyran ring carrying an additional C_6-side chain. An AB quartet at 6.66

and 5.63 ($J = 10$ Hz each) was assigned to the olefinic protons of the pyran ring. Enhancement of the signal at δ6.82 (H-7) on irradiation of the methoxy signal at δ4.0 showed that the methoxy group was at C-8 while irradiation of the aryl methyl at δ2.32 enhanced the signal of H-4 at δ7.13 thus locating the position of the aryl methyl at C-3. Hence the structure of murrayamine B was **280**.

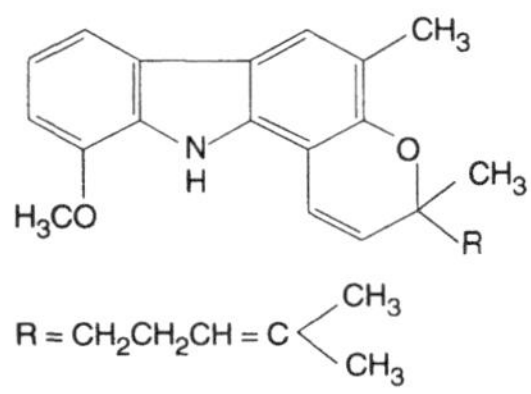

(280)

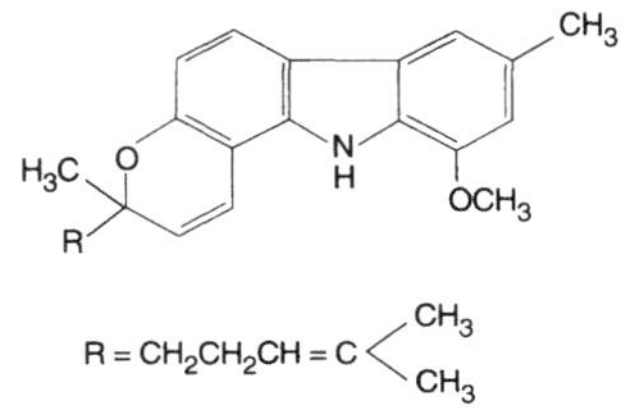

(281)

13. Murrayamine C

Murrayamine C (**281**), $C_{24}H_{27}NO_2$ (M^+361.2037), $[\alpha]_D \pm 0°$, was isolated from *Murraya euchrestifolia* as a syrup (*115*). Its UV and ir data suggested that it was a 1,7-dioxygenated carbazole derivative like murrayamine B. The ^{1}H NMR spectrum showed the presence of an aryl methyl group (δ2.49), an aromatic methoxy group (δ3.97), aromatic protons at H-4 (δ7.3) and H-2 (δ6.65) which showed enhancements when the aromatic methyl signal at δ2.47 was irradiated, thus confirming that H-4 and H-2 were unsubstituted. Irradiation of the methoxy signal enhanced the signal at δ6.65 showing that the methoxy group was at C-1. The NMR system also contained signals at δ7.70 (H-5) and 6.69 (H-6) suggesting that the pyran ring was attached to C-7 and C-8. Hence the structure (**281**) was proposed for murrayamine C.

14. Murrayanol

Murrayanol (**282**), $C_{24}H_{29}NO_2$ (M^+363.2196), m.p. 161°, was isolated from *M. koenigii* (*94*). Its UV [λ_{max} 208, 236, 296, 216, 316, 328 with logε 4.66, 4.69, 4.88, 4.25 (sh), 4.11 (sh), 4.62] and ir spectra showed it to be a phenolic carbazole alkaloid. The ^{1}H NMR spectrum had signals of a geranyl side chain at δ1.57, 1.63, 1.88 (3 Me groups), 3.62 (2H, *d*, $J = 6.7$ Hz), 5.02–5.12 (1H, *brs*), 2.06–2.09 (4H, *m*), 5.33 (1H, *m*), H-5 (δ7.74, *s*), H-4 (δ7.72, 1H, *d*, $J = 8.6$ Hz), H-6 (δ6.80, *s*), and H-3 (δ6.83, 1H, $J = 1.6$ Hz), aromatic C-Me at C-6 (δ2.38) and of an aromatic methoxy group at C-7 (δ3.91), which was also suggested by mass, spectral and ^{13}C NMR data. The alkaloid on cyclisation of the geranyl chain gave isomahanine (**277**). Hence structure **282** was assigned to murrayanol.

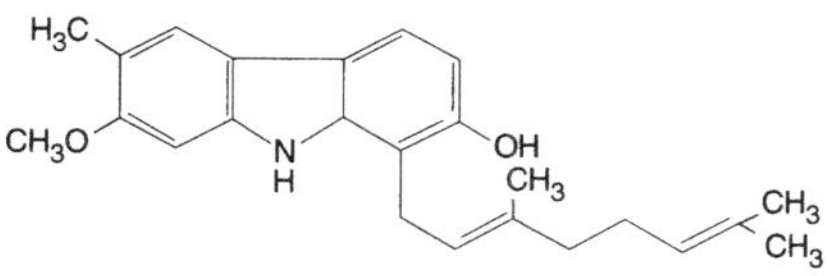

(**282**)

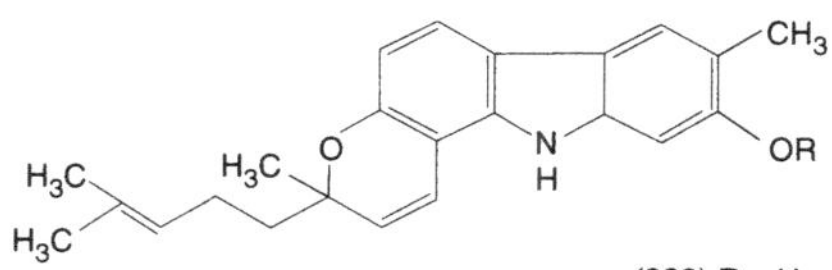

(**283**) R = H
(**283A**) R = CH_3

15. Pyrayafoline D

Pyrayafoline D (**283**), $C_{23}H_{23}NO_2$ (M^+347.1861), $[\alpha]_D$ $CH_3OH \pm 0°$ was obtained from *M. euchrestifolia* (*44*) as a pale brown powder. Its UV [λ_{max} 222, 238, 268 (sh), 296, 331, 342 nm] and ir spectra (ν_{max} 3600, 3450, 3380 cm^{-1}) and the high intensity mass spectral peak at m/z 365/264 showed that it was a phenolic pyranocarbazole, a conclusion supported by the ^1H NMR spectrum which indicated that the pyran contained a C_{10}-unit like that of mahanimbine (*10*). Presence of the C_6-carbon chain in the 2′ position geminal to the methyl group was appeared from the ^1H NMR signals [1.75 (2H, *m*), 2.15 (2H), 5.70 (1H, *t*), 1.57 (3H, *s*) and 1.65 (3H, *s*)]. In O-methylpyrayafoline D (**283A**) obtained by methylation of pyrayafoline D, the H-4 and H-1 signal was enhanced by irradiation of the aromatic C-Me and methoxy groups which indicated that the methyl and methoxy group were attached to C-3 and C-2 respectively, thus establishing structure **283**.

16. Murrayaquinone C

Murrayaquinone C (**284**), $C_{24}H_{27}NO_3$ (M^+377), m.p. 158–159°, obtained from *M. euchrestifolia* (*28*) contained a 1,4-benzoquinone system on the basis of its UV (λ_{max} 233, 267, 360 nm) and ^{13}C NMR spectra. The ^1H NMR spectrum established the presence of a murraquinone B skeleton with a geranyl unit substituted for the dimethylallyl chain of murrayaquinone B, i.e. formula **284**, also supported by a mass spectral peak at m/z 254. The position of the methoxy group at C-7 and that of the geranyl chain at C-8 were further supported by NOE exeriments.

17. Murrayaquinone D

Murrayaquinone D (**285**), $C_{23}H_{25}NO_3$ (M^+362), m.p. 164–168°, was isolated along with murrayaquinone E (*28*). The physical properties (UV, ir, ^{13}C NMR spectra showed that it was the phenol **285**. This was confirmed by methylation to **284** with diazomethane.

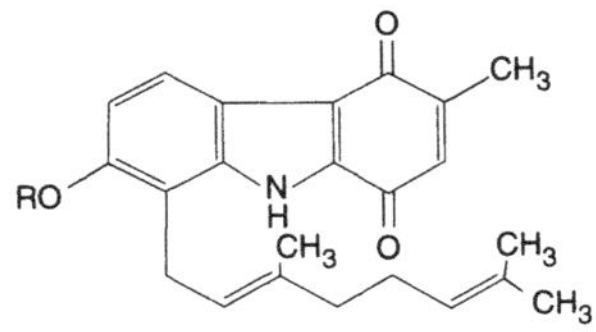

(**284**) R = CH_3 (**285**) R = H

18. Pyrayafoline E

Pyrayafoline E (**286**), $C_{23}H_{25}NO_2$ (M^+347), was obtained as a pale brown oil from *M. euchrestifolia* (*45*). The similarity of its UV, ir and ^1H NMR spectra to those of pyrayafoline B indicated the presence of a phenolic pyranocarbazole system like those in other pyrayafolines. The high intensity mass spectral peak at m/z 264 (M^+-83) characteristic of the carbazolopyrilium ion **287** showed that a C_6 unit was attached to the carbon of the pyran ring like that in mahanimbine (*10*), a unit also evident from the ^1H NMR signals at δ1.65 and δ1.57 (3H, each), 1.70 (2H), 2.13 (2H, *m*) and 5.10 (1H, *t*, $J = 7.3$) thus leading to structure (**286**).

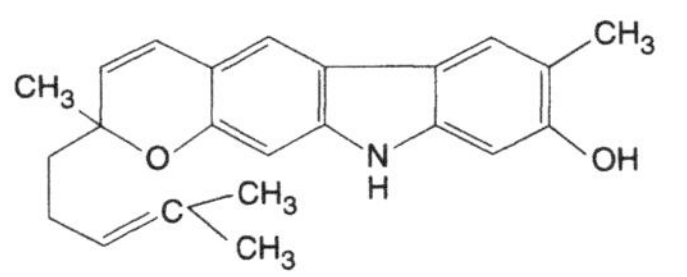

(**286**)

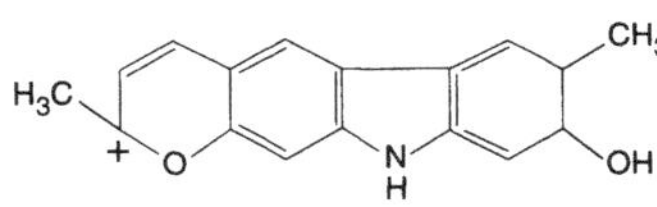

(**287**) m/z (M - 83)

19. Mukoenine B

Mukoenine B (**288**), $C_{23}H_{25}NO_2$ (M^+347), was isolated from *Murraya koenigii* by Ito *et al.* (*49*) as a colorless oil. Its UV [λ_{max} 222, 238, 284 (sh), 294, 331, 342 (sh), 360 (sh)] and ir spectra (3470 (br) 1700, 1640, 1610 cm^{-1}) indicated a phenolic carbazole skeleton while the ^1H NMR spectrum showed the presence of a formyl group (δ9.83), a hydroxyl group (δ11.60) and a deshielded H-4 proton (δ8.07) whose shift showed that the formyl group was at C-3. Since the hydroxyl proton was hydrogen bonded it was at C-2. The presence of 4 doublets at δ8.04

(H-5), 7.18 (H-6), 7.33 (H-7) and 7.44 (H-8) showed that ring A was unsubstituted, hence the geranyl chain, evident from the ^{1}H NMR spectrum [δ3.4, 5.0, 3.67, 2.10, and 1.91, 1.61, 1.57 (three singlets for methyl groups)] and from mass spectral peaks at m/z 224 and 278 (M^+-C_9H_5, M-C_5H_9) was at C-1, thus leading to structure **288** for mukoenine B.

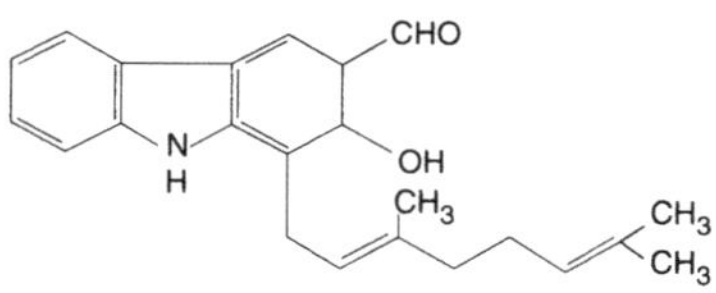

(**288**)

iv) Dimeric Carbazole Alkaloids from Higher Plants

Furukawa *et al.* (*27*) isolated the first dimeric carbazole alkaloid murrafoline from *Murraya euchrestifolia* thus substantiating the earlier prediction by Roy *et al.* (*96*) that such dimers should occur naturally on the basis of their work on the biomimetic hydroxylation of 3-methylcarbazole. In addition to dimeric alkaloids with C_{26}, C_{31} and C_{41} skeletons, many of which have been discussed previously (*13*), more recently alkaloids with C_{36} ($C_{23}+C_{13}\cdot C_{18}+C_{18}$) and C_{46} ($C_{23}+C_{23}$) skeletons have been discovered. The new dimeric alkaloids are discussed in the following section.

a) C_{26}-Alkaloids

1. Indole Dimer

Indole dimer (**217**), $C_{27}H_{30}N_2O_2$ (M^+414), was isolated from *Murraya gleni* by Kumar *et al.* (*67*). Its UV (λ_{max} 222, 271, 284 nm) and ir spectra (ν_{max}, 3400, 1730 cm^{-1}) indicated that it was an indole derivative containing an ester function while the mass spectrum suggested that it contains two monomeric units of mass 231 ($C_{14}H_{17}NO_2$) and of mass 184 ($C_{13}H_{14}N$). The ^{13}C NMR and ^{1}H NMR spectra confirmed that the ester was COOMe group [δ3.33 (*s*), 51.3 (-OMe), 177.2 (CO)] while the two NMR singlets at δ7.98 (N-H), 6.83 (H-2) and a doublet at δ5.20 ($J = 9.1$ Hz) indicated that the second indole nitrogen was substituted by a -CHCH-group. The presence of two sets of -$CH_2\cdot CH_2\cdot CHCH_3$ moieties [δ1.85–2.95 (*m*, 10H), 0.98 (*s*, 3H), 1.19 (*s*, 3H)] in the dimer was confirmed by H-H HomoCOSY, decoupling and long range coupling experiments suggesting that one of the $-CH_2CH_2CHCH_3$ units must carry the COOMe group while the other cyclized towards C-2 of the monoxygenated indole unit and then bound to the nitrogen of the second indole unit, thus constituting a 1,2-disubstituted 1,2,3,4-tetrahydrocarbazole structure **217**. The stereochemistry of the substituents on the tetrahydrocarbazole ring was found to be *trans*-diequatorial. Interestingly this dimer is the only alkaloid from higher plants which contains a methyl group on C-2 of a tetrahydrocarbazole ring. The five carbon chain in the indole skeleton provides circumstantial evidence for the genesis of carbazoles from 3-prenylated indoles.

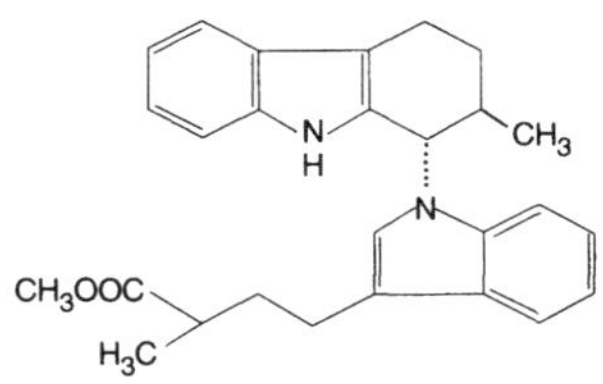

(217)

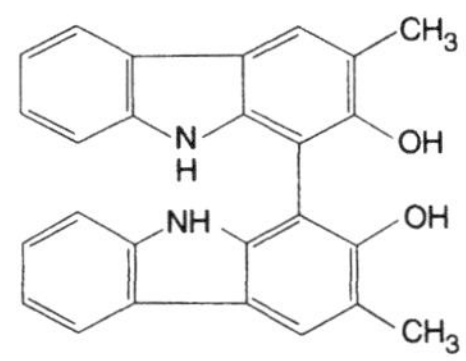

(289)

2. *Bis*-2-hydroxy-3-methylcarbazole

Bis-2-hydroxy-3-methylcarbazole (**289**), $C_{26}H_{20}N_2O_2$ (M^+392), isolated from *M. euchrestifolia* (*49*) had UV [λ_{max} 214, 237, 260 (sh), 304, 336] and ir spectra (δ_{max} 3550, 3450, 1610) characteristic of a carbazole. In the mass spectrum the presence of a peak at m/z 196 suggested that the alkaloid contained two carbazole units each containing hydroxy and methyl substituents. The ^{1}H NMR spectrum showed that it had two aryl methyls, two four spin protons (δ8.03, 7.22, 7.31, 7.23) and D_2O exchangeable hydroxyls (δ7.64, 5.27). Irradiation of the C-methyl signal enhanced the signal of H-4. The absence of signals due to H-1 and H-1′ suggested that the linkage between the two carbazole units was through the C-1 and C-1′ carbons thus leading to **289**.

3. Bismurrayaquinone A

Bismurrayaquinone A (**290**), $C_{26}H_{16}N_2O_4$ (M^+420), was obtained as an orange powder from *Murraya koenigii* (L.) Spreng (*49*). The UV [λ_{max} 228, 240, 252, 260 (sh), 288 (sh), 291, 330, 343 nm] and ir spectrum showed that it was a quinonoid carbazole derivative. The ^{1}H NMR spectrum showed the presence of two sets of 4-spin proton signals assignable to two ortho substituted rings with each set exhibiting signals similar to the spectrum of murrayaquinone A (3-methylcarbazole-1:4-quinone). Due to the absence of signals for H-2 and H-2′ structure **290** was assigned to bismurrayaquinone A.

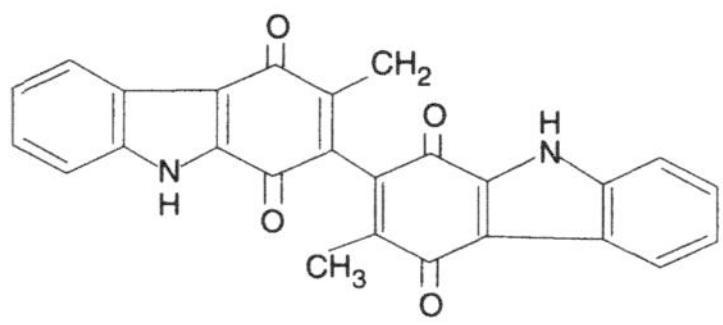

(290)

4. Chrestifoline A

Chrestifoline A (**291**), $C_{28}H_{24}N_2O_2$ (M^+420), obtained as an oil from *M. euchrestifolia* (*40*) along with chrestifoline B and C, had spectral properties which showed that the bisalkaloid consisted of a 1-methoxy-3-methylenecarbazole and a 1-methoxy-3-methylcarbazole unit. Irradiation of the protons of the benzylic methylene group enhanced the H-2′ signals of one unit and the H-4 signal of the other. A mass spectral peak at m/z 210 indicated presence of a unit containing methoxy and methylene groups on the aromatic rings of the alkaloid. Irradiation of the methoxy signal at δ4-0 resulted in enhancement of the H-2 signal at δ6.94 while irradiation of the aryl methyl caused enhancement of the H-2′ signal and the signal of the benzylic methylene. Conversely irradiation of the benzylic methylene at δ4.78 caused enhancement of the signal of the aromatic methyl and of H-5′ (δ8.08). From these data the structure of chrestifoline A was assigned as **291**.

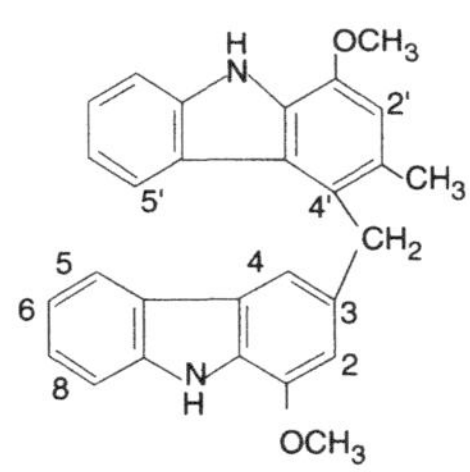

(291)

5. Chrestifoline D

Chrestifoline D (**292**), $C_{28}H_{22}N_2O_3$ (M^+434), was obtained from *M. euchrestifolia* (*46*) as a colorless oil. The molecular formula, and spectroscopic properties suggested that it was the formyl analog of bismurrayafoline A (**293**). This was confirmed by its synthesis from bismurrayafoline A by oxidation with DDQ.

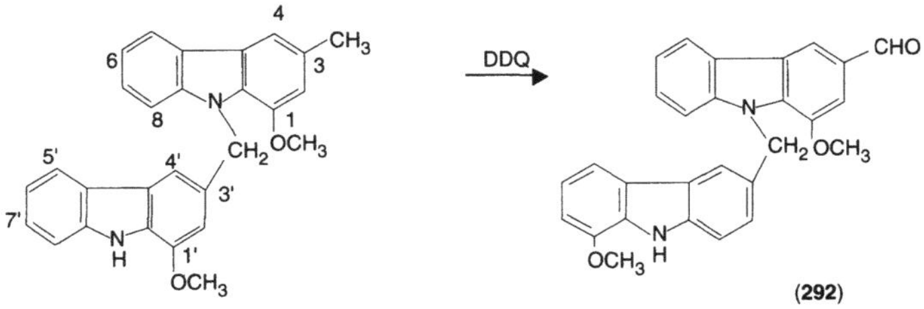

(293) (292)

6. Murrastifoline A

Murrastifoline A (**294**), $C_{28}H_{24}N_2O_2$ (M^+420), was isolated from *M. euchrestifolia* (*40*), as a colorless oil. The UV spectrum showed it to be a carbazole derivative. The ^{1}H-^{1}H COSY spectrum showed the presence of methoxy (δ3.57–4.11) and aromatic C-methyl groups (δ2.52–2.56), signals of two H-4 (δ7.56–7.63) and two H-2 protons. The ^{1}H NMR spectrum also indicated the presence of an unsubstituted ring A in one of the carbazole units. The positions of the methoxyl at C-1 and the methyl at C-3 of this unit were established by NOE spectrometry. The mass spectral fragments at m/z 210 and 211 represented a murrayafoline A unit, a common fragment in all alkaloids of the murrastifoline group. Other signals in the ^{1}H NMR spectrum were those of an aromatic C-methyl (δ2.48) and an aromatic methoxy group while signals of H-4′ (δ7.54), H-1′ and H-2′ were readily discernible together with signals of meta-coupled H-5′ (δ8.10, *d*, *J*-1.2) and H-8 (δ7.67, *d*, $J = 7.90$). Because of the latter and the absence of an H-6′ signal in the second carbazole unit the two monomeric units were thought to be linked through N and C-6, thus leading to structure **294**.

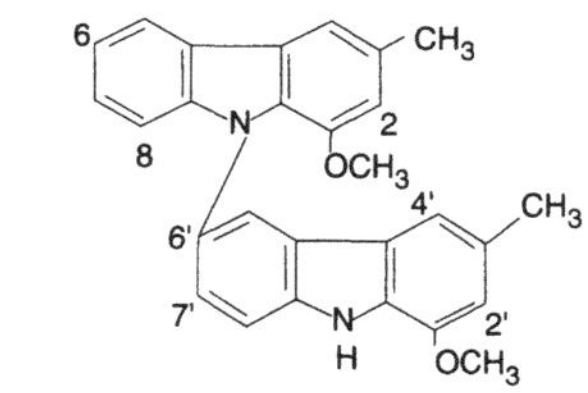

(294)

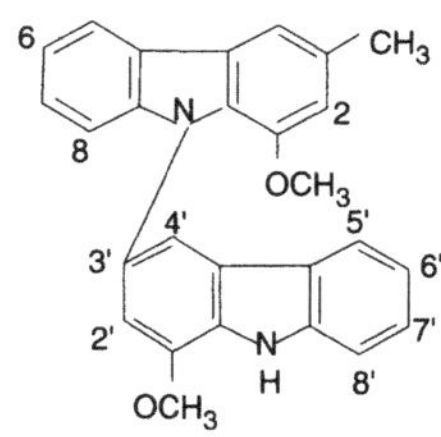

(295)

7. Murrastifoline B

Murrastifoline B (**295**), $C_{27}H_{22}N_2O_2$ (M^+406), from *M. euchrestifolia* (*40*) had UV and ir spectra similar to those of murrastifoline A, but the ^{1}H NMR spectrum, unlike that of murrastifoline A, showed the presence of two four spin aromatic ring systems. H-4 and H-4′ both were meta coupled. Irradiation of the methoxy signal at δ4.03 caused enhancement of the H-2 signal at δ8.7, H-5 and H-5′ both in the two monomeric units were ortho and meta coupled. These data suggested that the constituent carbazole units were linked through the nitrogen of one unit and C-3 of the other, with structure **295** being compatible with the evidence.

8. Murrastifoline F

Murrastifoline F (**296**), $C_{28}H_{24}N_2O_2$ (M^+420), was isolated from *Murraya koenigii* as a colorless oil by Ito *et al.* (*49*). Its UV [λ_{max} 227, 243, 253 (sh), 282, 292, 332, 346 nm] and ir spectra (ν_{max} 3480 cm^{-1}) were those of a carbazole. The ^{1}H NMR spectrum exhibited singlets of two aryl methyls (δ2.55, 2.05) and two methoxy groups (δ4.10, 3.33) while two four spin proton systems in the ^{1}H-^{1}H COSY experiments indicated the presence of two unsubstituted A-rings. Irradiation of the methyl at C-3 (δ2.55) caused enhancement of signals at H-4 (δ7.67) and H-2 (δ6.67) while irradiation of the second aromatic methyl at δ2.05 and the methoxy signal at δ4.10 enhanced the H-2′ signal at δ6.80. Based on these data, it was concluded that the dimeric alkaloid consisted of two 1-methoxy-3-methyl-carbazole (murrafoline A) units linked through the carbazole nitrogen of one unit to C-4′ of the second unit thus leading to structure **296**.

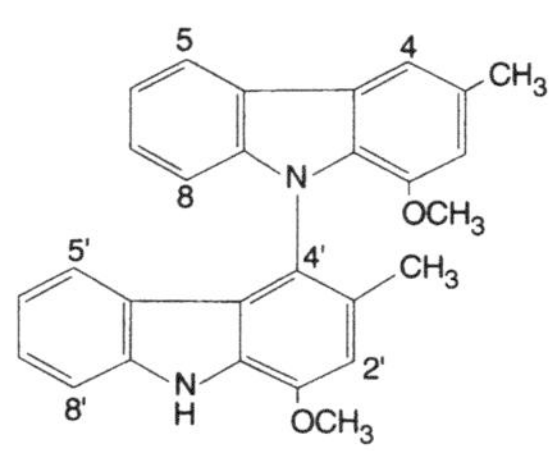

(296)

b) C_{31}-Alkaloids

1. Chrestifoline B

Chrestifoline B (**297**), $C_{32}H_{28}N_2O_2$ (M^+472), was obtained as an oily substance from *M. euchrestifolia* (*40*) together with its congeners chrestifoline A and C. Mass spectral peaks at m/z 262, 248 and 210 suggested that the alkaloid was constituted of girinimbine and murrafoline units. This was also supported by the ^{1}H NMR spectrum in which the signals of a 2′:2′-dimethyl-Δ^3-pyran system were evident [δ1.36 (6H, *s*), 6.9, 5.53, 1H each, *d*, $J = 9.9$ Hz each]. The H-4 signal was enhanced by irradiation of the aryl methyl signal at δ2.97 while irradiation of the benzylic methylene caused enhancement of the signal of H-4′ (δ6.9), H-2′ (δ6.90) and H-8 (δ7.3). Hence the benzylic methylene was placed on C-3′ and linked to the nitrogen of the pyranocarbazole fragment on in **297**.

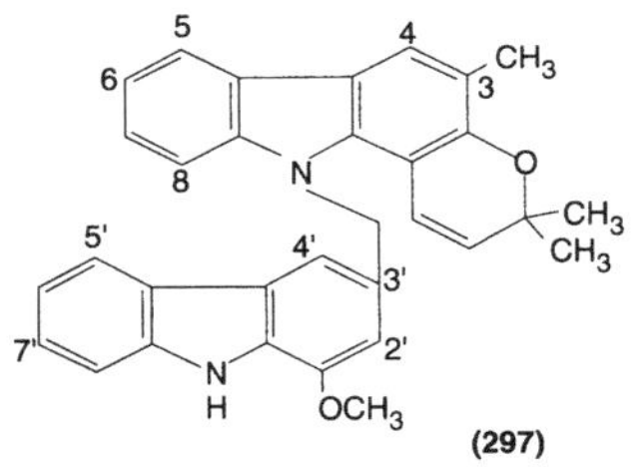

(297)

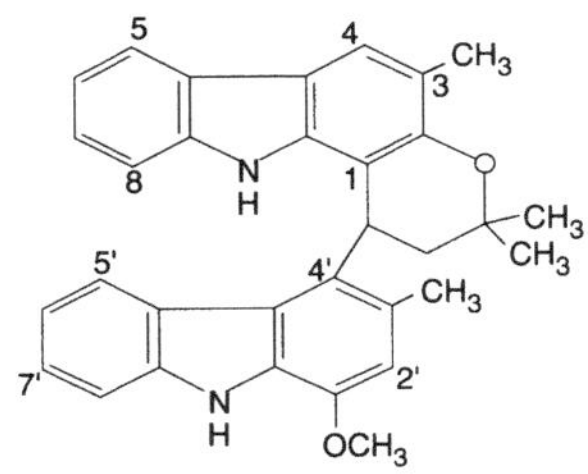

(298)

2. Murrafoline G

Murrafoline G (**298**), $C_{32}H_{30}N_2O_2$ (M^+474), was isolated from *M. euchrestifolia* (*29*). Its UV [λ_{max} 279, 243, 254 (sh), 284, 293, 302, 318 (sh)] and ir spectrum suggested that it was a carbazole. A mass spectral peak at m/z 264 suggested presence of a dihydrogirinimbine unit while a peak at m/z 210 could be assigned to the presence of a 1-methoxy-3-methylcarbazole fragment. The H-4 signals of the second unit and the signal of the benzylic protons of the dihydropyranocarbazole unit were missing which indicated that the two monomeric units were linked through C-4′ and the benzylic carbon of the dihydropyranocarbazole unit thus leading to formula **298**.

3. Murrastifoline D

Murrastifoline D (**299**), $C_{32}H_{30}N_2O_3$ (M^+490), $[\alpha]_D \pm 0°$ ($CHCl_3$), was obtained from *M. euchrestifolia* (*40*) as a colorless oil. Mass spectral peaks at m/z 211 and m/z 280 (base peak) suggested that the alkaloid contained a carbazole unit carrying a methoxy and a C-methyl group like those of other murrastifolines. The fragment of m/z 280 could be described to a C_{18} fragment with substitution. The ^{1}H NMR spectrum revealed the presence of geminal dimethyls on carbon linked to an oxygen (δ1.55, δ1.42), a *dol* (δ4.38, $J = 5.9$, 9.2 Hz) coupled to a hydroxylic proton (δ4.48, *d*, $J = 5.9$ Hz) and a signal at δ7.2 (1H, *d*, $J = 9.2$ Hz) assigned to H-4″ attached to a benzylic carbon attached to the N-atom of the carbazole unit carrying the methoxy and methyl groups (a murrayafoline A unit). The *trans* relationship of the substituents at C-3″ and C-4″ was ascertained from the coupling constants (9.2 Hz) and the absence of an NOE between these methine protons. Singlets at δ7.87 and 7.56 (H-4′, H-4) and two doublets at δ7.98 and 7.88 (H-5′, H-5″) together with overlapping methyl signals suggested the absence of substitution in rings A and A′ of the two carbazole units, thus leading to structure **299**.

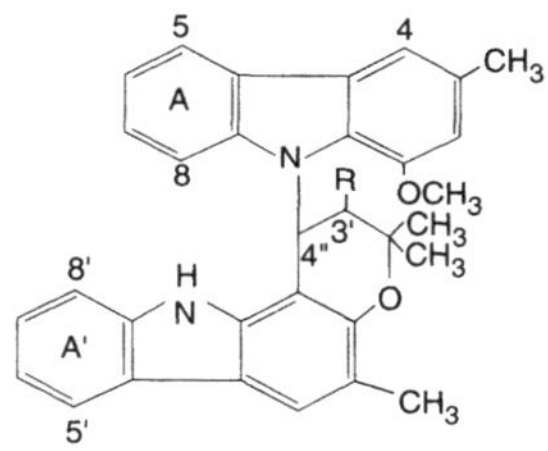

(299) R = OH **(300)** R = H

4. Murrastifoline E

Murrastifoline E (**300**), $C_{32}H_{30}N_2O_2$ (M^+474.2280), $[\alpha]_D -5.7°$ ($CHCl_3$), from *M. euchrestifolia* (*42*) had UV and ir spectral characteristics of a carbazole derivative. Mass spectral peaks at m/z 211 and 264 (base peak) arising from two constituent carbazole units suggested the presence of one carbazole unit with a methoxy and methyl substituent as in other murrafolines together with a dihydrogirinimbine unit. The ^{1}H NMR spectrum showed the presence of H-4 (δ7.58, *s*) and H-4′ (δ7.89, *s*) protons and two pairs of the four spin system which indicated the presence of unsubstituted A and A′ rings in the two constituent carbazole units.

The ^{1}H NMR and ^{13}C NMR spectra showed the presence of two aromatic methyls (δ2.56 and 2.42; δ17.56, 22.11) and one aryl methoxyl (δ4.11, 56.89) as well as the signal of one -NH proton. Signals at δ7.42 (1H, *dd*, $J = 7.3$, 11.0 Hz), 2.23 (2H, *m*), 1.50 and 1.48 (3H, *s* each) indicated the presence of a 4″-substituted 2,2-dimethyldihydropyran system which was confirmed by ^{13}C-NMR spectrometry. The chemical shift of an aromatic singlet at δ7.42 and presence of eleven sp^2 carbon doublets (δ50.59) in the ^{13}C NMR spectrum indicated that the two carbazole nuclei were linked through the benzylic carbon (C-4″) and an

N-atom thus leading to structure **300**, a deoxygenated murrastifoline D, a result further confirmed by NOE spectrometry.

c) C_{36}-Alkaloids

1. Chrestifoline C

Chrestifoline C (**301**), $C_{37}H_{36}N_2O_2$ (M^+540) $[\alpha]_D - 5.6°$ $CHCl_3$, a colorless oil, was obtained from *Murraya euchrestifolia* (*40*). The ^1H-^1H COSY spectrum evidenced two sets of four aromatic protons like those in murrastifoline C. The ^1H NMR spectrum showed signals at δ7.81 (H-4′), 2.30 (Ar-CH_3), 1.34 (methyl attached to a quaternary carbon bearing an oxygen), 6.95 and 5.53 (1H, *d*, $J = 9.9$ Hz) indicating presence of a double bond in a pyran ring, as well as a C_6H_{11} side chain [δ5.0 (1H, *t*, $J = 7.3$ Hz), 2.04 (2H, *m*), 1.68 (2H, *m*), 1.68 (2H, *m*), 1.54 (3H, *s*), 1.45 (3H, *s*) indicating the presence of a 2″-methyl, 2″-dimethylhomoallylic pyran system as in mahanimbine (10). Hence chrestifoline C contained a mahanimbine unit (m/z 331) and another unit responsible for a mass spectral peak at m/z 210 which could arise from a 1-methoxy-3-methylenecarbazole fragment. Irradiation of aromatic methyl at δ2.30 enhanced signals at δ7.81 (H-4′) while irradiation of the benzylic methylene at δ5.80 enhanced H-4, H-2, H-8′ and H-4″ signals. Thus structure **301** was assigned to chrestifoline C.

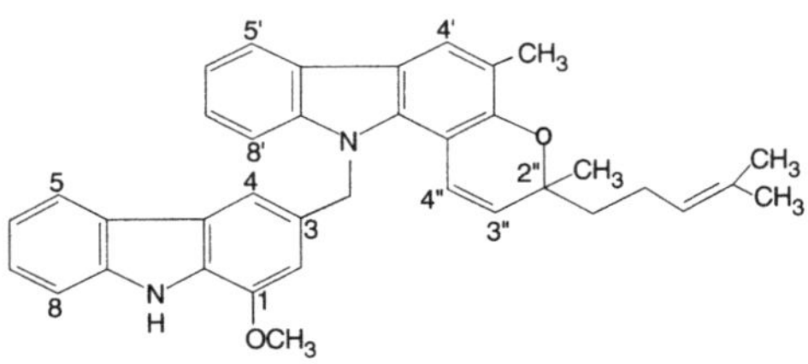

(**301**)

2. Murrafoline H

Murrafoline H (**302**), $C_{36}H_{34}N_2O_2$ (M^+526.2618), was obtained from *M. euchrestifolia* (*29*) as a colorless oily racemate. Its UV [λ_{max} 216, 239, 256 (sh), 281, 292, 304, 331, 345] and ir spectra (ν_{max} 3450 cm^{-1}) showed it to be a carbazole alkaloid. A mass spectral peak at m/z 248.1074 of a carbazolopyrilium ion like that produced by girinimbine suggested that the alkaloid contained a girinimbine and a dihydrogirinimbine unit. Irradiation of the benzylic proton of the dihydrogirinimbine unit enhanced H-5′ while irradiation of the methyl group at C-3 enhanced H-4. The H-5′ signal (δ7.95) was meta coupled while H-7′ was both meta and orthocoupled and H-8′ was ortho coupled. Thus the linkage between the two monomeric units was through C-6′ of the girinimbine unit and the benzylic carbon of the dihydrogirinimbine unit as in structure **302**.

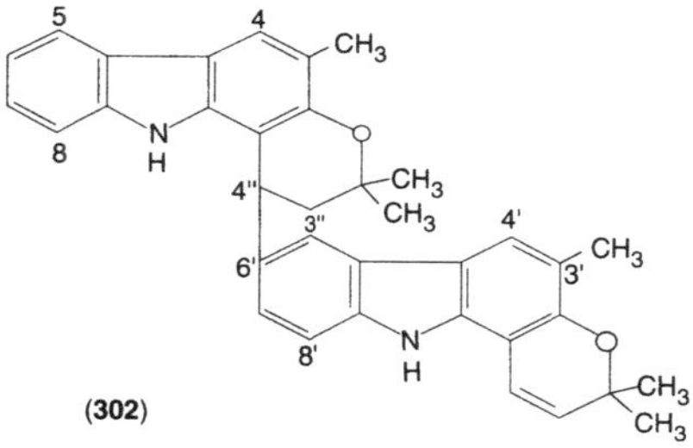

(302)

3. Murrastifoline C

Murrastifoline C (**303**), $C_{37}H_{36}N_2O_2$ (M^+540–2775), $[\alpha]_D$ 0°, was obtained from *M. euchrestifolia* (*40*) as a colorless oil. The UV and ir spectra resembled those of the dimeric alkaloids of the murrastifoline group. Mass spectral fragments at m/z 330 (base peak) and m/z 211 suggested that the dimer contained a mahanimbine unit (m/z 330) in addition to the 1-methoxy-3-methyl fragment. The presence of two sets of four spin systems in murrastifoline C and information from ^{1}H-^{1}H COSY experiments suggested that both rings A and A′ of the constituent carbazole units were unsubstituted. A mass spectral peak at m/z 457 (M^+-83) indicated scission of a C_6H_{11} unit like that present in mahanimbine. Signals of the benzylic methylene at δ6.13 and a deshielded H-4′ proton (δ7.36) together with the signal of one NH-proton showed that the benzylic methylene at C-3′ was linked to the nitrogen of the murrayafoline A unit. Hence the alkaloid was formulated as **303**.

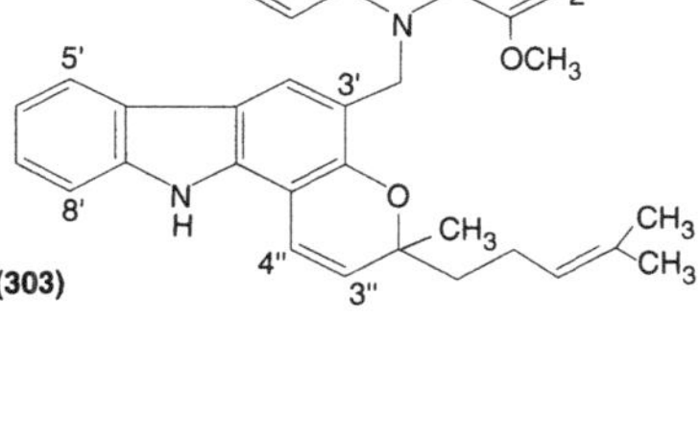

(303)

4. Bis-7-hydroxygirinimbine A

Bis-7-hydroxygirinimbine A (**304**), $C_{36}H_{32}N_2O_4$ (M^+556), a pale yellow oil from *M. euchrestifolia* (*116*), had a UV spectrum similar to that of girinimbine suggesting presence of a pyranocarbazole chromophore. The ^1H NMR signals showed the presence of aryl methyl groups, two quaternary methyl groups, two pairs of A, B type doublets ($J = 8.55$, $9.6\,Hz$) and aromatic protons. The ^1H NMR spectrum also suggested that it contained two 7,8-disubstituted girinimbine units. Methylation afforded a O-dimethyl derivative **304A**. NOE difference spectra showed enhancements of the H-4 and H-6 signals of each monomer units indicating that the methyl group was at C-3 and methoxy was at C-7 in each unit of **304**. Evidently the linkage between the monomeric units was between C-8 and C-8′. This conclusion was also supported by the HMBC spectrum of the O-methyl derivative **304A**.

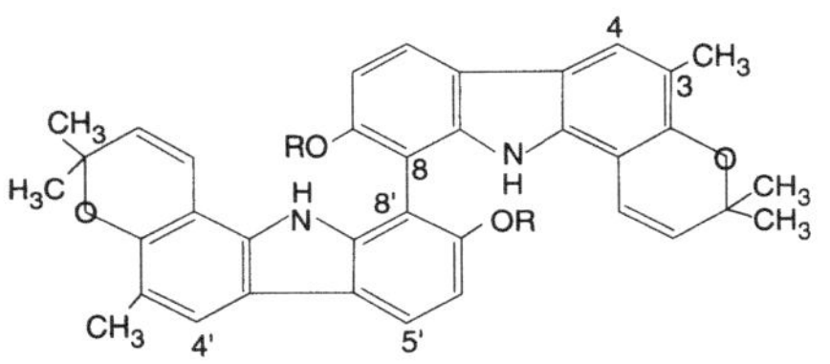

(**304**) R = H
(**304A**) R = CH_3

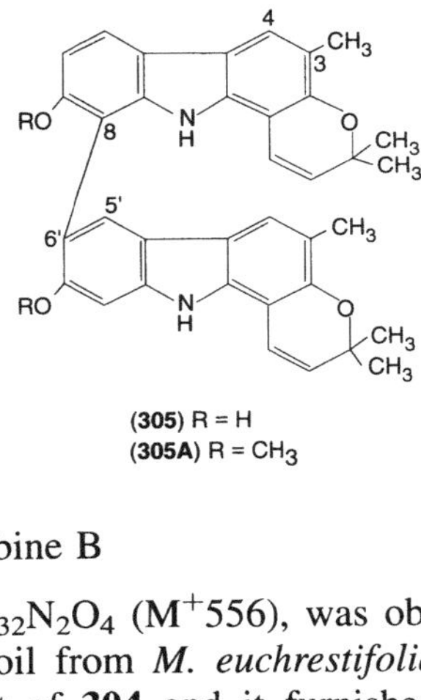

(**305**) R = H
(**305A**) R = CH_3

5. Bis-7-hydroxygirinimbine B

Bis-7-hydroxygirinimbine B (**305**), $C_{36}H_{32}N_2O_4$ (M^+556), was obtained together with **304** as a pale yellow oil from *M. euchrestifolia* (*116*). Its UV spectrum was similar to that of **304** and it furnished a dimethyl ether $C_{38}H_{36}N_2O_4$, **305A**, on methylation. Like **304**, the ^1H NMR and ^{13}C NMR spectra were suggestive of the presence of hydroxygirinimbine units, in which methyl and hydroxyl groups were at positions 3, 3′ and 7, 7′ respectively. NOE experiments on dimethyl ether **305A** indicated that the two girinimbine units, unlike in congener **304**, were linked through C-8 of the first and C-6′ of the second unit, thus leading to structure **305**.

6. Murranimbine

Murranimbine (**306**), $C_{36}H_{34}N_2O_2$ (M^+526), obtained as an oily racemate from *M. euchrestifolia* (*41*) had UV and ir spectra of a carbazole whose mass spectral peak at m/z 263.13 showed it to be a dimeric alkaloid. The ^{1}H NMR spectrum indicated the presence of two aryl methyls (δ2.30, 2.32), four geminal methyls at δ0.60, 1.54, 1.44, 1.60 attached to two carbons on oxygen and one D_2O exchangeable NH signal (δ10.55) which suggested that one of the carbazole nitrogens was involved in a linkage with the second unit. The presence of two sets of four spin aromatic systems and two deshielded signals at δ7.65 and 8.00 exhibiting long range coupling with an aromatic methyl in the H-H COSY spectrum indicated presence of an unsubstituted aromatic A-ring and H-4 in each carbazole nucleus. The spectrum further showed signals at δ2.74 (1H, *dd*, $J = 2.9, 5.1$ Hz, H-11′), 4.07 (1H, *m*, H-12), 2.21 (1H, *dd*, $J = 5.1$, 12.8, H-11a), 2.8 (H-11b obscued by the solvent signal) and a doublet at 6.12 ($J = 2.9$ Hz, H-12′) characteristic of a benzylic methine proton attached to a nitrogen atom indicating that the alkaloid contained two girinimbine units. The mutual attachment of the two rings was deduced by analysing the cross peaks resulting from three-bond correlations in the HMBC spectrum. Thus the benzylic methine proton attached to the N-atom (H-12′ at δ_H 6.12) was correlated with a benzylic methine carbon (C-12, δ_C 32.43) in the upper unit and with the quaternary carbon (C-10′, δ_C 80.13) in the lower unit, the latter being correlated with the benzylic methine proton (H-12, δ_H 4.07) in the upper unit. The methine proton (H-11′ at δ_H 2.74) coupled with the benzylic protons of H-12 and H-12′ at δ_H 6.12 and 4.07 had a three bond relationship with an aromatic carbon (C-1, δ_C 103.74) which in turn was correlated with the methylene protons (H-11 at δ_H 2.21) in the upper unit. Again the methylene carbon C-11 (δ_C 37.10) was related to the geminal dimethyl at δ_H 1.44 and 1.60 thus leading to structure **306**. The relative stereochemistry of **306** was deduced by NOE spectrometry; irradiation of H-12′ enhanced the signals of H-11′, H-12 and N-H, irradiation of H-11′ enhanced the signals of H-12 and H-12′ while irradiation of H-12 enhanced the signals of H-12′ and H-11′.

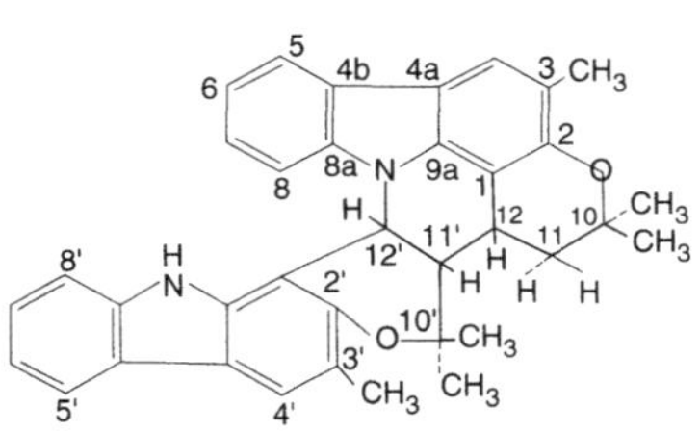

(306)

d) C_{46}-Alkaloid

1. Bismahanine

Bismahanine (**307**), $C_{46}H_{48}N_2O_4$ (M^+692), was isolated as a colorless oil from *M. euchrestifolia* (*49*). Its UV [λ_{max} 223, 245, 297, 331, 341 (sh), 359 (sh)] and ir (ν_{max} 3530, 3540, 1610 cm^{-1}) spectra showed it to be a carbazole derivative with hydroxyl functions. A peak at m/z 347 indicated that it was composed of two identical mass units. A peak at m/z 263 seemed to be that of a carbazolopyrilium ion like that obtained from girinimbine or mahanimbine but with an additional hydroxyl while a peak at m/z 347 could have arisen from a mahanimbine unit, also with an additional hydroxyl. The presence of carbazole skeletons having methyl-substituted pyran rings with 1,1-dimethyl-1-butene side chain was indicated by the ^1H NMR spectrum which showed signals at δ5.12 (1H, *t*, $J = 6.8$ Hz), 5.08 (1H, *t*, $J = 6.8$ Hz), 2.15 (4H, *m*), 1.77 (2H, *m*), 1.71 (2H, *m*) together with six methyl singlets and two pairs of AB type signals at δ6.67 and 5.71 (each 1H, *d*, $J = 9.8$ Hz) and δ6.48, 5.55 (each 1H, *d*, $J = 9.8$ Hz). The ortho coupled H-5 and H-6 protons of one unit appeared at δ7.84 and 6.95 (each 1H, *d*, $J = 8.3$ Hz), singlets at δ7.60 and 7.54 long range coupled to aryl methyls at δ2.33 and 2.31 were assigned to H-4 and H-4′ protons respectively while the other two singlets at δ7.89 and 7.16 were assigned to H-5′ and H-8′. The close similarity of these ^1H NMR data with that of bis-7-hydroxygirinimbine B (**305**) except for the signals due to the two side chains led to structure **307** for the dimer.

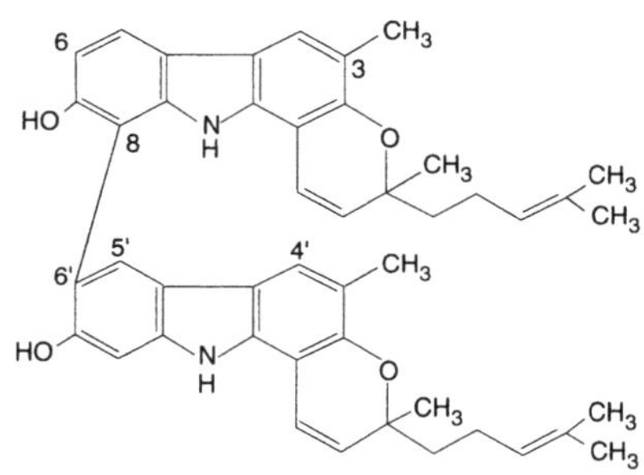

(307)

2. Bismurrayafoline C

Bismurrayafoline C (**308**), $C_{46}H_{52}N_2O_4$ (N^+696), was obtained as a pale yellow oil from *M. euchrestifolia* (*45*). Its UV spectrum [λ_{max} 226, 235, 260, 260 (sh), 285 (sh), 311, 335 (sh)] showed it to be a 1-oxygenated carbazole. The ^1H NMR spectrum exhibited signals of an arylmethyl group (δ2.48, *s*) ortho coupled H-5 and H-6 (δ7.69 and 6.85) and deshielded H-4 (δ7.85) and was very similar to that of murrafoline B except for the lack of an H-2 signal and signals due to a geranyl side chain while the presence of only half the number of proton signals expected from the molecular formula suggested that the dimer contained two completely symmetrical monomeric carbazole unts.

Irradiation of the aromatic methyl signal at δ2.48 enhanced the signals of H-4 and H-4′ while irradiation of the benzylic signal of the side chain at δ3.42 enhanced the signals of an allyl methyl at δ1.42 and N-H at δ7.55. In the dimethyl ether **309** irradiation of one methoxy signal at δ3.90 enhanced a doublet at δ6.85 (H-6, H-6′) suggesting that oxygens were at C-7 and C-7′ while irradiation of another methoxy signal at δ3.40 enhanced an N-H signal at δ7.85 and an aryl methyl singlet at δ2.52 located on another carbazole unit. Irradiation of this last enhanced a signal at δ7.85 (H-4, H-4′) and a methoxy signal at δ3.40 located on another carbazole unit. These results lead to formulation of bismurrayafoline C as **308**.

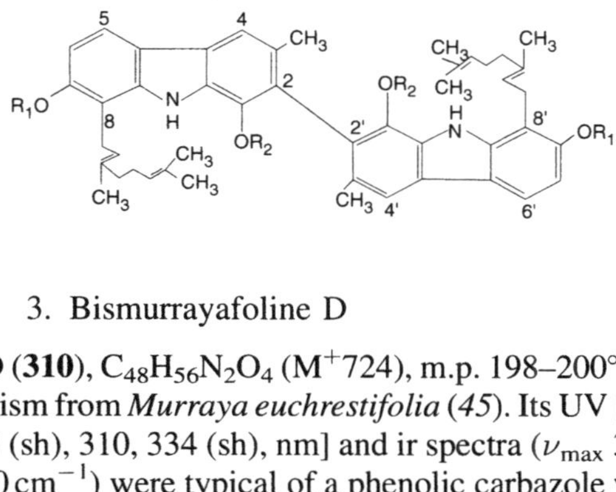

(**308**) $R_1 = R_2 = H$
(**309**) $R_1 = CH_3 = R_2$
(**310**) $R_1 = CH_3$; $R_2 = H$

3. Bismurrayafoline D

Bismurrayafoline D (**310**), $C_{48}H_{56}N_2O_4$ (M^+724), m.p. 198–200°, was obtained as colorless prism from *Murraya euchrestifolia* (*45*). Its UV [λ_{max} 224, 238, 260 (sh), 285 (sh), 310, 334 (sh), nm] and ir spectra (ν_{max} 3600, 3480, 3425, 3380, 1620 cm^{-1}) were typical of a phenolic carbazole while mass fragments at m/z 239 (M^{2+}-C_9H_{15}) and 293 (M^{2+}-C_5H_9) showed the presence of a geranyl chain. The ^1H NMR spectrum was very similar to that of **308** except for the presence of an additional methoxy signal at δ3.89. Methylation with diazomethane gave a dimethyl ether identical with dimethyl **309** from bismurrayafoline C. Irradiation of the signal at δ3.89 enhanced a doublet at δ6.85 (H-6, H-6′) indicating the location of the methoxy group at C-7, C-7′ thus leading to structure **310** for bismurrayafoline D.

B. Alkaloids from Lower Plants

Work on alkaloids of microbial and marine organisms has contributed interesting information to knowledge of chemistry and biology of carbazole alkaloids. Some of the results were described in earlier reviews (*11*, *13*).

Additional alkaloids including some not reported in previous reviews are described in this section. The first alkaloid from lower plants, also the first indolocarbazole alkaloid, staurosporine (**224**), was reported from *Streptomyces staurosporeus* in 1977 (*86a*). It contained two nitrogenous fragments while some others isolated subsequently contained only one such unit.

i) Alkaloids from Microbial Sources

a) Alkaloids Built on a Carbazole Skeleton (Single Nitrogen)

1. Aflavazole

Aflavazole (**222**), $C_{28}H_{35}NO_2$ (M^+417.1694), m.p. 156–60°, was isolated from *Aspergillus flavus* (*109*). The UV spectrum [λ_{max} 218, 243, 263, 297, 327, 341 nm with logε 15300, 17300, 7550, 7600, 1300, 1600] indicated that it was a carbazole and the degree of unsaturation in the formula suggested that it was a hexacyclic carbazole base, while similarity of its physical properties to those of aflavinine (**311**) as well as ^{1}H NMR, COSY, and INEPT spectral analysis indicated that the carbazole nucleus was 2,3,4-trisubstituted and a fragment matching the C11–C20 portion of the aflavinine ring system was attached to it (*110, 111*). The spectrum further showed the presence of an isolated unit and an aromatic methyl benzylically coupled to a broad singlet at δ7.12.

Irradiation of the proton signal at C-11 enhanced carbon singlets at C-3, 10 and 23 (δ_C119.80, 126.40) indicating direct attachment of C-11 to carbazole C-10, a conclusion supported by the downfield shift of H-11 at δ64.18. The protons of the aromatic methyl on C-26 correlated with three carbon signals at δ_C126.40, 135.13 and 110.25 (C-23, 24, 25), an observation locating the methyl group *meta* to C-11 and *ortho* to H-25. Irradiation of one of the signals at δ2.89 (H-22) enhanced the carbon signals at δ135.80, 126.40, 135.13 (C-10, 23, 24) indicating linkage of C-22 to C-23 while irradiation of another proton at δ2.15 (H-21) enhanced signal of C-19 and C-20 (δ_C71.47, 45.23) thus leading to structure **222** for aflavazole while the relative stereochemistry shown in (**222**) was assigned by analogy to that of aflavinines (*111*).

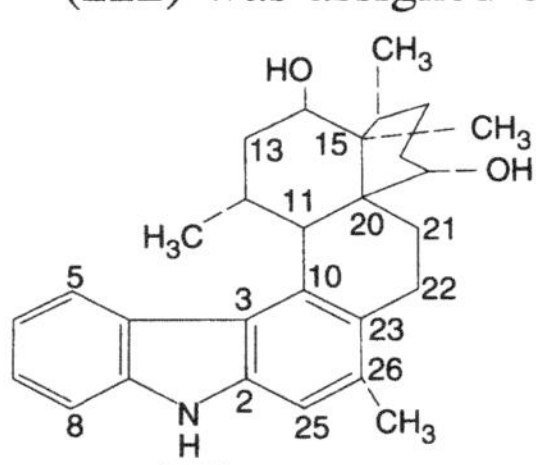

(**222**)

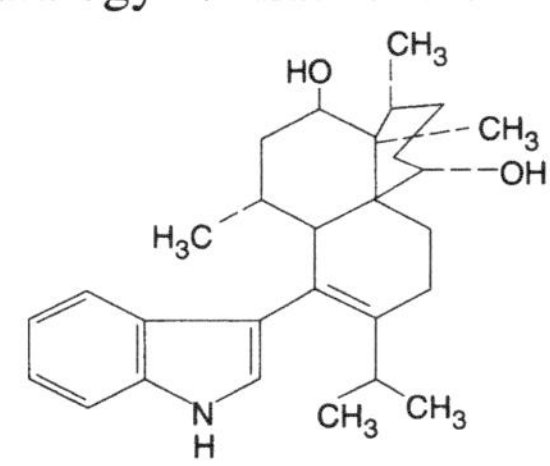

(**311**)

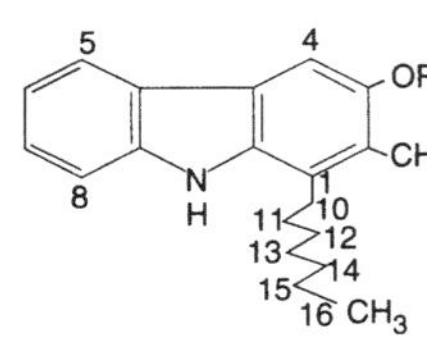

(**110**) R = H

(**110A**) R = CH_3

2. Carazostatin

Carazostatin (**110**), $C_{20}H_{25}NO$ (M^+295) m.p. 149.50°, was isolated from *Streptomyces chromofuscus* DC 118 (*60*). Its UV (λ_{max} 218, 235, 254, 266, 303, 342 nm with logε 153.000, 141.000, 141.000, 60000, 83000, 2000) and ir spectra (ν_{max} 3460, 3360 cm^{-1}) suggested that it was a 3-hydroxycarbazole. The ^{1}H NMR spectrum indicated presence of an unsubstituted ring A as well as signals of H-4 (δ7.3, *s*) and an aromatic methyl group (δ2.38). The signals of the protons of the n-heptyl chain appeared at δ2.87, 1.65, 1.46, −1.30 (8H, *m*) and 0.91 (terminal methyl, *t*, $J = 7.0$ Hz). The aromatic methyl group was assigned to C-2 as in many carbazoles of microbial origin while the alkyl chain was at C-1. Methylation of carazostatin gave a dimethyl derivative **110A** whose ^{1}H NMR spectrum had signals of an aromatic methoxyl at δ4.05 and an N-methyl group at δ3.9. Hence carazostatin was formulated as 1-heptyl-2-methyl-3-hydroxycarbazole (**110**) and its dimethyl derivative as **110A**.

3. Carquinostatin A

Carquinostatin A (**312**), $C_{21}H_{23}NO_3$ (M^+338.1757) m.p. 144–45°, isolated as reddish powder from *Streptomyces exfoliatus* 2419-SVT2 (*102*), had a UV [(λ_{max} 230 (32, 200), 267 (29, 700), 425 (5400)] and ir spectrum [(ν_{max} 3450, 3220, 1660 (sh), 1640 (sh), 1615 cm^{-1})] characteristic of a carbazole nucleus with quinone carbonyl functions, while the ^{13}C-chemical shift of C-9a (δ146.4) and long range coupling of one quinone carbonyl carbon (C-3) at δ183.8 to an aromatic methyl (H-13) suggested the presence of an ortho quinone system. The ^{1}H NMR spectrum together with ^{1}H-^{13}C long range coupling proved that a 1,2,4-trisubstituted benzene ring with a dimethylallyl residue at C-6 was fused by way of the pyrrole ring to a 1,2-quinone with a methyl and a 2-hydroxypropyl residue at contiguous position as shown in **312A**. Hence **312** was the structure of carquinostatin A, the absolute stereochemistry at C-11 having been determined by the Mosher method (*19a*).

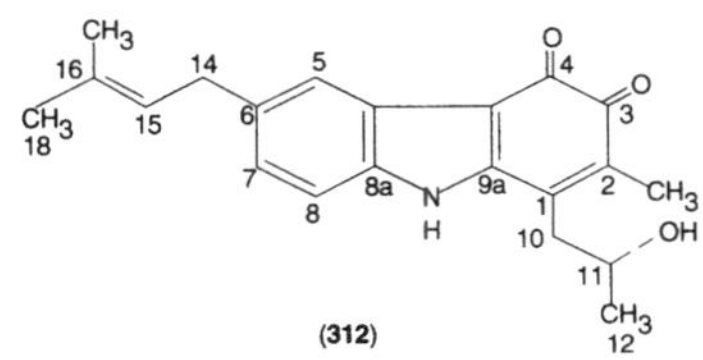

(**312**)

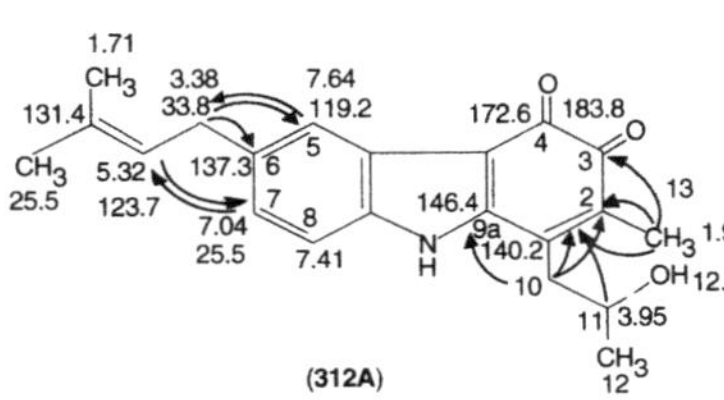

(**312A**)

:shows ^{1}H-^{13}C long-range couplings

b) Indolocarbazoles

i) Alkaloids Built on an Indolocarbazole Skeleton (Two Nitrogens)

Staurosporine (**224**), the first member of the indolocarbazole group of alkaloids, was isolated in 1977 (*86a*). Since then there has been substantial progress in the chemistry and biology of such alkaloids. The alkaloids are listed in alphabetic order below.**

1. Antitumor Compound AT 2433-A1

The antitumor compound AT 2433-A1 (**313**), $C_{34}H_{35}N_4O_9CL$ (M^+678 by FD), isolated as a yellow solid from *Actinomadura melliaura* ATCC 39691 (*33*) had a UV spectrum (λ_{max} 200, 235, 283, 316, 395 nm with logε 30781, 39934, 34307, 45562 and 3865) which suggested the presence of an indolocarbazole chromophore like that of rebeccamycin. The ir spectrum showed the presence of -NH-, hydroxyl and cyclic imide functions (ν_{max} 3120, 3365, 1750, 962 cm^{-1}). The ^1H NMR spectrum showed the presence of a N-CH_3 function (δ3.25) instead of the imide present in rebeccamycin (**116**). Both H-5 and H-13 were deshielded (δ9.27 and 9.18) due to the presence of the imide carbonyl function. Isotopic clusters in the mass spectrum indicated the presence of one chlorine atom. The ^1H and ^{13}C NMR spectrum showed the presence of a glucopyranoside fragment whose C-4′ methoxy (δ3.68) and C-1′ anomeric methine proton signals (δ6.91) together with the ^{13}C NMR data of C-1′, C-5′ and 4′-OCH_3 were very close to those of rebeccamycin. Location of the N-glycoside bond at N-12 was readily deduced by the downfield shift to 6.85 ppm of the anomeric proton (H-1′) compared with that of rebeccamycin (6.97 ppm) which was further supported by the shift of C-1′ at 81.7 ppm typical for an N-glycoside linkage. The downfield shift of C-6′ to 66.0 ppm compared with that of rebeccamycin suggested a further substitution at this position while the lack of the triplet at 5.32 ppm of the 6′-OH found in rebeccamycin indicated attachment of the aminosugar moiety at this position.

The structure of the aminosugar was elucidated by analyzing the MS, the ^1H NMR (taken at 110°C) and the CD spectrum of its dibromobenzoate methyl 2,4-dideoxy-4-N-methylaminopentapyranoside-3,4-dibromobenzoate, whose ^1H NMR spectrum showed the configuration at

** In consideration of the *Chemical Abstract* conventions for the tri-cyclic carbazole system, the numbering of carbon atoms in the aglycone fragments of the indolocarbazoles has been modified from that reported by the various authors (*11*).

C-1′ and equatorial substitution of the pyranose ring at C-3″ and C-4″ thus leading to structure (**313**).

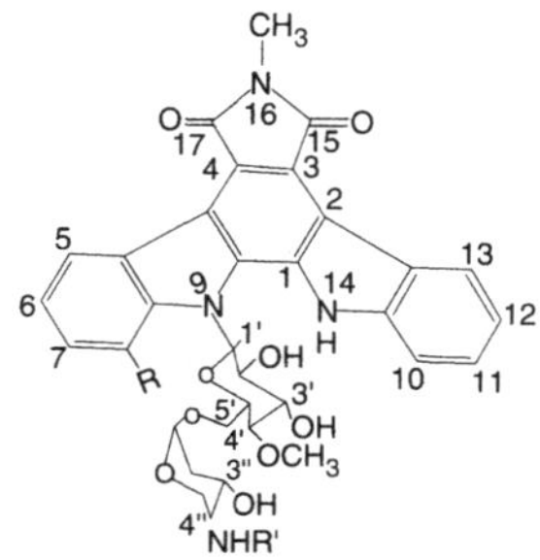

(**313**) R = Cl, R′ = CH_3
(**314**) R = Cl, R′ = H
(**315**) R = H, R′ = CH_3
(**316**) R = H, R′ = H

2. Antitumor Compound AT 2433-A2

Like antitumor compound At 2433-A1, antitumor compound AT 2433-A2 (**314**), $C_{33}H_{33}N_4O_9Cl$, (M^+664), was obtained from *Actinomadura melliaura* ATCC 39691 (*33*) as a yellow solid. Its structure was deduced by comparing the ^{1}H NMR spectrum and other physical properties with those of AT 2433-A1 which showed that the aglycone part was identical with that of AT 2433-A1 while the aminosugar part of AT 2433-A2 was N-demethylated as compared with that of AT 2433-A1. Hence it had structure **314**.

3. Antitumor Compound AT 2433-B1

Antitumor compound AT 2433-B1 (**315**), $C_{34}H_{36}N_4O_9$ (M^+644) was obtained as a yellow solid from *Actinomadura melliaura* ATCC 39691 (*33*). Its structure was deduced from the ^{1}H NMR and other physical data. Since the aglycone part of AT 2433-B1 lacked the chlorine atom at C-9, there were shifts in the proton and carbon spectra as compared with the spectra of AT 2433-A1 and AT-2433-A2, while the aminosugar fragment of AT-2433-B1 was identical with that of AT 2433-A1. Hence the structure of AT-2433-B1 was **315**.

4. Antitumor Compound AT 2433-B2

Antitumor compound AT 2433-B2 (316), $C_{33}H_{34}N_4O_9$ (M^+630), was also obtained from *Actinomadura melliaura* ATCC 39691 (*33*) as a yellow solid. Its structure was deduced by comparing the ^{1}H NMR and other physical data with those of AT 2433-A1 and its congeners. The aglycone part was identical with that of AT 2433-B1 while the sugar fragment was identical with that of AT 2433-A2, thus leading to structure **316**.

5. Arcyriaflavin B

Arcyriaflavin B (**54**), $C_{20}H_{13}N_3O_3$ (M^+343) m.p. 350°, was isolated together with the indolic pigments arcyriarubin B (**154**) and arcyriarubin C (**155**) from the fruiting bodies of the slime mould *Arcyria denudata* (*103*). Its UV spectrum was similar to that of staurosporine (**224**) suggesting the presence of an indolocarbazole skeleton. Structures of arcyriarubin B and arcyriarubin C provided clues for the structure of the arcyriaflavins. Thus ir bands at 1750 and 1705 cm^{-1} and the ^{13}C NMR signals at δ175.3 and δ129.3 suggested the presence of a maleimide unit in arcyriaflavin B. The ^{1}H NMR signals of H-2 and H-2′ in arcyriarubin B were absent showing the presence of an indolocarbazole skeleton in **54**. Signals of H-5 and H-13 (δ8.95) in the NMR spectrum showed that the maleimide fragment was fused to the carbazole skeleton at the 3,4 positions. Conversion of arcyriarubin B to arcyriaflavin B on treatment with H_2SO_4 then established the structure **54**. The structure has been confirmed by synthesis (see p. 147–148, 160).

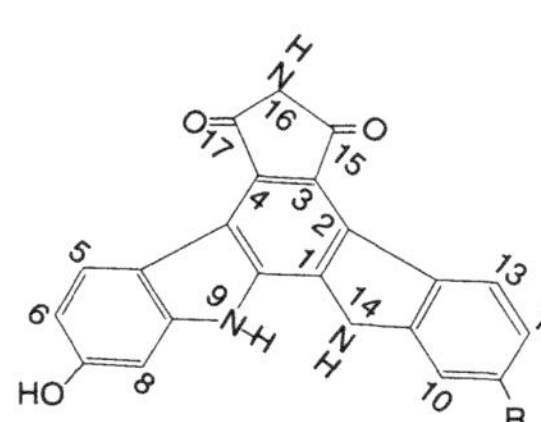

(**54**) R=H
(**155**) R=OH

6. Arcyriaflavin C

Arcyriaflavin C (**156**), $C_{20}H_{13}N_3O_4$ m.p. 350°, was isolated from *Arcyria denudata* along with arcyriaflavin-B. Like its congener (**54**) its UV spectrum resembled that of staurosporine. The presence of the maleimide moiety in arcyriaflavin C was detected from its ir band at 1705 cm^{-1} and the ^{13}C NMR signal at δ129.2. Arcyriaflavin C lacked the H-2 and H-2′ signals of arcyriarubin C showing that the two indolic units, like those in arcyriaflavin B, were linked through the 2- and

2′-positions. The deshielded H-5 and H-13 signals (δ8.93) suggested that the maleimide fragment was fused to the carbazole skeleton at the 3,4-position. Arcyriaflavin C was obtained from arcyriarubin C (**155**) on treatment with H_2SO_4.

7. Protein Kinase C Inhibitor K-252a

Protein kinase C inhibitor K-252a (**317**), $C_{27}H_{21}N_2O_5$ (M^+467), m.p. 262–273°, $[\alpha]_D^{CHCl_3} -23°$, was isolated from *Nocardiopsis sp.* K-252 and K-290 (**119**). The UV and ir spectra indicated that it was an indolocarbazole with NH, OH, ester and amide functions. ^{1}H and ^{13}C NMR spectra showed the presence of one secondary amide and one carbomethoxy group, one tertiary hydroxyl, one quaternary carbon, one methine, two methylenes, one methyl and eight sp^2 carbons. Decoupling revealed the presence of two 1,2-disubstituted benzenes, a γ-lactam and a sugar moiety while ^{13}C NMR spectrometry showed the presence of six quaternary carbons of which the γ-lactam ring contained two (δ119.5, 132.9) and two benzene rings contained one each (δ114.6, 115.8). These four quaternary carbons together with the remaining quaternary carbons (δ128.3, 123.9 connecting the sugar moiety through heteroatom) constituted another aromatic ring system. Irradiation of methylene protons at δ5.04 and 5.00 enhanced the signal of H-5 at δ8.05 indicating the position of methylene at C-13, thus leading to the aglycone of **317** which was identical with that of staurosporine (**224**).

COLOC experiments determined the structure of the sugar moiety as a 2-deoxyfuranoside, with a carbomethoxy group attached to the C-3′, a methyl to C-4′, C-1′ linked to N-9 and C-4′ to N-14. Structure **317** with the 3′-configuration shown in the formula was finally established by X-ray crystallography.

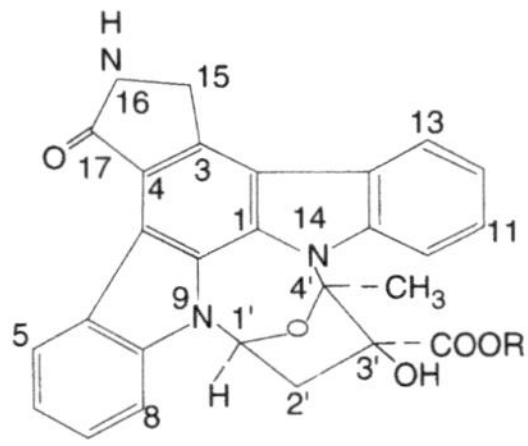

317 K-252a R=CH_3
318 K-252b R=H

8. Protein Kinase C Inhibitor K-252-b

Protein kinase C inhibitor K-252b (**318**), $C_{29}H_{19}N_3O_5$, m.p. 262–266° (dec.), $[\alpha]_D^{DMF} + 97$, isolated from *Nocardiopsis sp.* K-252 and

K-290 (*119*) had UV and ir spectra similar to those of protein kinase C inhibitor K-252a. Comparison of the ^{1}H NMR and ^{13}C NMR spectra of protein kinase C inhibitors K-252b and K-252a indicated that K-252a was the methyl ester **318** of K-252b. This was confirmed by hydrolysis to K-252b.

9. Protein Kinase C Inhibitor K-252c

Protein kinase C inhibitor K-252c (**319**), $C_{20}H_{13}N_3O$ (M^+371.1079), m.p. 300° was also isolated from *Nocardiopsis sp.* K-252 and K-290 (*119*). The presence of an indolocarbazole chromophore was evident from the UV spectrum [λ_{max} 230, 238, 246, 257, 287, 320, 331, 341, 358 nm with logε 37000, 34000, 28000 (sh), 29000 (sh), 86000, 16000, 20000, 16000, 1100]. The ^{1}H NMR spectrum showed signals of two NH-protons (δ11.46 and 11.38) and a secondary amide (δ8.49) whereas the indole nitrogens were unsubstituted. Drastic hydrolysis of K-252a furnished **319** which was thus conformed as the aglycone of **317**.

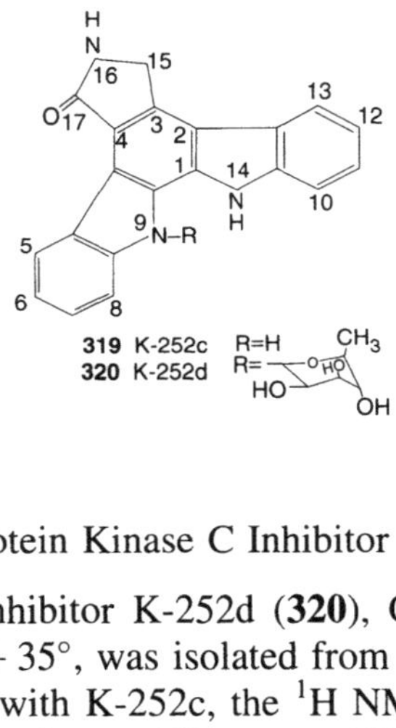

319 K-252c
320 K-252d

10. Protein Kinase C Inhibitor K-252d

Protein kinase C inhibitor K-252d (**320**), $C_{26}H_{23}N_3O_5$, m.p. 240–245° (dec.), $[\alpha]_D^{CH_3OH}$ + 35°, was isolated from *Nocardiopsis sp.* K-252 and K-290. Compared with K-252c, the ^{1}H NMR spectrum of K-252d exhibited the signal of only one NH proton (δ11.68) instead of the two NH signals (δ11.46, 11.38) of K-252c as well as protons assigned to a sugar moiety in addition to those of the aglycone. This suggested that **320** was an N-glycoside of K-252c. The high resolution electron impact (EI) mass spectrum showed that the composition of the sugar moiety was $C_6H_{14}O_4$. Hydrolysis of K-252d gave rhamnose and K-252c. Detailed analysis of the ^{1}H and ^{13}C NMR spectral data confirmed the structure as 9-N-(α-L-rhamnopyranose)-K-252c.

References, pp. 224–230

11. Rebeccamycin

Rebeccamycin (**116**), $C_{27}H_{21}N_3O_7Cl_2$ ($M^+ + 1$, 570), m.p. 326–330° (decomp.), $[\alpha]_D^{THF} + 131°$, from *Nocardia aerocoligenes* (*85*) had UV [λ_{max} 238, 256 (sh), 293 (sh), 314, 362 (sh) nm] and ir (ν_{max} 3418, 3350, 1752 cm^{-1}) absorption showing it to be an indolocarbazole with hydroxyl and amide carbonyl groups. It furnished a tetracetate (**116A**). The ^1H NMR spectrum of the parent compound exhibited signals of an amide NH (δ11.37) and an indole NH (δ10.30), as well as the signals of deshielded H-5 and H-13 aromatic protons and of a sugar unit. Deshielding of the H-5 and H-13 signals was attributed to a phthalimide function whereas in the case of staurosporine (**224**) which contains only one lactam carbonyl only H-5 is deshielded. The chemical shift of H-1′ of rebeccamycin and its tetracetate suggested that the sugar residue whose presence was indicated by the NMR spectrum might be attached to a carbon or nitrogen atom. Structure **116** was finally established by X-ray crystallography (*85*). The absolute configuration was however established eventually only by synthesis (see p. 155–156).

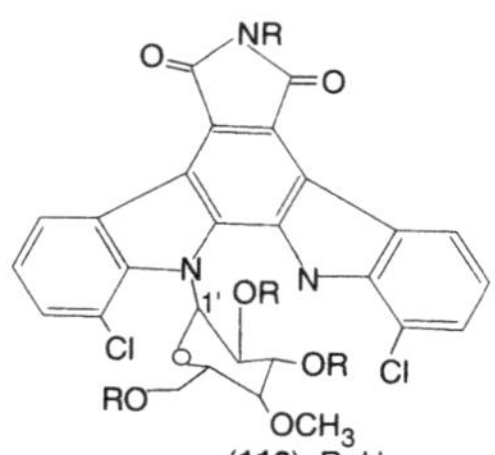

(**116**) R=H
(**116A**) R=OAc

12. Staurosporine

Staurosporine (**224**), $C_{29}H_{26}N_4O_3$ (M^+466), m.p. 270° (dec.) $[\alpha]_D^{MeOH} + 35°$, was isolated from *Streptomyces staurosporeus* Anaya D, Takahashi and Omura sp. nov. (*86a*). Its UV (λ_{max} 243, 263 (sh), 292, 322 (sh), 335, 356, 372 nm) and ir (ν_{max} 3200, 3500 cm^{-1}) spectra showed the presence of an aromatic system with hydroxyl and amino groups. Structure and stereochemistry of the alkaloid were established by X-ray crystallography of its methanol solvate (*32*). Subsequently Meksuriyen and Cordell (*76*) made definite assigments for the signals in the ^1H and ^{13}C NMR spectra. Several syntheses of staurosporine have also confirmed the structure (**224**) assigned to it (see pp. 157–160).

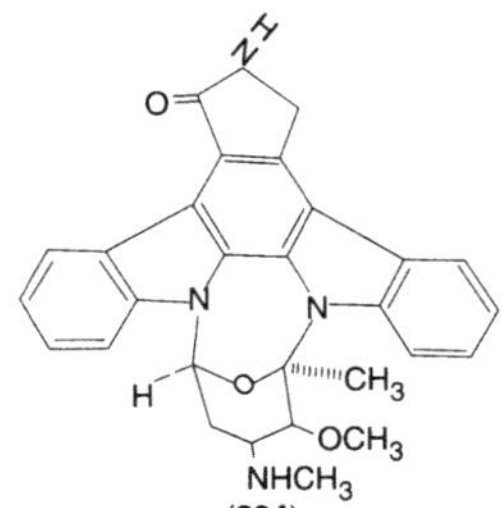

(**224**)

13. Tan-1030 A

Tan 1030-A (**321**), $C_{27}H_{22}N_4O_4$ $[(M+H)^+$ 467], m.p. 290–95° (dec.), was isolated from *Streptomyces* sp. C-71799 by Tanida *et al.* (*107, 112*) and thought it to be an indolocarbazole alkaloid with NH, OH and amide functions from its UV [λ_{max} 233, 344, 263 (sh), 275 (sh), 289, 329 (sh), 333, 252, 369 nm with logε 29400, 28000, 31300, 4200, 71000, 13400, 17700, 12000, 13400] and ir spectra (ν_{max} 3430, 1680 cm^{-1}). The ^1H and ^{13}C NMR spectra were very similar to those of staurosporine as they indicated the presence of an amide group, nineteen sp^2 carbons, a quaternary carbon, two methines, two methylenes, one methyl and one methoxy group. On hydrogenation, the alkaloid furnished an amine **322**, $C_{27}H_{24}N_4O_3$ (M^+452) $[\alpha]_D+90.4°$ which gave an N-acetyl derivative **327A**, $C_{29}H_{26}N_4O_4$ (M^+474).

Studies using modern NMR techniques provided further information about the structure, ^1H-^{13}C COSY and correlation spectroscopy via long range coupling (COLOC) substantiated the presence of a γ-lactam, two 1,2-disubstituted benzene rings, one of which had two quaternary carbons and a methoxy methine carbon, the methoxy carbon (δ58.24) of which coupled with a methine proton (H-4′, δ4.73) and vice versa. An oxime function was detected by long range coupling. The oxime carbon was correlated with methine protons H-4′ and H-1′ and a $CHCH_2$ group and the methoxy methine group were connected through the oxime function, thus forming a part of a six membered ring. The downfield shifts of the methine (C-1′, δ82.17) and the quaternary carbon (C-5′, δ96.16) showed that these two carbons were bonded to heteroatoms (N, N, O). The results of COLOC experiments with $J = 12.5$ Hz (*81a*) showed that ten quaternary carbons were connected. The three remaining quaternary carbons appeared at δ114.98, 114.02 and 128.03 in refocussed INEPT experiments (*81a*). The next COLOC experiments with $J = 4.2$ Hz showed that two quaternary carbons C-4a (δ114.98) and C-3a (δ114.02) were coupled with H-5 (δ9.31) and H-13 (δ87.98) respectively, while the downfield shifts of two quaternary C-14a and C-9a at δ139.82 and δ136.01 indicated that these carbons were bonded to two nitrogen atoms. Irradiation of the methylene protons (H-15) enhanced an amide NH proton (H-16, δ8.58) and H-13 while irradiation of the methyl (δ2.47, H-6′) and the methine (δ7.04, H-1′) enhanced the H-10 and H-8 signals.

From this it was evident how the lactam function, the disubstituted benzene rings and the hexacyclic sugar fragment were connected. Irradiation of the oxime proton at δ10.45 enhanced methoxy (δ3.3) and methyl protons (δ2.47, H-6′) which settled the stereochemistry of the oxime group. Thus Tan-1030 A was assigned structure **321**.

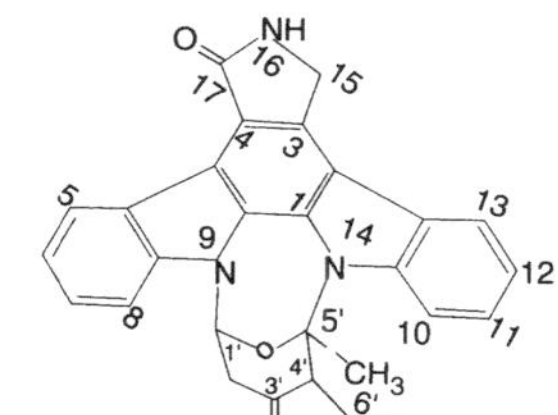

(321) Tan 1030, R= =NOH
(322) R=NH$_2$
(322A) R=NHCOCH$_3$

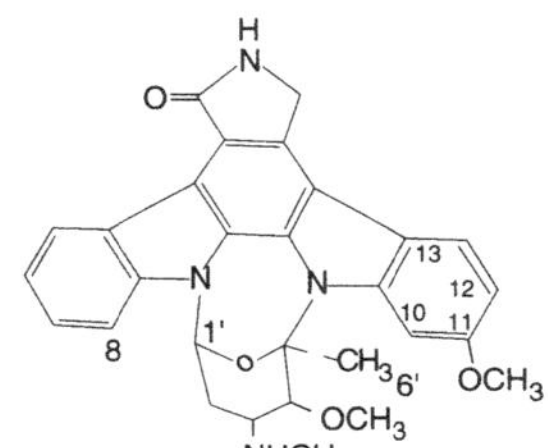

(323) Tan-999

14. Tan-999

Tan-999 (**323**), $C_{29}H_{28}N_4O_4$ (M^+496), m.p. 221° (dec.), $[\alpha]_D + 42°$ (DMF), accompanied Tan-1030 A from *Nocardiopsis dassonville* (*112*, *106*). Its UV and ir spectra were similar to those of Tan-1030 A and staurosporine which suggested the presence of an indolocarbazole skeleton with a lactam function. Comparison of the ^{1}H NMR spectrum with that of staurosporine showed that the former, in addition to one more methoxy group than staurosporine, had a 1,2,4-trisubstituted instead of one of the two 1,2-disubstituted benzene rings of staurosporine. The position of the additional methoxy group was determined by refocussed INEPT and COLOC experiments (*81a*). The COLOC spectrum ($J = 4.2$ Hz) showed a correlation between a quaternary carbon (δ157.49, C-11), the methoxy protons (δ3.95) and H-13. A cross peak between the methoxy protons and two aromatic protons (H-12 and H-10) was detected from the NOESY data. Irradiation of the methyl signal at δ2.33 (H-6′) enhanced H-10 indicating that the methoxy group was substituted at C-11. The NOESY spectrum also showed a correlation between H-15 (δ4.96) and H-13 and that of H-1′ and H-8. Hence structure **323** was assigned to Tan-999.

15. UCN-01

UCN-01 (**324**), $C_{28}H_{26}N_4O_4$ (M^+466), m.p. 245–250° (decom.) $[\alpha]_D^{MeOH}$ + 132.0°, was isolated from *Streptomyces* sp. (*104*). The presence of an indolocarbazole skeleton was evident from the UV and ir data. The ^{13}C spectrum showed the presence of one amide group, eighteen sp^2 carbons, eight aromatic protons, one quaternary carbon, four methines, one methoxy group, one N-methyl group, one methylene group and one methyl group, the spectrum being very similar to that of staurosporine except for the absence of the signal at δ45.3 (*t*, C-15). Analogously in the ^1H NMR spectrum the signal of δ4.95 (2H, *s*) of the 15-methylene of staurosporine was replaced by signals of D_2O exchangeable proton at δ6.44 (1H, *d*, $J = 9.8$ Hz) and a second signal at δ6.39 (1H, *dd*, $J = 9.8$, 1.1 Hz). Irradiation of an imino proton (H-16) at δ8.72 (1H, *d*, $J = 1.1$ Hz) enhanced the signals at δ6.39. Hence the signals at δ6.44 and δ6.39 were assigned to the 15-hydroxyl and 15-methine protons, respectively leading to structure **324** for UCN-01.

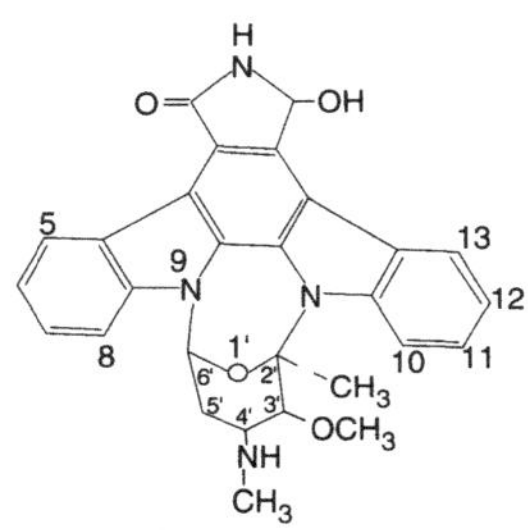

(**324**) UCN-01

ii) Alkaloids from Marine Sources

Hyellazole and 6-chlorohyellazole, the first two alkaloids from a marine source *Hyella caespitosa*, were described in a previous review (*13*). Since then several other alkaloids have been reported from marine sources.

1. 1-Methylcarbazole

1-Methylcarbazole (**325**), $C_{13}H_{11}NO$ (M^+ 181.0811) m.p. 120°, was isolated from *Tednia ignis* (*23*). In the ^1H NMR spectrum the H-1 signal of carbazole was replaced by a methyl signal of an aromatic methyl group (δ2.56).

References, pp. 224–230

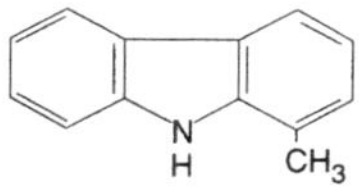
(325)

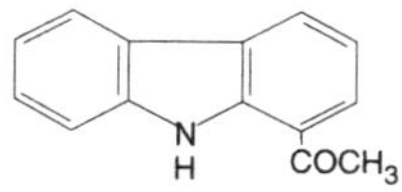
(326)

2. 1-Acetylcarbazole

1-Acetylcarbazole (**326**), $C_{14}H_{11}NO$ (M^+ 209.0811) was isolated from the same source (*23*). The ir spectrum showed the presence of an NH and an acetyl function (ν_{max} 3444, 1671 cm^{-1}) while the ^1H NMR spectrum contained all signals of carbazole systems except H-1, with C-1 being substituted by an acetyl group (δ2.8, 3H, *s*) as also shown by the ^{13}C NMR spectrum.

3. Aldose Reductase Inhibitors

From a marine sponge, *Dictyodendrilla* sp., Sato *et al.* (*100*) isolated three pyrrolocarbazole alkaloids **327**, **328** and **329**. Compound **327**, $C_{43}H_{29}N_2O_{12}SNa^+$, [(M + H)m/z. 821.1479], m.p. > 300°, exhibited characteristic UV [λ_{max} 475 (1900), 391 (10,700), 324 (22100), 288 (23100), 229 (45800) nm] and ir (ν_{max} 3396, 1760, 1575, 1512, 1383, 1262 cm^{-1}) spectra. The ^1H NMR spectra contained two methylene signals at δ2.37 (2H, *m*), 2.97 (2H, *m*) ABC-type aromatic signals at δ5.74 (1H, *d*, $J = 8.6$ Hz, H-10), 6.61 (1H, *dd*, $J = 7.9$, 8.6, Hz H-9), 7.15 (1H, *d*, $J = 7.9$ Hz, H-8) and unresolved broad signals of protons while the ^{13}C NMR spectrum exhibited signals of two sp^3 carbons at δ36.8 (*t*) and 47.5 (*t*), a carbonyl carbon at δ181.6 (*s*) and many unassignable aromatic carbons. Chemical and enzymatic hydrolysis of (**327**) gave **330** whose ^1H NMR spectrum displayed significant upfield shifts of the H-8 (δ6.54) and H-10 (δ5.43) signals, suggesting that the position of the sulfate of **327** was at C-7 *ortho* to H-8 and *para* to H-10. The structure of **327** was finally established by X-ray crystallographic analysis of an acetate **330A** which showed that the aromatic ring at C-1 and the carbazole ring were nearly perpendicular to each other thus explaining the high-field resonances of H-10 and H-8 as being due to the aromatic ring current at C-1. Thus **327** possessed a unique pyrrolo-(2,3-C)-carbazole spirolactone system while **328** was the corresponding acid. The asymmetric centers C-2 of both compounds were apparently racemic.

Compound **329**, $C_{43}H_{33}N_2O_{11}S$ [(M + H)m/z 785.1826], m.p. >300°, had ^1H NMR and UV spectra similar to those of **327**, thus locating a sulfate at C-7. On treatment with trifluoroacetic acid both **329** (at room temperature) and **330** (under refluxing conditions) furnished an identical substance **329A** containing a carbomethoxy group (ν_{max} 1691) presence of the latter being supported by long range coupling between the methoxy protons (δ3.17) and a carboxyl carbon (δ171.3) in the HMBC. Correlations between the C-1″ methylene (δ2.98), C-2 (δ151.4) and C-3a (δ157.4) indicated the change in hybridization of the C-2 aminoketal carbon (δ106.4) of **330** from sp^3 to sp^2 in **329A**. Irradiation of the C-1″ methylene signal enhanced the signals of all *ortho* protons of the aromatic rings attached to C-4 and C-2″ indicating a *cis* arrangement of these substituents. Hence (**329**) could be formulated as shown.

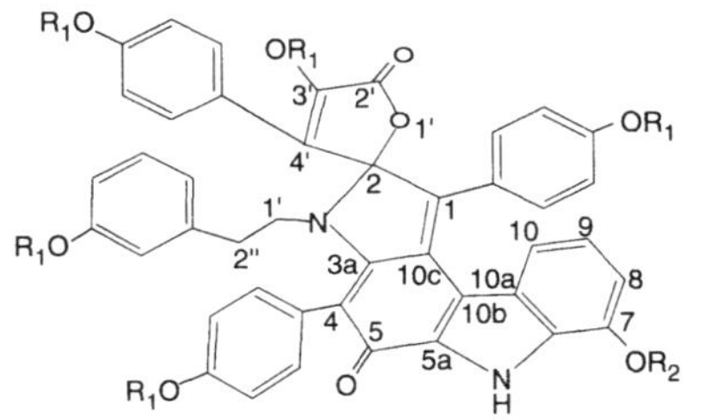

(**327**) $R_1 = H; R_2 = SO_3Na$
(**328**) $R_1 = H; R_2 = SO_3H$
(**330**) $R_1 = R_2 = H$
(**330A**) $R_1 = R_2 = OAc$

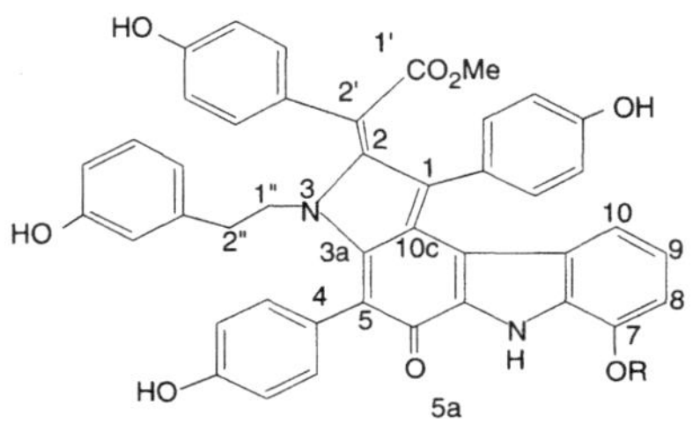

(**329**) $R = SO_3H$
(**329A**) $R = H$

References

1. Bhattacharyya P, Chakraborty DP (1987) Carbazole Alkaloids. In: Herz W, Grisebach H, Kirby GW (eds) Fortschr Chem Organ Naturstoffe, Vol 52, p 159. Springer-Verlag, Wien, New York

2. Birch AJ, Leipa AJ, Stephenson GR (1979) Organometallic Compounds in Organic Synthesis: Some Tricarbonyl (Cyclohexadienyl) Iron Cations and Nitrogen Containing Nucleophils. Tetrahedron Lett 3565

3. Broka CA (1996) Eur Pat EP 695–755, US Pat 286043/1994 [(CA **124**, 28951 (1996)]

4. Bush JA, Long BH, Catino JJ, Brander WT, Tomita K (1987) Production and Biological Activity of Rebeccamycin, a Novel Antitumour Agent. J Antibiot **40**: 668

5. Carruthers W, Evans N (1974) Photocyclisation of Stryrylindole. Synthesis of 1-, 2-, 3- and 4-Methyl-11H-benzo [a] carbazoles. J Chem Soc Perkin 1, 421

6. Chaichatipyuth C, Pummangure S, Naowseran K, Thanyavuthi D, Anderson J, McLaughlin JL (1988) Two New Bioactive Carbazole Alkaloids from the Root Bark of *Clausena harmandiana*. J Nat Prod **51**: 1285

7. Chakraborty A, Nath AC, Khasnobis S, Chakraborty M, Kored Y, Harigaya Y, Komiyana K (1997) Carbazole Alkaloids from *Murraya koenigii*. Phytochemistry **46**: 751
8. Chakraborty A, Chowdhury BK, Bhattacharyya P (1995) Clausenol and Clausenine, Two Carbazole Alkaloids from *Clausena anisata*. Phytochemistry **40**: 295
9. Chakraborty A, Saha C, Poddar G, Chowdhury BK, Bhattacharyya P (1995) Carbazole Alkaloid with Antimicrobial Activity from *Clausena heptaphylla*. Phytochemistry **38**: 787
10. Chakraborty DP (1977) Carbazole Alkaloids. In: Herz W, Grisebach H, Kirby GW (eds) Fortschr Chem Organ Naturstoffe, Vol 34, p 299. Wien-New York: Springer
11. Chakraborty DP (1993) Chemistry and Biology of Carbazole Alkaloids. In: Cordell GA (ed) Alkaloids, Vol 44, p 257. New York: Academic Press
12. Chakraborty DP, Barman BK, Bose PK (1965) On the Constitution of Murrayamine, A Carbazole Derivative Isolated from *Murraya koenigii* Spreng. Tetrahedron **21**: 681
13. Chakraborty DP, Roy S (1991) Carbazole Alkaloids III. In: Herz W, Kirby GW, Steglich W, Tamm Ch (eds) Fortschr Chem Organ Naturstoffe, Vol 57, p 71. Wien-New York: Springer
14. Chesire MK, Cranwell PK, Falshaw FF, Floyd AJ, Haworth RD (1967) Humic Acid-II, Structure of Humic Acids. Tetrahedron **23**: 1669
15. Chunchatprasert L, Rao KRN, Shanon PV (1992) A New Synthetic Route to Pyrrolo-[3, 2-b]-carbazoles, 1H-Benzofuro-[3, 2-f]-indole and 1H-[1] Benzo-thieno-[2,3 f-] indoles. J Chem Soc Perkin 1: 1779
16. Chunchatprasert L, Shanon PVR (1994) Further Observations on Novel Products from Acid Catalysed Indole-Pyrrole Condensation. Formation of Pyrrolo-[2, 3-b]. carbazole. J Chem Soc Perkin 1: 1965
17. Clemo GR, Perkin Jr WH, Robinson R (1927) Strychnine and Brucine, Part IV. J Chem Soc 1589
18. Clemo GR, Perkin Jr WH (1924) Vinyl Derivatives Specially of Carbazoles and Tetrahydro Carbozole and Their Behavior with Acids. J Chem Soc **115**: 1804
19. Clive DLG, Et Kin N, Joseph T, Lown JW (1993) Synthesis of Carbazomycin B. J Org Chem **58**: 2442
19a. Dale JA, Mosher HS (1973) Nuclear Magnetic Resonance Enantiomer Reagents. Configurational Correlations *via* Nuclear Magnetic Resonance, Chemical Shifts of Diastereomeric Mandelet, O-Methyl Mandelet and α-Methoxy-α-trifuromethyl-phenylacetate (MTPA) Esters. J Amer Chem Soc **95**: 512
20. Danheiser RL, Brisbois RG, Kowalczyk JJ, Miller RK (1990) An Annulation Method for the Synthesis of Highly Substituted Polycyclic Aromatic and Heteroaromatic Compounds. J Amer Chem Soc **112**: 3093
21. De Silva SO, Snieckus V (1974) The Witting Synthesis and Photo-cyclodehydrogenation of 3-(2-Aryl)-vinyl Indole Derivatives. Can J Chem **52**: 1294
22. De Silva SO, Snieckus V (1974) Photochemical Synthesis of Benzo-[c]-carbazole and Pyridocarbazoles. Synthesis 254
23. Dillman RL, Cardellina JH (1991) Aromatic Secondary Metabolites from the Sponge *Tednia ignis*. J Nat Products **51**: 1056
24. Erdtman H (1969) In: Harborne JB, Swain T (eds) Perspectives in Phytochemistry, p. 107. New York: Academic Press
25. Fiebig NJM, Soejarts DD, Kinghorn AD (1985) Koenoline, a Further Cytotoxic Carbazole Alkaloid from *Murraya koenigii*. Phytochemistry **24**: 3041
26. Ferris RM, Tang FLM, Chang KJ, Russel A (1986) Evidence that the Potential Antipsychotic Agent Rincazole (BW234U) is Specific Competitive Antagonistic of Sigma Sites in Brain. Life Science **38**: 2329

27. Furukawa H, Wu TS, Kuch C (1985) Structure of Murrafoline-B and -C, New Binary Carbazole Alkaloids from *Murraya euchrestifolia*. Chem Pharm Bull **33**: 2613
28. Furukawa H, Wu TS, Kuch C (1985) Chemical Constituents of *Murraya euchrestifolia* Hayata. Structure of Novel Carbazoquinones and Other New Carbazole Alkaloids. Chem Pharm Bull **33**: 4132
29. Furukawa H, Ito C, Wu TS, Mephail AT (1993) Structural Elucidation of Murrafolines, Six Novel Binary Carbazole Alkaloids Isolated from *Murraya euchrestifolia*. Chem Pharm Bull **41**: 1249
30. Furukawa H, Ito C, Yogo M, Wu TS (1985) Structure of Murrayastine, Murrayaline and Pyrayafoline, Three New Carbazole Alkaloids from *Murraya euchrestifolia*. Chem Pharm Bull **33**: 1320
31. Furukawa H, Ito C, Yogo M, Wu TS (1986) Synthesis of Pyrafoline B. Chem Pharm Bull **34**: 2672
32. Furusaki A, Hashiba N, Matsumoto T, Hirano A, Iwai Y, Omura S (1978) X-Ray Crystal Structure of Staurosporine: A New Alkaloid from a Streptomyces Strain. J Chem Soc Chem Comm: 800
33. Golik J, Doyle TW, Krishnan B, Dubay G, Matson JA (1989) AT 2433-A1, AT 2433-A2, AT 2433-B1, and AT 2433-B2, Novel Antitumour Compounds Produced by *Actinomadura melliaura*, II. Structure Determination. J Antibiot **42**: 1785
34. Gould ST, Tamay N, Meville CR, Cone MC (1994) Revised Structures for Kinamycin Antibiotics: 5-Diazobenzo-(b)-fluorenes Rather than Benzo-(b)-carbazole Cyanamides. J Amer Chem Soc **116**: 2207
35. Graede C, Glazer C (1872) Über Carbazol. Chem Ber **5**: 12
36. Hibino S, Tonari A, Chosi T, Suigino E (1993) New Approach to Antibiotic Carbazomycins: The Formal Total Synthesis of Carbazomycins-A and -B. Heterocycles **35**: 441 (CA.119,49705Y, 1993)
37. Hino T, Nagawa M (1988) In: Brossi A (ed) 'The Alkaloids', Vol 34, p 1. San Diego: Academic Press
38. Hughes I, Raphael RA (1983) Synthesis of Arcyriaflavin B. Tetrahedron Lett **24**: 1441
39. Husson HP, Thal C, Potier P, Wenkert E (1970) Photocyclisation of 1-(α-Indolyl)-2(β-pyridyl)-acrylonitrile. J Org Chem **35**: 442
40. Ignatiaias I, Dorbas M, Appino A (1986) Identification of Carbazole Derivatives in a Hydrotreated Gas Oil by Gas Chromatography and Gas Chromatography-Mass Spectrophotometry. J Chromatography **366**: 251
41. Ito C, Furukawa H (1991) Structure of Murranimbine, A Novel Dimeric Carbazole Alkaloid from *Murraya euchrestifolia*. Chem Pharm Bull **39**: 1355
42. Ito C, Furukawa H (1990) New Carbazole Alkaloid from *Murraya euchrestifolia*. Chem Pharm Bull **38**: 1548
43. Ito C, Ohta H, Tan HTW, Furukawa H (1996) Constituents of *Clausena excavata*, Isolation and Structural Elucidation of Seven New Carbazole Alkaloids and a New Coumarin. Chem Pharm Bull **44**: 2231
44. Ito C, Nakagawa M, Wu T-S, Furukawa H (1991) New Carbazole Alkaloids from *Murraya euchrestifolia*. Chem Pharm Bull **39**: 1668
45. Ito C, Nakagawa M, Wu T-S, Furukawa H (1991) New Carbazole Alkaloids from *Murraya euchrestifolia*. Chem Pharma Bull **39**: 2525
46. Ito C, Okahana N, Wu T-S, Wang ML, Lai JS, Khon CS, Furukawa H (1992) New Carbazole Alkaloids from *Murraya euchrestifolia*. Chem Pharm Bull **40**: 230
47. Ito C, Katsuno S, Ohta H, Omura M, Kajiura I, Furukawa H (1997) Constituents of *Clausena excavata*. Isolation and Structure Elucidation. Chem Pharm Bull **45**: 48

48. Ito C, Wu T-S, Furukawa H (1990) Novel Binary Carbazole Alkaloids from *Murraya euchrestifolia*. Chem Pharm Bull **38**: 1143
49. Ito C, Thoyama Y, Omura M, Kajiura I, Furukawa H (1993) Alkaloidal Constituents of *Murraya koenigii.* Isolation and Structural Elucidation of Novel Binary Carbazoloquinone and Carbazole Alkaloids. Chem Pharm Bull **44**: 2096
50. Jackson PM, Moody CJ, Mortimer RJ (1991) Synthesis and Electro-chemical Properties of Naturally Occurring Free Radical Scavenger Carazostatin. J Chem Soc Perkin 1: 2941
51. Kadiri A, Fadouch M, Benali B, Boucetta A (1994) The Effects of Complexation by Hydrogen Bonds and N-Ethylation on UV Absorption Spectra of Carbazoles. Spectrochim Acta **50A**: 851 (C.A. 1994, **120**: 3222638)
52. Kaneda M, Kitahara T, Yamasaki K, Nakamura S (1990) Biosynthesis of Carbazomycin B, II: Origin of Whole Carbon Skeleton. J Antibiotics **43**: 1623
53. Kaneko T, Wong H, Okamoto KT, Clardy J (1986) Two Synthetic Approaches to Rebeccamycin. Tetrahedron Lett **26**: 4015
54. Kano S, Sugino E, Shibuya S, Hibino S (1981) A New Synthesis of Antitumour 6-H-Pyrido-[4, 4-b]-Carbazole Alkaloid, Elipticine. J Org Chem **46**: 2979
55. Kano S, Sugino E, Shibuya S, Hibino S (1987) Synthesis of Carbazole Alkaloids Hyellazole and 6-Chlorohyellazole. J Org Chem **46**: 3856
56. Kapil RS (1971) In: Manske RHF (ed) 'The Alkaloids', Vol 12, p 273. New York-London: Academic Press
57. Kase H, Ewahasi K, Matsuda Y (1986) K252a, A Potent Inhibitor of Protein Kinase C from Microbial Origin. J Antibiot **39**: 1019
58. Kato S, Kawai H, Kawasaki T, Toda Y, Urata T, Hayakawa Y (1989) Studies of Free Radical Scavenging Substances from Microorganisms, I: Carazostatin. A New Free Radical Scavenger Produced by *Streptomyces chromofuscus* DC 118. J Antibiot **42**: 1879
59. Kawasaki T, Sakamoto M (1994) Electrocyclisation for Synthesis of Carbazoles and Their Annulated Compounds. J Indian Chem Soc **71**: 443 and references cited therein
60. Kazano S, Mosahiro, Kanno F, Yoichi H, Harno S (1986) Structure of Carazostatin, A Neutral Cell Promoting Substance Produced by Streptomyces. J Antibiot **39**: 1059
61. Kinoshita J, Tatara S, Sankawa U (1985) Structures of Paniculadins A and B, Novel Phenylindoles from *Murraya paniculata*. Chem Pharm Bull **33**: 1770
62. Kinoshita J, Tatara S, Yamasaki K, Nakamura S (1990) Biosynthesis of Carbazomycin B, III. J Anbibiot **43**: 1623
63. Knolker H-J (1995) Transition Metal-mediated Synthesis of Carbazole Derivatives. In: Advances in Nitrogen Heterocycles **1**: 173. JAI Press Inc and references cited therein
64. Knolker H-J, Bauermeister M (1989) The Total Synthesis of Carbazole Antibiotic Carbozomycin B and an Improved Route to Carbazomycin A. J Chem Soc Chem Comm 1468
65. Knolker H-J, Bauermeister M (1990) Iron-Mediated Total Synthesis of the Cytotoxic Carbazole Koenoline and Related Alkaloids. J Chem Soc Chem Comm 664
66. Knolker H-J, Wolpert M (1997) Transition Metal Complexes in Organic Synthesis, Part 36, Cyclization of Tricarbonyl Iron Complexes by Oxygen to 4a,9a-Dihydro-9H-carbazoles: Applications to the Synthesis of Mukonine, Mukonidine and Pyrido-[3, 2, 1-J]-carbazoles. Tetrahedron Lett **38**: 533
67. Kumar V, Wickramasinghe DBM, Jacobson U (1990) Indole Dimer from *Murraya gleni.* Tetrahedron Lett **31**: 5217

68. Kumar V, Reisch J, Wickramasinghe A (1989) Glycomaurin and Glycomaurol, New Carbazole Alkaloids from *G. mauritiana* Bark. Aust J Chem **42**: 1375
69. Kusano G, Sibano M, Idoji M, Minoura K (1993) A New Alkaloid 2-Hydroxy-7-methyl-9H Carbazole from *Cimicifuga simplex*. Heterocycles **136**: 2367
70. Ljunggren BO, Lundberg K (1985) *In vivo* Phototoxicity of Nonsteroidal Anti-inflammatory Drugs Evaluated by the Mousetail Technique. Photodermatology **2**: 377 (C.A. 104 1986 81699a)
71. Li WS, Mechesney JD, El-Feraly FG (1991) Carbazole Alkaloids from *Clausena lansium*. Phytochemistry **30**: 343
72. Luk KL, Stern L, Weigel M, O'Brien RA, Spirit N (1983) Isolation and Identification of 'Diazepam'-like Compounds from Bovine Urine. J Nat Prod **46**: 852
73. Magnus PD, Sear NL (1984) Indole-2, -3 Quinodimethanes, Synthesis of Selectively Protected Derivatives of Fused Dimeric Indole Alkaloid Staurosporine. Tetrahedron **40**: 2795
74. Martin T, Moody CJ (1985) Synthesis of Carbazole Alkaloid Murrayaquinone B. J Chem Soc Chem Commun: 1391
75. Matson JA, Claridge C, Bush JA, Titus J, Bradner WT, Doyle TW, Horan AC, Patel M (1989) AT 2433-A1, AT 2433-A2, AT 2433-B1 and AT 2433-B2, Novel Antitumour Antibiotic Compounds Produced by *Actinomadura melliaura*, Taxonomy, Fermentation, Isolation and Biological Properties. J Antibiot **42**: 1547
76. Meksuriyen D, Cordell GA (1988) Biosynthesis of Staurosporine 2. Incorporation of Tryptophan. J Nat Prod **51**: 893
77. Mitra AR (1974) PhD Thesis, Calcutta University
78. Mithani S, Weeratunga G, Taylor NJ, Dimitrierienko GI (1994) The Kinamycins are Diazofluorens and not Cyanocarbazoles. J Amer Chem Soc **116**: 2209
79. Moody CJ, Shah P (1989) Diels-Alder Reactivity of Pyrano-[3, 4-b]-indole-3-ones, Synthesis of Carbazoles Carbazomycin-A, -B and Hyellazole. J Chem Soc Perkin 1, 2463
80. Morioka H, Ishahara M, Shibai H, Suzuki T (1985) Agri Biol Chem **49**: 1959
81. Morioka H, Shibai H, Yokagawa Y, Ishahara M, Kida T, Suzuki T (C.A. **104** 1986, 18649u)

81a. Morris GA, Freeman R (1979) Enhancement of Nuclear Magnetic Resonance Signals by Polarization Transfer. J Amer Chem Soc **101**: 760

82. Mudry CA, Frasca AR (1974) Photochemical Reductions of 2,3-Diphenyl Indoles. Tetrahedron **30**: 2983
83. Nagadjui BT, Ayafor JF, Sondongam BL, Connolly JD (1989) Quinolone and Carbazole Alkaloids from *Clausena anisata*. Phytochemistry **28**: 1517
84. Nakanishi S, Matsuda Y, Iwahasi K, Kase H (1986) K-252a, b, c and d, Potent Inhibitors of Protein Kinase C from Microbial Origin. Antibiot **39**: 1066
85. Nettleton DE, Doyle TW, Krishnan B, Matsumoto GK, Clardy J (1985) Isolation and Structure of Rebeccamycin–A New Antitumour Antibiotic from *Nocardia aerocoligenes*, Tetrahedron **26**: 4011
86. Omura S, Nakagawa A, Yamada H, Hata T, Furusaki A, Watanabe T (1971) Structure of Kinamycin C and Structural Relations Among Kinamycin A, B, C, D. Chem Pharm Bull **19**: 2428

86a. Omura S, Iwai Y, Hirano A, Nakagawa A, Awaya J, Suchiya HT, Takahashi Y, Masuma R (1977) A New Alkaloid AM-2282 of Streptomyces Origin: Taxonomy, Fermentation, Isolation and Preliminary Characterisation. J Antibiot **30**: 275

87. Patterson AM, Capbell LT (1940) The Ring Index P 229 No 1675. New York: Reinhold

88. Pearce CJ, Doyle TW, Forenza S, Lem KS, Schroeder DS (1988) The Biosynthetic Origin of Rebeccamycin. J Nat Prod **51**: 937
89. Pindur U, Erfauion-Ardoust H (1989) Indole-2,3-quinodimethanes and Stable Cyclic Analogues for Regio- and Stereo-controlled Synthesis of (b) Annealated Indoles. Chem Rev **89**: 1681
90. Pindur U, Adam R (1991) 1,6-Electrocyclisation Reaction of Acceptor Substituted 2,3-Vinylindoles to Functionalised Carbazole. Heterocycles **31**: 581
91. Pointek JA, Wang RY (1986) Acute and Subchronic Effects of Rincazole (B234U) a Potential Antipsychotic Drug on A9 and A10 Dopamine Neurons in the Rat. Life Sci **39**: 651
92. Ramesh K, Kapil RS (1986) Synthesis of Murrayafoline B and Murrayaquinone-B. Indian J Chem **25B**: 462
93. Rice LM, Scot KB (1970) 3-Substituted 1,2,3,4-Tetrahydrocarbazoles. J Medicin Chem **13**: 308
94. Reisch JO, Goz O, Wickramsinghe A, Baudra Herath HKT, Henkel G (1992) Carbazole Alkaloids from Seeds of *Murraya koenigii*. Phytochemistry **31**: 2877
95. Reppe W, Stark W, Voss A (1934) Complex Polymerization Products (CA **28**: 32556)
96. Roy S, Guha R, Ghosh S, Chakraborty DP (1982) Biomimetic Hydroxylation Studies on Carbazole Alkaloids. Ind J Chem **21B**: 617
97. Ruangrungi N, Ariyapraycon J, Lang GL, Organ HG (1990) Three New Carbazole Alkaloids Isolated from *Murraya siamensis*. J Nat Prod **53**: 946
98. Sarstedt B, Winterfeldt E (1983) Reactions with Indole Derivatives, XLVIII. Heterocycles **20**: 469
99. Saroja B, Srinivasan PC (1986) Synthesis and (4+2) Cyclo Additions of N-(Phenylsulphonyl-3)-vinyl-indole. Synthesis: 748
100. Sato A, Morishita P, Shiraki T, Yoshioka S, Horikoshi H, Kaswano H, Hanzawa H, Hata T (1993) Aldose Reductase, Inhibitors from Marine Sponge *Doctyodendri* sp. J Org Chem **58**: 7632
101. Sezaki M, Sasaki T, Nakazawa T, Takeda U, Iwata M, Watanabe T, Koyama M, Kai F, Shomura T, Kojima N (1985) A New Antibiotic SF-2370 Produced by *Actinomadura* sp. SF-2370. Antibiot **38**: 1439
102. Shinya K, Tanaka M, Furihata K, Hayakawa Y, Seto H (1993) Structure of Carquinostatin A, A New Cell Protecting Substance Produced by *Streptomyces exfoliatus*. Tetrahedron Lett **34**: 4943
103. Steglich W, Steffan R, Kopanski L, Echardt G (1980) Archyriaflavin-B, -C, Angew Chem Int Engl **19**: 459
104. Takahashi I, Kobayashi E, Asano K, Yoshida M, Nakano H (1987) UCN-01, A Selective Inhibitor of Protein Kinase C from Streptomyces. J Antibiot **40**: 1782
105. Takeya K, Itoigawa M, Furukawa H (1989) Triphasic Inotropic Response of Muscle to Murrayaquinone. Isolated from Rutaceae. Eur J Pharmacol **169**: 137
106. Tamaoki T, Nomoto H, Takahasi I, Kato Y, Morimoto N, Tomita P (1986) Staurosporine, A Potent Inhibitor of Phospholipid/Ca^{++} Dependent Protein Kinase. Biochem Biophys Res Commun **135**: 397
107. Tanida S, Takizawa M, Takahashi T, Tsubotani S, Harda S (1989) Tan-999 and Tan-1030A. New Indole Carbazole Alkaloids with Macrophage Activating Properties. J Antibiot **42**: 1619
108. Taylor MR, Westwood SA (1993) An Investigation of Drug Detection in Equine Urine Samples by Modern TLC. J Planar Chromatography and TLC **6**: 415
109. Te Paske MR, Gloer JB, Wicklow DT, Dowd PF (1990) Aflavazole: A New Antiinsectant Carbazole Metabolite from Sclerotia of *Aspergllius flavus*. J Org Chem **55**: 5299

110. Te Paske MR, Gloer JB, Wicklow DT, Dowd PF (1989) An Antiviral Carbazole Alkaloid from Sclerotia of *Aspergillus tubingensis.* J Org Chem **54**: 4743
111. Te Paske MR, Gloer JB, Wicklow DT, Dowd PF (1989) Structure of Tubingensin B, A Cytotoxic Carbazole Alkaloid from Sclerotia of *Aspergillus tubingensis.* Tetrahedron **45**: 4961
112. Tsubotani S, Tanida S, Harda S (1991) Structure Determination of Indolocarbazole Alkaloids by NMR Spectroscopy. Tetrahedron **47**: 3565
113. Wada A, Hirai S, Hanaoka M (1994) Novel Synthesis of Quinazolinequinone and Carbazoloquinone Through Anionic Cycloaddition: Its Application to Murrayaquinone A. Chem Pharm Bull **42**: 416
114. Weinreb SM, Garigipati RS, Gainor JA (1984) Natural Product Synthesis via Cycloaddition with N. Sulfinyl Dienophiles. Heterocycles **21**: 309
115. Wu T-S (1991) Murrayamine A, B, C and (+)-Mahanine, Carbazole Alkaloids from *Murraya euchrestifolia.* Phytochemistry **30**: 1048
116. Wu T-S, Wang M-L, Lai J-S, Ito C, Furukawa H (1991) Binary Carbazole Alkaloids from *Murraya euchrestifolia.* Phytochemistry **30**: 1054
117. Yang W, Jiang T, Davis PT, Acosta D (1991) In vitro Metabolism and Toxicity Assessment of N-Methyl Carbazole in Primary Culture Rat Hepatocytes. Toxicology **68**: 217 (C.A. **115** 225969, 1991)
118. Yamasaki K, Kaneda M, Watanabe K, Ueki Y, Ishamari K, Nakamura S, Nomi R, Yoshida N, Nakajima T (1983) New Antibiotics Carbazomycin A and B, III. J Antibiot **36**: 552
119. Yasuzawa T, Iada T, Yoshida M, Hirayama N, Takahashi M, Shirahata K, Sano H (1986) The Structures of Novel Protein Kinase C. Inhibitors K-252a, -b, -c, and -d. J Antibiot **39**: 1072
120. Yogo M, Ito C, Furukawa H (1991) Synthesis of Some Carbazoloquinone Alkaloids and Their Analogues, *Facile pelladium* Assisted Intramolecular Ring Closure of Arylamino-1,4-Benzoquinones to Carbazole-1,4-quinones. Chem Phar Bull **39**: 328

Author Index

Page numbers printed in *italics* refer to References

Subject Index

Springer-Verlag and the Environment

MIX
Papier aus verantwortungsvollen Quellen
Paper from responsible sources
FSC® C105338

If you have any concerns about our products,
you can contact us on
ProductSafety@springernature.com

In case Publisher is established outside the EU,
the EU authorized representative is:
Springer Nature Customer Service Center GmbH
Europaplatz 3, 69115 Heidelberg, Germany

Printed by Libri Plureos GmbH
in Hamburg, Germany